Lecture Notes in Mathematics

Edited by A. Dold and B. Eckmann

1027

Morisuke Hasumi

Hardy Classes on Infinitely Connected Riemann Surfaces

Springer-Verlag
Berlin Heidelberg New York Tokyo 1983

Author

Morisuke Hasumi
Department of Mathematics, Ibaraki University
Mito, Ibaraki 310, Japan

AMS Subject Classifications (1980): 30F99, 30F25, 46J15, 46J20, 31A20, 30D55

ISBN 3-540-12729-1 Springer-Verlag Berlin Heidelberg New York Tokyo
ISBN 0-387-12729-1 Springer-Verlag New York Heidelberg Berlin Tokyo

Printing and binding: Beltz Offsetdruck, Hemsbach/Bergstr.
2146/3140-543210

PREFACE

The purpose of these notes is to give an account of Hardy classes on infinitely connected open Riemann surfaces and some related topics. Already in this Lecture Notes series we have a beautiful monograph "Hardy Classes on Riemann Surfaces" by Maurice Heins, which appeared in print in 1969. It is therefore natural that our stress should now be placed on some new advances we have seen during subsequent years.

As generally recognized, Hardy classes made their debut in the literature in 1915, when G. H. Hardy discussed the mean growth of functions analytic on the unit disk in his paper [15]. The theory of these very useful classes of functions was laid its foundation in the work of Hardy himself, J. E. Littlewood, F. and M. Riesz, G. Szegö among others. And we now have a large and still growing amount of literature in this area. Speaking roughly, Hardy classes have been studied most intensively in the case of the unit disk from the cradle for its importance as well as simplicity. The case of finitely connected surfaces, planar or not, has drawn much attention and enjoyed considerable progress in recent years.

Opposed to this, our knowledge seems to be relatively small in the case of infinitely connected surfaces. The classical theory of Hardy classes deals mostly with the unit disk and does not have much direct bearing on our present problem. From 1950's downwards, functional-analytic methods have found their successful applications in the field of complex function theory including Hardy classes and the abstract Hardy class theory thus created has grown to form the core of the newly-born theory of function algebras, as evidenced from Gamelin's book [10] for example. Nevertheless, the case of infinitely connected surfaces, as I understand it, still lies beyond the reach of the new theory and needs an independent study as in the case of polydisks and balls. The structure of general Riemann surfaces is not yet very well known and so we should begin with this basic question: "For which class of infinitely connected open Riemann surfaces can one get a fruitful extension of the classical theory of Hardy classes?" In the present notes we will try to give an answer, partial at least, to this question.

Our idea for attacking the problem is very simple and says that any nice surface should carry an ample family of holomorphic functions. But what do we mean by this? The candidate we wish to put forward here as most promising is the class of Riemann surfaces of Parreau-Widom type (abbreviated to "PWS" in the following). The definition will be given in Chapter V. This class was first introduced by M. Parreau [52] in 1958 and also, perhaps independently, by H. Widom [70] in 1971 from very different motives. They used different definitions, which later turned out to be essentially the same. The main reason why we are interested in this class of surfaces is expressed in the following fundamental result of Widom:

(A) An open Riemann surface R is of Parreau-Widom type if and only if every unitary flat complex line bundle over R has nonconstant bounded holomorphic sections.

Moreover, surfaces of this kind inherit many other nice properties of the unit disk or finitely connected surfaces. We list most relevant results in the following, where R denotes a PWS.

(B) Every positive harmonic function on R has a limit along almost every Green line issuing from any fixed origin in R.

(C) The Dirichlet problem on the space of Green lines on R for any bounded measurable boundary function has a unique solution, which converges to the given boundary function along almost every Green line.

(D) Almost every Green line ℓ on R issuing from any fixed origin O converges to a point, say b_ℓ, in the Martin boundary Δ of R and the correspondence $\ell \to b_\ell$ is measure-preserving with respect to the Green measure on the one hand, and the harmonic measure on Δ for the point O on the other. Furthermore, the usual solution of the Dirichlet problem for R with any bounded measurable boundary function on Δ converges to the data along almost every Green line.

The statement (D) means in particular that the Brelot-Choquet problem (see Brelot [5]) has a completely affirmative answer for any PWS. It seems indeed that PWS's form the first general class of surfaces of infinite genus for which this problem has a positive solution. Moreover, the statement (D) can be refined so as to have the following:

(E) Stolz regions with vertex at almost every point in Δ can be defined and the boundary behavior of analytic maps from R can be analysed in detail as in the case of the unit disk.

It is possible to generalize the Cauchy-Read theorem to PWS's. The theorem consists of two statements, whose refinements are called, respectively, the direct Cauchy theorem--(DCT) for short--and the inverse Cauchy theorem. Concerning these, we first have:

(F) The inverse Cauchy theorem holds for any PWS.

The converse of this statement is also true. Namely, we have:

(G) If R is a hyperbolic Riemann surface and if the set $H^\infty(R)$ of bounded holomorphic functions on R separates the points of R, then the inverse Cauchy theorem holds if and only if R is a PWS.

Thus the inverse Cauchy theorem almost characterizes surfaces of Parreau-Widom type. On the other hand, the direct Cauchy theorem--an utmost refinement of the Cauchy integral formula--is not always valid. As we shall see, there exist PWS's of infinite genus for which the direct Cauchy theorem holds, while there exist planar PWS's for which the direct Cauchy theorem fails.

The corona problem for PWS's can be studied with some interesting results. First we have:

(H) Every PWS R can be embedded homeomorphically as an open subset in the maximal ideal space of the Banach algebra $H^\infty(R)$.

This, however, is all one can say about an arbitrary PWS. The corona problem for general PWS's has a negative answer, contrary to the expectation one might have. A curious fact in this connection is that a PWS can satisfy the corona theorem without satisfying the direct Cauchy theorem. These two things look like independent. Finally, we state another fact:

(I) If a PWS R is regular in the sense of potential theory, then $H^\infty(R)$ is dense in the space of all holomorphic functions on R in the topology of uniform convergence on compact subsets of R.

We have been talking about relevant properties of PWS's to be discussed in these notes. Apart from PWS's, this book contains a detailed account of the classification problem of plane regions in terms of Hardy classes. As it turns out, almost all that happen in the category of Riemann surfaces can happen in the category of plane regions. Intuitively, PWS's are surfaces which are good in some sense or other. So our last result shows that plane regions can be as ill-behaved as one can imagine.

In writing these notes I have been led by the feeling that the surfaces of Parreau-Widom type probably form the widest family of well-behaved surfaces as far as the theory of Hardy classes is concerned. It is hoped that our description in the following pages will justify this feeling somehow or other. One thing I wish to note is that our study is not a mere adaptation of the existing knowledge of Riemann surfaces. It also aims at finding some new facts which are neither too general nor too special. The present notes are not complete in any sense but just reflect the author's personal interest in the field.

At all event I hope that our effort would help not only extend the theory of Hardy classes but also increase our knowledge of Riemann surfaces in general.

The prerequisites for reading these notes are the fundamentals of advanced complex function theory and some knowledge of functional analysis. As for the function theory, we assume that the reader has some acquaintance with the facts to be found in Chapters I, II, III, V of Ahlfors and Sario [AS] and also in the first four chapters of the book [CC] by Constantinescu and Cornea. As for the functional analysis, Chapter 1 of Hoffman [34] may be useful, if not sufficient.

We now comment on the contents of the present notes. The function-theoretic prerequisites are sketched in Chapter I without proof. In order to deal reasonably with Hardy classes on multiply-connected open surfaces, we rely on two concepts: multiplicative analytic functions and Martin compactification. These are explained in Chapters II and III, respectively. Chapter IV contains preliminary observations on Hardy classes, where the boundary behavior is our principal concern. The main body of this book begins in Chapter V. There, the definition of surfaces of Parreau-Widom type is given after Widom. After showing, by means of regularization, that this definition is essentially equivalent to Parreau's, we present a detailed proof of Widom's fundamental theorem mentioned in (A) above. In Chapter VI we discuss the Dirichlet problem on the space of Green lines (see (B), (C)) and then solve the Brelot-Choquet problem for PWS's (see (D)). Two types of Cauchy theorems--direct and inverse--on PWS's form the main theme of Chapter VII. There, the statement (F) is established by using Green lines and, as an application, it is shown that H^∞ is a maximal weak-star closed subalgebra of L^∞ on the Martin boundary. Next, the direct Cauchy theorem (DCT) is precisely stated. We prove a weaker version of (DCT), which is valid for any hyperbolic Riemann surface, together with some application. On the other hand, (DCT) itself fails sometimes. But this cannot be seen until we know something about invariant subspaces, which are studied in Chapter VIII. In Chapter VIII we classify closed H^∞-submodules of L^p--(shift-)invariant subspaces of L^p--on the Martin boundary of a PWS. Corresponding to the known results for the case of the unit disk, we consider two principal types of invariant subspaces, which are called doubly invariant and simply invariant, respectively. As for doubly invariant subspaces the situation is rather simple for any PWS. But the so-called Beurling type theorem for simply invariant subspaces is not always valid for PWS's. It is proved in fact that the

Beurling type theorem is valid if and only if (DCT) is. In this connection, examples in Chapter X may be interesting. We give there three types of construction: the first defines PWS's of infinite genus (of Myrberg type) for which (DCT) holds; the second yields a family of planar PWS's for which (DCT) fails but the corona theorem holds; and the third gives PWS's for which the corona theorem is false. In the same chapter we also prove the statements (H) and (I). In Chapter IX we first prove the statement (G), which characterizes PWS's among hyperbolic Riemann surfaces, and then collect a couple of conditions on PWS's equivalent to (DCT). Finally in Chapter XI we solve Heins' problem concerning classification of plane regions by using Hardy classes. The statement (E) will not be proved but just sketched in Chapter VI. There are three appendices and a list of references, which is by no means exhaustive.

My interest in the subject treated here was first aroused while I was visiting the University of California at Berkeley in 1962-64. For this I am indebted to Professors E. Bishop, H. Helson and J. L. Kelley. It was C. Neville's thesis [45], that led me in 1972 to a serious study of Hardy classes on infinitely connected Riemann surfaces. I would thus like to thank Professor L. A. Rubel, who showed me the thesis right after its completion. I also owe thanks to Professor L. Carleson, who invited me to the Mittag-Leffler Institute during its Beurling Year in 1976-77, when I was able to deepen my knowledge of Hardy classes. The primitive version of these notes was written then. New discoveries have made the notes expand subsequently. Most of the main chapters were rewritten for a series of lectures I gave on the present topics at Tokyo Metropolitan University during the week of July 5, 1982. Particular thanks are due to Professors M. Sakai and S. Yamashita, who organized the lectures and made some valuable suggestions. I have much benefited by very helpful remarks from Professors Z. Kuramochi and H. Widom and Dr. M. Hayashi, to whom I wish to express my appreciation.

I would like to dedicate this book to Professor Zirô Takeda, who is my first teacher on the research level and whose inspiration and encouragement have remained with me as fresh as ever.

Mito, Ibaraki
July, 1983

Morisuke Hasumi

CONTENTS

PREFACE iii

CHAPTER I. THEORY OF RIEMANN SURFACES: A QUICK REVIEW

§1. Topology of Riemann Surfaces 1
1. Exhaustion 1
2. The Homology Groups 2
3. The Fundamental Group 3
§2. Classical Potential Theory 4
4. Superharmonic Functions 4
5. The Dirichlet Problem 5
6. Potential Theory 7
§3. Differentials 9
7. Basic Definition 9
8. The Class Γ and its Subclasses 11
9. Cycles and Differentials 12
10. Riemann-Roch Theorem 14
11. Cauchy Kernels on Compact Bordered Surfaces 17
Notes 22

CHAPTER II. MULTIPLICATIVE ANALYTIC FUNCTIONS

§1. Multiplicative Analytic Functions 23
1. The First Cohomology Group 23
2. Line Bundles and Multiplicative Analytic Functions . . . 28
3. Existence of Holomorphic Sections 31
§2. Lattice Structure of Harmonic Functions 33
4. Basic Structure 33
5. Orthogonal Decomposition 36
Notes 38

CHAPTER III. MARTIN COMPACTIFICATION

§1. Compactification 39
1. Definition 39
2. Integral Representation 40
3. The Dirichlet Problem 43

§2. Fine Limits 49
4. Definition of Fine Limits 49
5. Analysis of Boundary Behavior 50
§3. Covering Maps 57
6. Correspondence of Harmonic Functions 57
7. Preservation of Harmonic Measures 59
Notes 63

CHAPTER IV. HARDY CLASSES

§1. Hardy Classes on the Unit Disk 64
1. Basic Definitions 64
2. Some Classical Results 66
§2. Hardy Classes on Hyperbolic Riemann Surfaces 73
3. Boundary Behavior of H^p and h^p Functions 73
4. Some Results on Multiplicative Analytic Functions . . . 74
5. The β-Topology 75
Notes 82

CHAPTER V. RIEMANN SURFACES OF PARREAU-WIDOM TYPE

§1. Definitions and Fundamental Properties 83
1. Basic Definitions 83
2. Widom's Characterization 85
3. Regularization of Surfaces of Parreau-Widom Type 86
§2. Proof of Widom's Theorem (I) 90
4. Analysis on Regular Subregions 90
5. Proof of Necessity 95
§3. Proof of Widom's Theorem (II) 99
6. Review of Principal Operators 99
7. Modified Green Functions 102
8. Proof of Sufficiency 111
9. A Few Direct Consequences 117
Notes 118

CHAPTER VI. GREEN LINES

§1. The Dirichlet Problem on the Space of Green Lines 119
1. Definition of Green Lines 119
2. The Dirichlet Problem 121
§2. The Space of Green Lines on a Surface of Parreau-Widom Type 124
3. The Green Star Regions 124
4. Limit along Green Lines 129

§3. The Green Lines and the Martin Boundary 132
5. Convergence of Green Lines 132
6. Green Lines and the Martin Boundary 135
7. Boundary Behavior of Analytic Maps 140
Notes 143

CHAPTER VII. CAUCHY THEOREMS

§1. The Inverse Cauchy Theorem 144
1. Statement of Results 144
2. Proof of Theorem 1B 145
§2. The Direct Cauchy Theorem 151
3. Formulation of the Condition 151
4. The Direct Cauchy Theorem of Weak Type 152
§3. Applications 155
5. Weak-star Maximality of H^∞ 155
6. Common Inner Factors 156
7. The Orthocomplement of $H^\infty(d\chi)$ 157
Notes 159

CHAPTER VIII. SHIFT-INVARIANT SUBSPACES

§1. Preliminary Observations 160
1. Generalities 160
2. Shift-Invariant Subspaces on the Unit Disk 162
§2. Invariant Subspaces 167
3. Doubly Invariant Subspaces 167
4. Simply Invariant Subspaces 169
5. Equivalence of (DCT_a) 177
Notes 178

CHAPTER IX. CHARACTERIZATION OF SURFACES OF PARREAU-WIDOM TYPE

§1. The Inverse Cauchy Theorem and Surfaces of Parreau-Widom Type 179
1. Statement of the Main Result 179
2. A Mean Value Theorem 183
3. Proof of the Main Theorem 187
§2. Conditions Equivalent to the Direct Cauchy Theorem 198
4. General Discussion 198
5. Functions $m^p(\xi,a)$ and (DCT) 200
Notes 207

CHAPTER X. EXAMPLES OF SURFACES OF PARREAU-WIDOM TYPE

§1. PWS of Infinite Genus for Which (DCT) Holds 208

1. PWS's of Myrberg Type 208

2. Verification of (DCT) 213

§2. Plane Regions of Parreau-Widom Type for Which (DCT) Fails . 215

3. Some Simple Lemmas 215

4. Existence Theorem 217

§3. Further Properties of PWS 221

5. Embedding into the Maximal Ideal Space 221

6. Density of $H^\infty(R)$ 223

§4. The Corona Problem for PWS 227

7. (DCT) and the Corona Theorem: Positive Examples 227

8. Negative Examples 229

Notes 233

CHAPTER XI. CLASSIFICATION OF PLANE REGIONS

§1. Hardy-Orlicz Classes 234

1. Definitions 234

2. Some Basic Properties 235

§2. Null Sets of Class N_Φ 238

3. Preliminary Lemmas 238

4. Existence of Null Sets 247

§3. Classification of Plane Regions 253

5. Lemmas 253

6. Classification Theorem 256

7. Majoration by Quasibounded Harmonic Functions 260

Notes 261

APPENDICES

A.1. The Classical Fatou Theorem 262

A.2. Kolmogorov's Theorem on Conjugate Functions 267

A.3. The F. and M. Riesz Theorem 269

References 272

Index of Notations 276

Index 278

CHAPTER I. THEORY OF RIEMANN SURFACES: A QUICK REVIEW

Some basic results in the theory of Riemann surfaces are collected here for our later reference. They are stated without proof but most of them can be found together with complete proofs either in Ahlfors and Sario, Riemann Surfaces or in Constantinescu and Cornea, Ideale Ränder Riemannscher Flächen, referred to as [AS] or [CC] below.

§1. TOPOLOGY OF RIEMANN SURFACES

We refer to [AS] for most basic definitions concerning Riemann surfaces, which will not be given here, e.g. conformal structure, local variable, parametric disk, intersection number, etc. In what follows, R denotes a Riemann surface. Unless otherwise stated, all Riemann surfaces are assumed to be connected.

1. Exhaustion

1A. Every connected open set in R is called a (sub-)region (or domain) in R. Every region in R is supposed to have the conformal structure induced from that of R. A region D in R is called a regular region in R if it is relatively compact, the boundary ∂D of D in R consists of a finite number of nonintersecting analytic curves, and $R \setminus D$ has no compact components.

Theorem. If R is an open Riemann surface, then there exists an increasing sequence $\{R_n\}_{n=1}^{\infty}$ of regular regions in R such that $\mathrm{Cl}(R_n)$ is included in R_{n+1} for each $n = 1, 2, \ldots$ and $R = \bigcup_{n=1}^{\infty} R_n$. ([AS], Ch. II, 12D)

Any sequence of regular regions R_n in R having this property is called a regular exhaustion of R.

1B. Existence of a regular exhaustion shows that every Riemann surface R admits a locally finite covering consisting of parametric disks and hence that R can be regarded as a polyhedron, i.e. a triangulated surface, which we denote by $K = K(R)$ ([AS], Ch. I, 46A).

Theorem. $K = K(R)$ is an orientable polyhedron. It is a finite polyhedron if and only if R is a compact or compact bordered surface.

1C. Let K be an orientable polyhedron. A finite subcomplex P of K is called a canonical subcomplex if (i) P is a polyhedron and (ii) every component of $K \setminus P$ is infinite and has a single contour. An increasing sequence $\{P_n\}_{n=1}^{\infty}$ of canonical subcomplexes of K is called a canonical exhaustion of K if K is the union of P_n and if every border simplex of P_{n+1} does not belong to P_n.

Theorem. Let R be an open Riemann surface. Then the polyhedron $K = K(R)$ has a subdivision which permits a canonical exhaustion. ([AS], Ch. I, 29A)

By a canonical (sub-)region in R we mean a regular region D such that every component of $R \setminus D$ has a single contour.

Corollary. Every open Riemann surface R admits a regular exhaustion $\{R_n\}_{n=1}^{\infty}$ consisting of canonical subregions.

2. The Homology Groups

2A. Let R be a Riemann surface and $H_1(R)$ the 1-dimensional singular homology group of R ([AS], Ch. I, 33B). By 1C the surface R can be regarded as an orientable polyhedron, which we denote by $K = K(R)$. Let $H_1(K)$ be the 1-dimensional homology group of this polyhedron K ([AS], Ch. I, 23D). Every oriented 1-simplex in K is regarded as a singular 1-simplex on the surface R. This gives rise to an isomorphism of $H_1(K)$ onto $H_1(R)$, which is called the canonical isomorphism ([AS], Ch. I, 34A).

2B. Thus the properties of the group $H_1(R)$ can be obtained by looking into $H_1(K)$. A finite or infinite sequence of cycles in K, labeled in pairs by a_i, b_i, is called a canonical sequence if $a_i \times a_j = b_i \times b_j = 0$, $a_i \times b_i = 1$ and $a_i \times b_j = 0$ for $i \neq j$, denoting by $a \times b$ the intersection number of 1-chains a and b ([AS], Ch. I, 31A).

If R is a compact surface, then K is a finite orientable polyhedron and there exists a canonical sequence a_i, b_i, $i = 1,\ldots, g$, which forms a basis for $H_1(K)$. Hence, $\dim H_1(R) = \dim H_1(K) = 2g$.

The number g is called the genus of the surface R.

If $\overline{R}$ is a compact bordered surface with q contours $c_0, \ldots, c_{q-1}$, $q \geqq 1$, then there exists in K a canonical sequence a_i, b_i, $i = 1, \ldots, g$, which, together with all but one contours, forms a basis of $H_1(K)$. Hence, $\dim H_1(\overline{R}) = \dim H_1(K) = 2g + q - 1$ ([AS], Ch. I, 31D). The number g is again the genus of $\overline{R}$.

2C. Finally let R be an open surface, so that K is an orientable open polyhedron. Let D be a regular subregion of R. Since the closure $Cl(D)$ of D is regarded as a compact bordered surface, it defines a finite polyhedron $K(Cl(D))$. By applying a suitable subdivision to K if necessary we may assume that $K(Cl(D))$ is a subcomplex of K. So we have a natural homomorphism of $H_1(K(Cl(D)))$ into $H_1(K)$. Since no components of $R \setminus D$ are compact, we see that the homomorphism is in fact an isomorphism and therefore that $H_1(K(Cl(D)))$ is identified with a subgroup of $H_1(K)$. Turning to the region D itself, we have

Theorem. Let D be a regular subregion of R. Then the group $H_1(D)$ is regarded as a subgroup of $H_1(R)$ by identifying every 1-chain in D with one in R.

In case D is a canonical subregion of R, it is seen moreover that $H_1(D)$ is a direct summand of the free abelian group $H_1(R)$. See [AS], Ch. I, §§31-32 for a detailed discussion.

3. The Fundamental Group

3A. Let $\mathcal{O}$ be a point in R, which is held fixed. The fundamental group $F_{\mathcal{O}}(R)$, referred to the "origin" $\mathcal{O}$, is defined to be the multiplicative group of homotopy classes of closed curves issuing from $\mathcal{O}$ ([AS], Ch. I, 9D). Every closed curve γ from $\mathcal{O}$ can be considered as a singular 1-simplex. We thus get a natural homomorphism of $F_{\mathcal{O}}(R)$ onto $H_1(R)$, which takes the homotopy class of γ to its homology class.

Theorem. Under the natural homomorphism the homology group $H_1(R)$ is isomorphic with the quotient group $F_{\mathcal{O}}(R)/[F_{\mathcal{O}}(R)]$ of $F_{\mathcal{O}}(R)$ modulo the commutator subgroup $[F_{\mathcal{O}}(R)]$ of $F_{\mathcal{O}}(R)$. ([AS], Ch. I, 33D)

3B. By a character of an abstract group G we mean any homomorphism of G into the circle group (= the multiplicative group of complex numbers of modulus one). The set of all characters of G forms a group with respect to the pointwise multiplication of characters. The group thus obtained is called the character group of G and is denoted by G^*. The preceding theorem shows that $H_1(R)^*$ is isomorphic with $\mathcal{F}_0(R)^*$. Combined with the last remark in 2C we have

Theorem. Let D be a canonical subregion of an open Riemann surface R such that $0 \in D$. Then $\mathcal{F}_0(D)$ is a subgroup of $\mathcal{F}_0(R)$ and every character of $\mathcal{F}_0(D)$ is the restriction of some character of $\mathcal{F}_0(R)$.

§2. CLASSICAL POTENTIAL THEORY

4. Superharmonic Functions

4A. Let D be a region in a Riemann surface R. An extended real-valued function s on D is called superharmonic if (i) $-\infty < s(z) \leq +\infty$ on D and $s \not\equiv +\infty$; (ii) s is lower semicontinuous; and (iii) for every a in D there exists a parametric disk V with center a such that $Cl(V) \subseteq D$ and

$$s(a) \geq \frac{1}{2\pi}\int_0^{2\pi} s(re^{it})dt$$

for $0 < r \leq 1$, where V is identified with the open unit disk. A function s on D is subharmonic if $-s$ is superharmonic. A function on D is harmonic if it is both superharmonic and subharmonic.

4B. A collection S of superharmonic functions on D is called a Perron family if (i) for every pair of elements s_1, s_2 in S there exists an s_3 in S with $\min\{s_1(z), s_2(z)\} \geq s_3(z)$ for all z in D; (ii) for every s in S and every parametric disk V with $Cl(V) \subseteq D$, there exists an s' in S such that

$$s'(z) \leq \frac{1}{2\pi}\int_0^{2\pi} s(e^{it})\mathrm{Re}\left(\frac{e^{it}+z}{e^{it}-z}\right) dt$$

for all z in V; and (iii) S has a subharmonic minorant, i.e. there exists a subharmonic function s'' on D such that $s \geq s''$ for all s in S. A collection S of subharmonic functions on D is a Perron family if $-S = \{-s: s \in S\}$ is a Perron family of superharmonic functions. Perron families are very useful in constructing harmonic functions. The reason is given by the following

Theorem. Let S be a Perron family of superharmonic functions on D and set $u(z) = \inf\{s(z): s \in S\}$. Then u is a harmonic function on D. ([CC], p. 14)

4C. In particular, if a subharmonic function u on D has a superharmonic majorant, i.e. a superharmonic function s with $s \geq u$ on D, then the set S_u of all superharmonic majorants of u forms a Perron family and the minimum element in S_u, which exists by the preceding theorem, is the least harmonic function that majorizes u on D. This is called the least harmonic majorant of u and is denoted by LHM(u). Let u_1 and u_2 be harmonic on D. Then the pointwise maximum $\max\{u_1, u_2\}$ (resp. minimum $\min\{u_1, u_2\}$) is subharmonic (resp. superharmonic) on D. So, if $\max\{u_1, u_2\}$ (resp. $\min\{u_1, u_2\}$) has a superharmonic majorant (resp. a subharmonic minorant), then it has the least harmonic majorant (resp. the greatest harmonic minorant), which is denoted by $u_1 \vee u_2$ (resp. $u_1 \wedge u_2$).

5. The Dirichlet Problem

5A. Let D be a region in R, whose boundary ∂D in R is nonvoid. For every extended real-valued function f on ∂D, we denote by $\overline{S}(f;D)$ the set of superharmonic functions s on D such that (i) s is bounded below; (ii) $\liminf_{D \ni z \to \infty} s(z) \geq 0$, where ∞ denotes the point at infinity of the one-point compactification of R; and (iii) $\liminf_{D \ni z \to a} s(z) \geq f(a)$ for every a in ∂D. We set $\underline{S}(f;D) = -\overline{S}(-f;D)$. Suppose that both $\overline{S}(f;D)$ and $\underline{S}(f;D)$ are nonvoid and let $\overline{H}[f;D]$ (resp. $\underline{H}[f;D]$) be the pointwise infimum (resp. supremum) on D of functions in $\overline{S}(f;D)$ (resp. $\underline{S}(f;D)$). Since $s' \geq s''$ for every s' in $\overline{S}(f;D)$ and every s'' in $\underline{S}(f;D)$, both $\overline{S}(f;D)$ and $\underline{S}(f;D)$ are Perron families, so that $\overline{H}[f;D]$ and $\underline{H}[f;D]$ are harmonic on D and $\overline{H}[f;D] \geq \underline{H}[f;D]$. A function f on ∂D is called resolutive if both $\overline{S}(f;D)$ and $\underline{S}(f;D)$ are nonvoid and $\overline{H}[f;D] = \underline{H}[f;D]$. The common function is then denoted by $H[f;D]$ and is called the solution of the Dirichlet problem for D with the boundary data f.

Theorem. (a) If f_1, f_2 on ∂D are resolutive and α_1, α_2 are real numbers, then $\alpha_1 f_1 + \alpha_2 f_2$, $\max\{f_1, f_2\}$ and $\min\{f_1, f_2\}$ are also resolutive and satisfy the following, in which we write $H[\cdot]$ in place of $H[\cdot;D]$:

$$H[\alpha_1 f_1 + \alpha_2 f_2] = \alpha_1 H[f_1] + \alpha_2 H[f_2],$$

$$H[\max\{f_1, f_2\}] = H[f_1] \vee H[f_2], \quad H[\min\{f_1, f_2\}] = H[f_1] \wedge H[f_2].$$

(b) Let $\{f_n\}_{n=1}^{\infty}$ be a monotone sequence of resolutive functions on ∂D with limit $f = \lim_n f_n$. Then f is resolutive if and only if the sequence $\{H[f_n]\}_{n=1}^{\infty}$ converges. If this is the case, then $H[f] = \lim_n H[f_n]$.

(c) Let f be a resolutive function on ∂D. Let D' be a region included in D and define f' on $\partial D'$ by setting $f' = f$ on $\partial D \cap \partial D'$ and $= H[f;D]$ on $D \cap \partial D'$. Then, $H[f';D'] = H[f;D]$ on D'.
([CC], p. 22)

5B. A point a in ∂D is called <u>regular</u> if, for every bounded real-valued function f on ∂D,

$$\limsup_{\partial D \ni b \to a} f(b) \geqq \limsup_{D \ni z \to a} \overline{H}[f;D](z).$$

If f is continuous at a, then this inequality implies

$$\lim_{z \to a} \overline{H}[f;D](z) = \lim_{z \to a} \underline{H}[f;D](z) = f(a).$$

A point in ∂D which is not regular is called <u>irregular</u>. A positive superharmonic function s on D is called a <u>barrier</u> for a point a in ∂D if (i) $\lim_{z \to a} s(z) = 0$ and (ii) $\inf_{z \in D \setminus V} s(z) > 0$ for every neighborhood V of a.

<u>Theorem</u>. Let D be a region in R and let a be a point in ∂D. Then the following statements are equivalent:

(a) The point a is a regular boundary point of D.

(b) There exists a barrier for the point a with respect to the region D.

(c) There exist a neighborhood V of a and a positive superharmonic function u on $V \cap D$ such that $u(z) \to 0$ as $z \to a$.

If the connected component of ∂D containing a is a continuum, then a is a regular boundary point of D. ([CC], p. 23)

5C. A region D in R is called <u>hyperbolic</u> if it carries a non-constant positive superharmonic function.

<u>Theorem</u>. If D is a hyperbolic region in R, then every bounded continuous function on ∂D is resolutive. ([CC], p. 27)

5D. Let D be a hyperbolic region in R and let $C_b(\partial D,\mathbb{R})$ be the set of all real-valued bounded continuous functions on ∂D. This is a linear space over $\mathbb{R}$ under the usual addition and scalar multiplication of functions. It becomes a Banach space if we equip it with the supremum norm $\|f\|_\infty = \sup_{b\in\partial D} |f(b)|$. By the preceding theorem every f in $C_b(\partial D,\mathbb{R})$ is resolutive and thus determines a harmonic function $H[f;D]$ on D. In view of Th. 5A, the map $f \to H[f;D]$ is a linear map from $C_b(\partial D,\mathbb{R})$ into the space of bounded harmonic functions on D, which is order-preserving, i.e. $f \geq 0$ on ∂D implies $H[f;D] \geq 0$ on D. For every fixed point a in D the map $f \to H[f;D](a)$ is a bounded linear functional on $C_b(\partial D,\mathbb{R})$ with $|H[f;D](a)| \leq \|f\|_\infty$. By Riesz's representation theorem there exists a unique nonnegative regular Borel measure, ω_a^D, on ∂D such that

$$(1) \qquad H[f;D](a) = \int_{\partial D} f(b)\, d\omega_a^D(b)$$

for every f in $C_b(\partial D,\mathbb{R})$ with compact support. The inequality $H[1;D] \leq 1$ shows that $\omega_a^D(\partial D) \leq 1$. It may happen that the Dirichlet problem has only the trivial solution. In that case the measures ω_a^D vanish identically for all a in D. When the Dirichlet problem has nontrivial solutions, the measure ω_a^D is called the harmonic measure on ∂D at the point a (with respect to the region D).

Theorem. A function f on ∂D is resolutive if and only if it is ω_a^D-summable. If this is the case, then the equation (1) holds. ([CC], p. 28)

We now slightly extend the definition of $H[f;D]$. Namely, let f be an extended real-valued function defined on a subset of R including the boundary ∂D and let f_0 be the restriction of f to the set ∂D. We call f resolutive with respect to D if f_0 is resolutive in the sense described above, i.e. f_0 is ω_a^D-summable for some a in D. If this is the case, then we again use the symbol $H[f;D]$ to denote the solution $H[f_0;D]$ of the Dirichlet problem for the region D with the boundary data f_0.

6. Potential Theory

6A. In the rest of §2, R denotes a hyperbolic Riemann surface. Let $g_a(z) = g(a,z)$ be the Green function for R with pole a in R. This is characterized by the following

Theorem. For every a in R there exists a unique positive superharmonic function $z \to g_a(z)$ satisfying the conditions:

(A1) in any parametric disk V with center a there exists a harmonic function u on V such that $g_a(z) = -\log|z| + u(z)$ for all z in V, where V is identified with the open unit disk $\{|z| < 1\}$;

(A2) for every subregion D of R with $a \notin Cl(D)$ we have $g_a = H[g_a;D]$ on D. ([CC], p. 32)

6B. For any given finite positive regular Borel measure μ on R we set

$$U^\mu(z) = \int_R g(z,w)\, d\mu(w)$$

for z in R. We call it the potential generated by μ, unless it is everywhere infinite. So a potential U^μ is a positive superharmonic function on R, which is harmonic outside the closed support of μ. Moreover we have

Theorem. A potential U^μ is harmonic on an open set D in R if and only if D does not intersect the support of μ. Every potential is generated by a unique positive measure. ([CC], p. 35)

6C. A subset E of R is called a polar set if there exists a positive superharmonic function on R which is equal to $+\infty$ identically on E. A countable union of polar sets is again a polar set. We say that a condition holds quasi-everywhere (abbreviated to q.e.) on a subset A of R if there exists a polar set $E \subseteqq A$ such that the condition holds on A except at the points of E.

Theorem. Let u be a positive superharmonic function on R and U^μ a potential. If $U^\mu \leqq u$ q.e. on the closed support of μ, then $U^\mu \leqq u$ everywhere on R. ([CC], p. 37)

6D. As for the regularity of the Dirichlet problem we have the following

Theorem. Let D be a region in R such that $R \setminus D$ is not a polar set. Then the irregular boundary points of D form a polar set. ([CC], p. 42)

6E. For a subset E of R and a positive superharmonic function u on R we define the function $R[u;E]$ on R to be the pointwise infimum of positive superharmonic functions s on R such that $s(z) \geqq u(z)$ on E. We set $u_E^R(a) = \liminf_{z \to a} R[u;E](z)$ for all a in R. The function u_E^R (resp. $R[u;E]$) is called the balayaged function (resp. the reduced function) of u relative to E.

Theorem. Let E be a closed set in R, u a positive superharmonic function on R and $v = u_E^R$. Then the following hold:

(a) v is a superharmonic function on R, which is equal to $H[u;R \setminus E]$ on $R \setminus E$ and to u on E except at the irregular boundary points of $R \setminus E$.

(b) v is the smallest positive superharmonic function on R that is not smaller than u q.e. on the set E.

(c) If E is compact, then $v = U^\mu$ for some nonnegative measure μ on E, which is determined uniquely. ([CC], p. 43)

6F. Finally we state the decomposition theorem of F. Riesz:

Theorem. Every positive superharmonic function on R can be written uniquely as the sum of a nonnegative harmonic function and a potential on R. ([CC], p. 41)

§3. DIFFERENTIALS

7. Basic Definition

7A. Let R be a Riemann surface. A first (resp. second) order differential ω on R is defined as a collection of 1-forms $a_\alpha dx_\alpha + b_\alpha dy_\alpha$ (resp. 2-forms $c_\alpha dx_\alpha dy_\alpha$), one for each parametric disk V_α with local variable $z_\alpha = x_\alpha + iy_\alpha$, such that (i) a_α and b_α (resp. c_α) are complex-valued functions on $z_\alpha(V_\alpha)$ and (ii) for any two parametric disks V_α and V_β, respectively with local variables z_α and z_β, we have

$$a_\beta = a_\alpha \frac{\partial x_\alpha}{\partial x_\beta} + b_\alpha \frac{\partial y_\alpha}{\partial x_\beta}, \quad b_\beta = a_\alpha \frac{\partial x_\alpha}{\partial y_\beta} + b_\alpha \frac{\partial y_\alpha}{\partial y_\beta} \quad \left(\text{resp. } c_\beta = c_\alpha \frac{\partial(x_\alpha, y_\alpha)}{\partial(x_\beta, y_\beta)}\right)$$

on $V_\alpha \cap V_\beta$. Such a differential is expressed symbolically as $adx + bdy$ (resp. $cdxdy$). A differential ω on R is said to be of class C^k (or a C^k-differential), $k = 0, 1, \ldots$, if all the coefficients of ω are of class C^k.

Basic operations on first order differentials $\omega = adx + bdy$ are as follows:

(a) multiplication by numerical functions f: $f\omega = (fa)dx + (fb)dy$;

(b) complex conjugation: $\overline{\omega} = \overline{a}dx + \overline{b}dy$;

(c) conjugate differential: $\omega^* = -bdx + ady$;

(d) exterior differentiation: $d\omega = (\partial b/\partial x - \partial a/\partial y)dxdy$ when ω is of class C^1;

(e) exterior product: $\omega_1\omega_2 = (a_1b_2 - a_2b_1)dxdy$ if $\omega_j = a_jdx + b_jdy$ for $j = 1, 2$.

A first order C^1-differential ω is said to be closed if $d\omega = 0$, i.e. $\partial b/\partial x = \partial a/\partial y$. It is called exact if $\omega = df$ $(= (\partial f/\partial x)dx + (\partial f/\partial y)dy)$ for some function f of class C^2.

7B. Let γ be an arc on R. Here and in what follows, arcs and 1-chains are always assumed to be piecewise smooth. Let $\omega = adx + bdy$ be a first order C^1-differential on R. If the arc γ is included in a coordinate neighborhood V with local variable $z = x + iy$ and is defined by $t \to z(t) = x(t) + iy(t)$, $0 \leqq t \leqq 1$, then the integral $\int_\gamma \omega$ is defined by the formula

$$\int_0^1 (a(z(t))\frac{dx}{dt}(t) + b(z(t))\frac{dy}{dt}(t))dt,$$

the value being independent of both local variables and parametrization of the arc. If the arc is not included in a single coordinate neighborhood, then we divide γ into a finite number of subarcs γ_j so that each γ_j is in a single coordinate neighborhood and we set

$$\int_\gamma \omega = \sum_j \int_{\gamma_j} \omega,$$

the result being independent of the subdivision we choose. The integral of ω is then extended to any 1-chain by linearity.

The integration of a second order C^0-differential $\omega = cdxdy$ is defined similarly. Namely, let Δ be a singular 2-simplex in R. If it is included in a single coordinate neighborhood V with local variable $z = x + iy$ and is defined by a differentiable map $(t,u) \to z(t,u)$, $0 \leqq u \leqq t \leqq 1$, then we define

$$\iint_\Delta \omega = \iint_{0\leqq u\leqq t\leqq 1} c(z)\frac{\partial(x,y)}{\partial(t,u)}\,dtdu,$$

the value again being independent of the choice of local variables and parametrization. When Δ is not included in a single coordinate neighborhood, then we have only to apply a suitable subdivision to Δ as in the case of line integrals, the result being independent of subdivision.

The integral of ω over any singular 2-chain is then defined by linearity, where each simplex is assumed to be differentiable. With these definitions we have the Stokes formula

$$\iint_X d\omega = \int_{\partial X} \omega$$

for any first order C^1-differential ω and any singular 2-chain X.

8. The Class Γ and its Subclasses

8A. Let ω be a first order C^1-differential. Then $\omega\bar{\omega}^* = (|a|^2 + |b|^2)\,dxdy$ is a second order C^1-differential. We set

$$\|\omega\| = \iint \omega\bar{\omega}^* = \iint (|a|^2 + |b|^2)dxdy. \tag{2}$$

We denote by $\Gamma^1 = \Gamma^1(R)$ the totality of first order C^1-differentials ω on R with $\|\omega\| < \infty$. Then Γ^1 is a linear space, in which the function (2) defines a norm. Indeed, this norm is induced from the inner product

$$(\omega_1, \omega_2) = \iint \omega_1\bar{\omega}_2^* \tag{3}$$

for ω_1, ω_2 in Γ^1. For a set A of differentials we denote by A^* the set of conjugate differentials of elements in A. A differential ω in Γ^1 is called harmonic if both ω and ω^* are closed. It is called analytic if $d\omega = 0$ and $\omega^* = -i\omega$. The set of all harmonic (resp. analytic) differentials in Γ^1 is denoted by Γ_h^1 (resp. Γ_a^1). We denote by Γ_c^1 (resp. Γ_e^1) the class of closed (resp. exact) differentials belonging to the class Γ^1.

8B. Suppose that R is the interior of a compact bordered surface $\bar{R}$. Then we use the following notation: $\Gamma^1(\bar{R})$ is the set of first order differentials on $\bar{R}$ whose coefficients are of class C^1 up to the border. $\Gamma_c^1(\bar{R})$ and $\Gamma_e^1(\bar{R})$ have the obvious meaning. We say that ω belongs to the class $\Gamma_{c0}^1(\bar{R})$ if $\omega \in \Gamma_c^1(\bar{R})$ and $\omega = 0$ along the border ∂R and that ω belongs to $\Gamma_{e0}^1(\bar{R})$ if $\omega = df$ for some f in C^2 with $f = 0$ on ∂R. The class $\Gamma^1(\bar{R})$ is equipped with the inner product (3), which defines the orthogonality relation there.

Theorem. $\Gamma_{c0}^1(\bar{R})$ (resp. $\Gamma_{e0}^1(\bar{R})$) is the orthocomplement of $\Gamma_e^1(\bar{R})^*$ (resp. $\Gamma_c^1(\bar{R})^*$) in $\Gamma^1(\bar{R})$. ([AS], Ch. V, 5A)

8C. If R is an open Riemann surface, then we denote by $\Gamma^1_{c0} = \Gamma^1_{c0}(R)$ (resp. $\Gamma^1_{e0} = \Gamma^1_{e0}(R)$) the set of ω in Γ^1_c (resp. Γ^1_e) with compact support. In this case we have

Theorem. Γ^1_{c0} is orthogonal to $\Gamma^1_e{}^*$ and Γ^1_c is the orthocomplement of $\Gamma^1_{e0}{}^*$ in Γ^1. ([AS], Ch. V, 6C)

8D. We denote by Γ the completion of Γ^1 with respect to the norm (2), so that Γ is a Hilbert space. We define $\Gamma_e = \mathrm{Cl}(\Gamma^1_e)$, $\Gamma_{e0} = \mathrm{Cl}(\Gamma^1_{e0})$, $\Gamma_c = (\Gamma_{e0}{}^*)^\perp$, and $\Gamma_{c0} = (\Gamma_e{}^*)^\perp$, where the closure and the orthocomplementation are taken in the space Γ. We set $\Gamma_h = \Gamma_c \cap \Gamma_c{}^*$. The fundamental result concerning the class Γ_h is the following

Theorem. If $\omega \in \Gamma_c \cap \Gamma_c{}^*$, then there exists a ω_1 in Γ^1_h such that $\|\omega - \omega_1\| = 0$; namely, Γ_h and Γ^1_h can be identified. ([AS], Ch. V, 9A)

8E. Among basic formulae for orthogonal decomposition of Γ, we shall use the following later, in which $\dotplus$ denotes the orthogonal direct sum in the space Γ:

$$\Gamma = \Gamma_{e0} \dotplus \Gamma_c{}^* = \Gamma_{e0} \dotplus \Gamma_{e0}{}^* \dotplus \Gamma_h. \tag{4}$$

As a consequence, we have

$$\Gamma_c = \mathrm{Cl}(\Gamma^1_c). \qquad \text{([AS], Ch. V, 10C)} \tag{5}$$

9. Cycles and Differentials

9A. If a first order C^1-differential, say Ω, is defined in a neighborhood V of a point a in R except at a, then Ω is said to have a singularity at a. The singularity is called analytic (resp. harmonic) if Ω is analytic (resp. harmonic) in some deleted neighborhood of a. It is called removable if Ω is continued to the point a as a C^1-differential.

Take a parametric disk $V = \{|z| < 1\}$ about a point a in R and consider a holomorphic differential Ω on $V \setminus \{a\}$. So we have $\Omega = f dz$ for some holomorphic function f on $V \setminus \{a\}$, i.e. in terms of the local variable z we have

$$f(z) = \sum_{k=-\infty}^{\infty} c_k z^k, \qquad 0 < |z| < 1.$$

We set $\nu_a(\Omega) = \inf\{k\colon c_k \neq 0\}$ and call it the order of Ω at a. The order is independent of the choice of local variables. Moreover, the coefficient c_{-1}, the <u>residue</u> of Ω at a, is also independent of the local variables and is denoted by $\operatorname{Res}_a(\Omega)$. In fact, we have

$$\operatorname{Res}_a(\Omega) = c_{-1} = \frac{1}{2\pi i}\int_\gamma \Omega$$

for any cycle γ in V having winding number 1 with respect to the point a. The singularity at a is removable if and only if $\nu_a(\Omega) \geqq 0$. Analytic differentials with prescribed singularities may be constructed by means of the following fact: If Ω is a closed C^1-differential on R with a finite number of analytic singularities and is holomorphic off a compact subset of R, then there exists a unique harmonic differential τ with singularities such that $\tau - \Omega \in \Gamma_{e0}$ ([AS], Ch. V, 17D).

9B. We construct some differentials related to cycles on R. Let p, q be two distinct points in R and c an arc joining p to q. Suppose first that these p, q and c are contained in a single parametric disk $V = \{|z| < 1\}$. Set $\zeta_1 = z(p)$, $\zeta_2 = z(q)$, and choose positive numbers r_1, r_2 which satisfy $|\zeta_1|$, $|\zeta_2| < r_1 < r_2 < 1$. Let v be a single-valued branch of the function $\log\{(z-\zeta_2)/(z-\zeta_1)\}$ in the annulus $r_1 < |z| < 1$ and let e be a C^2-function on R such that $e \equiv 1$ on $\{|z| \leqq r_1\}$ and $\equiv 0$ on $R \setminus \{|z| < r_2\}$. Further we define a differential Ω by setting $\Omega = \{(z-\zeta_2)^{-1} - (z-\zeta_1)^{-1}\}dz$ for $|z| < r_1$, $= d(ev)$ for $r_1 \leqq |z| < 1$, and $= 0$ otherwise in R. Then we use the result in 9A to find a unique harmonic differential τ with $\tau - \Omega \in \Gamma_{e0}$. Thus, for any 1-cycle γ not passing p and q, we have

$$\int_\gamma \tau = 2\pi i(c \times \gamma). \qquad \text{([AS], Ch. V, 19C)}$$

If p, q and c are not necessarily contained in a single parametric disk, then we divide c into a finite number of subarcs c_j, $j = 1,\dots, n$, so that each c_j joins a point p_{j-1} to a point p_j within a parametric disk with $p_0 = p$ and $p_n = q$. We define Ω_j and τ_j for every j as above and set $\Omega = \sum_j \Omega_j$ and $\tau = \sum_j \tau_j$. Then τ is a harmonic differential with singularities only at p and q, and $\tau - \Omega \in \Gamma_{e0}$. For every 1-cycle γ not passing any of the p_j's, we have

$$\int_\gamma \tau = \int_\gamma \Omega = \sum_j \int_\gamma \Omega_j = \sum_j 2\pi i(c_j \times \gamma) = 2\pi i(c \times \gamma).$$

The differential τ is seen to depend only on the homology class of c.

We write $\tau(c)$ in place of τ. Further we set

$$\phi(c) = (\tau(c) + i\tau(c)^*)/2 \quad \text{and} \quad \psi(c) = (\overline{\tau(c)} + i\overline{\tau(c)}^*)/2.$$

Then, both $\phi(c)$ and $\psi(c)$ are analytic differentials on R and $\tau(c) = \phi(c) + \overline{\psi(c)}$. We see that $\phi(c)$ has only two singularities, simple poles at p and q with residues -1 and 1 respectively, and $\psi(c)$ is everywhere holomorphic.

The above construction works even in the case $p = q$ and then the resulting differentials $\phi(c)$ and $\psi(c)$ are everywhere holomorphic. We can extend our definition of $\tau(c)$, $\phi(c)$ and $\psi(c)$ to arbitrary 1-chains c by linearity. $\phi(c)$ has no singularities if and only if c is a cycle. Suppose that this is the case. Then $\phi(c) \in \Gamma_a$ and $\tau(c) \in \Gamma_h$. Moreover, since Ω has compact support and since $\tau(c) - \Omega$ belongs to Γ_{e0}, $\tau(c)$ belongs to $\Gamma_h \cap \Gamma_{c0}$ (= Γ_{h0}, by definition). By setting

$$\sigma(c) = \frac{1}{2\pi i}\tau(c),$$

we get the following

<u>Theorem</u>. If c is a cycle on R, then there exists a unique real harmonic differential $\sigma(c)$ in Γ_{h0} such that

$$(\omega, \sigma(c)^*) = \int_c \omega$$

for each ω in Γ_c. If c and γ are any cycles on R, then

(6) $$(\sigma(c), \sigma(\gamma)^*) = \int_\gamma \sigma(c) = c \times \gamma. \qquad \text{([AS], Ch. V, 20)}$$

10. <u>Riemann-Roch Theorem</u>

10A. Here we consider a compact Riemann surface R of genus g. Fix a canonical basis of 1-cycles A_n, B_n, $n = 1, \ldots, g$, for which all intersection numbers vanish except the cases $A_n \times B_n = -B_n \times A_n = 1$, $n = 1, \ldots, g$ (see 2B). By our observation in 9B $\sigma(A_n)$, $\sigma(B_n)$ are harmonic differentials and we have for all m, n

$$\int_{B_m} \sigma(A_n) = A_n \times B_m = \delta_{mn}, \quad \int_{A_m} \sigma(B_n) = B_n \times A_m = -\delta_{mn},$$

and

$$\int_{A_m} \sigma(A_n) = \int_{B_m} \sigma(B_n) = 0.$$

Theorem. (a) $\sigma(A_n)$ and $\sigma(B_n)$, $n = 1,\ldots, g$, form a basis for Γ_h and therefore $\dim \Gamma_h = 2g$.

(b) $\dim \Gamma_a = g$, and every ϕ in Γ_a is determined by its A-periods, i.e. $(\int_{A_1} \phi, \ldots, \int_{A_g} \phi)$. ([AS], Ch. V, 24A-B)

10B. A divisor D on R is a formal finite sum

(7) $$D = n_1 a_1 + \cdots + n_k a_k,$$

where a_j are points in R and n_j are integers. Such a divisor D is called a principal divisor if it comes from a nonzero meromorphic function f, i.e. $D = (f) = \sum_{a\in R} \nu_a(f)a$, where $\nu_a(f)$, the order of f at the point a, is defined to be $\inf\{k: c_k \neq 0\}$ with $f(z) = \sum_k c_k(z-a)^k$.

Let $\mathcal{D}$ (resp. $\mathcal{D}_0$) be the set of all divisors (resp. principal divisors) on R. $\mathcal{D}$ (resp. $\mathcal{D}_0$) is a free Abelian group written additively. The quotient group $\mathcal{D}/\mathcal{D}_0$ is called the divisor class group on the surface R. The degree, $\deg D$, of a divisor D of the form (7) is defined by $\sum_j n_j$. Since every nonconstant meromorphic function on R makes R a complete covering surface of the Riemann sphere, we see that $\deg D = 0$ for every D in $\mathcal{D}_0$ and therefore that $\deg$ is a function on the quotient group $\mathcal{D}/\mathcal{D}_0$.

We also define the divisor of a meromorphic differential α by the formula $(\alpha) = \sum_{a\in R} \nu_a(\alpha)a$, where $\nu_a(\alpha)$ was defined in 9A. If α' is another meromorphic differential, then α'/α is a meromorphic function on R so that $(\alpha') - (\alpha) = (\alpha'/\alpha) \in \mathcal{D}_0$. Thus, (α)'s belong to one and the same divisor class called the canonical class.

10C. A divisor D of the form (7) is called integral if $n_j \geq 0$ for all j. D_1 is a multiple of D_2 if $D_1 - D_2$ is integral. For every divisor D we denote by $L(D)$ the space of all meromorphic functions f on R such that (f) is a multiple of D. The complex dimension of $L(D)$ is denoted by $\dim D$, which depends only on the divisor class of D.

For every divisor D in $\mathcal{D}$ we denote by $\Omega(D)$ the space of meromorphic differentials α such that (α) is a multiple of D. If we take a fixed nonzero meromorphic differential α_0 on R, then α belongs to $\Omega(D)$ if and only if (α/α_0) is a multiple of $D - Z$ with $Z = (\alpha_0)$. Hence $\dim \Omega(D) = \dim (D - Z)$.

Theorem (Riemann-Roch). For every D in $\mathcal{D}$ and every canonical divsor Z we have

(8) $$\dim D = \dim(-D - Z) - \deg D - g + 1.$$

([AS], Ch. V, 27A)

10D. By setting $D = -Z$ in (8) and using the fact that $\dim \Omega(0) = \dim \Gamma_a = g$ (see Th. 10A), we have

(9) $$\deg Z = 2g - 2$$

for each canonical divisor Z. This, together with (8), implies

Theorem. There exists a meromorphic function f which has a simple pole at any prescribed point in R. We can prescribe the location of other poles of f as well. ([AS], Ch. V, 28B)

10E. To proceed further, we need

Theorem. Let J be any infinite subset of R. Then there exist distinct points $a_1, \ldots, a_g$ in J such that $\dim \Omega(a_1 + \cdots + a_g) = 0$.

Proof. Take any $a_1, \ldots, a_g$ from J. Since $L(-\sum_{j=1}^{s} a_j)$ contains constant functions, we have $1 \leqq \dim(-\sum_{j=1}^{s} a_j) = \dim \Omega(\sum_{j=1}^{s} a_j) + s - g + 1$ by (8) and therefore

(10) $$\dim \Omega\left(\sum_{j=1}^{s} a_j\right) \geqq g - s.$$

We now prove the theorem by induction. If $g = 0$, then there is nothing to prove, for $\Omega(0) = \Gamma_a = \{0\}$. Suppose that $g > 0$. We note first that

$$\dim(a - Z) = \dim \Omega(a) \leqq g - 1$$

for each a in R. In fact, if this were not the case for some a_0, then $g \leqq \dim(a_0 - Z) = \dim \Omega(a_0) \leqq \dim \Gamma_a = g$. By (8) $\dim(-a_0) = \dim(a_0 - Z) - \deg(-a_0) - g + 1 = 2$. Thus there would be a nonconstant meromorphic function f on R with only one pole--a simple pole--at the point a_0. As noticed in 10B, this implies that f would map R conformally onto the Riemann sphere and thus g should be zero, a contradiction.

Thus, for $s = 1$, we have $\dim \Omega(a_1) = g - 1$. If $g > 1$, then we take a nonzero ϕ_1 in $\Omega(a_1)$. Since the zeros of ϕ_1 is finite in number and J is infinite, we can find a_2 in J with $\phi_1(a_2) \neq 0$. Then $\dim \Omega(a_1+a_2) \leqq \dim \Omega(a_1) - 1 = g - 2$. Combined with (10), this gives us $\dim \Omega(a_1+a_2) = g - 2$. We have only to repeat this argument to finish the proof. □

10F. We add one more result, which is clear from the definitions.

Theorem. Let $\phi_1, \ldots, \phi_g$ be a basis of Γ_a. Take distinct points $a_1, \ldots, a_g$ in R and choose one parametric disk $\{V_j, z\}$ for each point a_j, $j = 1, \ldots, g$. Then $\dim \Omega(a_1+\cdots+a_g) = 0$ if and only if

$$\det (f_j(z_k)) \neq 0,$$

where $\phi_j = f_j dz$ and $z_k = z(a_k)$.

11. Cauchy Kernels on Compact Bordered Surfaces

11A. Let R be a compact Riemann surface of genus g and let $\{A_j, B_j : j = 1, \ldots, g\}$ be a canonical basis of 1-cycles on R. We set $R_0 = R \setminus \{\cup_{j=1}^{g} (A_j \cup B_j)\}$. R_0 can be assumed to have a canonical form, in which the single boundary contour is of the form

$$A_1 B_1 A_1^{-1} B_1^{-1} \cdots A_g B_g A_g^{-1} B_g^{-1},$$

where A_j^{-1} (resp. B_j^{-1}) is the curve obtained from A_j (resp. B_j) by giving the opposite orientation ([AS], Ch. I, 40D).

Let p, q be distinct points in R not on any A_j and B_j. Take an arc c in R_0 joining p to q and define an analytic differential $\phi(c)$ on R as in 9B. So $\phi(c)$ has singularities only at p and q. Then we use Th. 10A,(b) to find a unique differential ϕ_1 in Γ_a which has the same A-periods as $\phi(c)$. The properties of $\phi(c)$ noticed in 9B imply that $\phi(c) - \phi_1$ has only two poles--simple poles at p and q with resides -1 and 1, respectively--and has zero A-periods. $\phi(c) - \phi_1$ is seen to be independent of the choice of c and is denoted by $\phi_{p,q}$.

Next we take distinct points p, q, p', q' in R_0 and set

$$\phi = \phi_{p,q}, \quad \omega = \phi_{p',q'} \quad \text{and} \quad \Phi = \int^z \phi.$$

Let γ be a simple arc joining p to q within R_0 and not passing p' and q'. Then Φ is single-valued on $R_0 \setminus \gamma$ and therefore $\Phi\omega$ is an analytic differential on $R_0 \setminus \gamma$ with possible singularities only at p' and q'. Let Σ_0 be the total sum of residues (see 9A) of $\Phi\omega$

over the region $R_0 \setminus \gamma$. Denoting by z^+ (resp. z^-) the left-hand (resp. right-hand) bank of γ at the point z on γ, we have

$$2\pi i \Sigma_0 = \int_{\partial R_0} \Phi\omega + \int_\gamma (\Phi(z^+) - \Phi(z^-))\omega.$$

Since $\Phi(z^+) - \Phi(z^-) = 2\pi i$, the second term of the right-hand side is equal to $2\pi i \int_\gamma \omega$. We also see that $\int_{\partial R_0} \Phi\omega = 0$ because all the A-periods of ϕ and ω vanish. On the other hand, if we take any arc γ' joining p' to q' within $R_0 \setminus \gamma$, then we have

$$\Sigma_0 = \mathrm{Res}_{p'}(\Phi\omega) + \mathrm{Res}_{q'}(\Phi\omega) = \Phi(q') - \Phi(p') = \int_{\gamma'} \phi_{p',q'}.$$

So we have proved the following

Theorem. Let γ and γ' be nonintersecting arcs on R_0 with $\partial\gamma = q - p$ and $\partial\gamma' = q' - p'$. Then

$$\int_{\gamma'} \phi_{p,q} = \int_\gamma \phi_{p',q'}.$$

11B. Let $\bar{R}$ be a compact bordered surface with nonvoid boundary ∂R. We denote by $R' = R \cup \partial R \cup \tilde{R}$ the double of R, where $\tilde{R}$ is the conjugate of R (see [AS], Ch. II, 3E). Let g' be the genus of R'.

In order to define the Cauchy kernel for R, we take a nonconstant meromorphic function h on R', which is held fixed. Then, by Th. 10E, there exist g' points $a_1,\ldots, a_{g'}$ in $\tilde{R}$ such that the differential dh has neither zeros nor poles at $a_1,\ldots, a_{g'}$ and such that

$$\dim \Omega_{R'}(a_1+\cdots+a_{g'}) = 0,$$

where $\Omega_{R'}(\cdot)$ denotes the space $\Omega(\cdot)$ for the surface R' (see 10C). Let $\{A_j, B_j: j = 1,\ldots, g'\}$ be a canonical basis of 1-cycles on R' such that $R'_0 = R' \setminus \{\cup_{j=1}^{g'}(A_j \cup B_j)\}$ is a canonical form in the sense of 11A. We may also assume that R'_0 contains all the zeros and poles of dh as well as all the points $a_1,\ldots, a_{g'}$ in its interior. By use of Th. 10A we choose a basis $\{\phi_n: n = 1,\ldots, g'\}$ of $\Gamma_a(R')$ such that $\int_{A_m} \phi_n = \delta_{mn}$ for $m, n = 1,\ldots, g'$. We fix one parametric disk $\{V_j, z\}$ around each of the points $a_1,\ldots, a_{g'}$ and express ϕ_n in terms of the local variables chosen as $\phi_n = h_n(z)dz$ on each V_j. If we write $z_k = z(a_k)$ for $k = 1,\ldots, g'$, then we have, by Th. 10F, $\det (h_j(z_k)) \neq 0$. Since $dh(a_k) \neq 0$ for $k = 1,\ldots, g'$, we see that

$$\phi_j(a_k)/dh(a_k) = h_j(z_k)/\frac{dh}{dz}(z_k)$$

are well-defined and that the matrix

$$\left(\frac{\phi_j(a_k)}{dh(a_k)}\right)_{j,k=1,\ldots,g'}$$

is nonsingular. So a basis $\{\omega_n : n = 1,\ldots, g'\}$ of $\Gamma_a(R')$ is determined by the formulas

$$\phi_j = \sum_{k=1}^{g'} \frac{\phi_j(a_k)}{dh(a_k)}\,\omega_k \tag{11}$$

for $j = 1,\ldots, g'$. We now fix a point $q'' \in \tilde{R} \cap R_0'$ which is distinct from $a_1,\ldots, a_{g'}$. If we take three distinct points p, p', q in $R_0' \setminus \{q''\}$ and draw disjoint arcs γ' and γ'' within the region R_0' such that $\partial\gamma' = p - p'$ and $\partial\gamma'' = q - q''$, then Th. 11A implies that $\int_{\gamma'} \phi_{q'',q} = \int_{\gamma''} \phi_{p',p}$. Writing this in the form

$$\int_{p'}^{p} \phi_{q'',q} = \int_{q''}^{q} \phi_{p',p}, \tag{12}$$

we see that the function $F(p,q)$, defined by either side of this equation, is holomorphic in q in $R_0' \setminus \gamma'$ and holomorphic in p in $R_0' \setminus \gamma''$. It follows that $F(p,q)$ is holomorphic jointly in p and q when $p \neq q, q''$ and $q \neq p'$. Differentiation of F with respect to p shows that $\phi_{q'',q}(p)$ is locally holomorphic in q when $q \neq p, q''$.

Denoting by z and ζ local variables for p and q, respectively, we set $\phi_{q'',q}(p) = f(z,q'',\zeta)dz$ and also

$$\int_{A_j} \partial_q \phi_{q'',q} = \left(\int_{A_j} \frac{\partial}{\partial\zeta} f(z,q'',\zeta)d\zeta\right)dz.$$

Then, by using (12), we have

$$\begin{aligned}
\int_{p'}^{p}\left(\int_{A_j} \partial_q\phi_{q'',q}\right) &= \int_{p'}^{p}\left(\int_{A_j} \frac{\partial}{\partial\zeta} f(z,q'',\zeta)d\zeta\right)dz \\
&= \int_{A_j} \frac{\partial}{\partial\zeta}\left(\int_{p'}^{p} f(z,q'',\zeta)dz\right)d\zeta = \int_{A_j} \frac{\partial}{\partial\zeta}\left(\int_{p'}^{p} \phi_{q'',q}\right)d\zeta \\
&= \int_{A_j} \frac{\partial}{\partial\zeta}\left(\int_{q''}^{\zeta} \phi_{p',p}\right)d\zeta = \int_{A_j} \phi_{p',p} = 0,
\end{aligned}$$

the last equality sign being valid because all the A-periods of $\phi_{p',p}$ vanish. As p is arbitrary, we conclude that, for every fixed p in the region $R \cap R_0'$,

(13) $$\int_{A_j} \partial_q \phi_{q'',q} = 0.$$

A similar computation shows that

$$\int_{p'}^{p} \left(\int_{B_j} \partial_q \phi_{q'',q}\right) = \int_{B_j} \phi_{p',p}.$$

As we shall see below in 11C, an application of Th. 11A easily shows that

(14) $$\int_{B_j} \phi_{p',p} = -2\pi i \int_{p'}^{p} \phi_j$$

for $j = 1, \ldots, g'$. Thus we have arrived at the following:

(15) $$\int_{B_j} \partial_q \phi_{q'',q} = -2\pi i \phi_j$$

for $j = 1, \ldots, g'$.

We finally set

$$\omega(p,q) = \phi_{q'',q}(p) - \sum_{j=1}^{g'} \frac{\phi_{q'',q}}{dh}(a_j)\omega_j(p).$$

Since q'' and all a_j belong to $\tilde{R}$, $\phi_{q'',q}(a_j)$ is holomorphic in q everywhere on $\overline{R}$, while $\phi_{q'',q}(p)$ is holomorphic jointly in p and q on $\overline{R}$ except at $p \neq q$. Moreover, $p \to \phi_{q'',q}(p)$ has a simple pole with residue $+1$ at the point q. To finish the construction of the Cauchy kernel, we claim that, for every fixed p, the function $q \to \omega(p,q)$ is single-valued. To see this, we fix p in R_0' and compute the periods of the differential $q \to \partial_q \omega(p,q)$ along A_j and B_j. First, from (13) follows easily that the period along any A_j vanishes. Secondly, by use of (11), (14) and (15) we get

$$\int_{B_j} \partial_q \omega(p,q) = \int_{B_j} \partial_q(\phi_{q'',q}(p)) - \sum_{k=1}^{g'} \frac{\int_{B_j} \partial_q \phi_{q'',q}}{dh}(a_k)\omega_k(p)$$

$$= -2\pi i\left(\phi_j(p) - \sum_{k=1}^{g'} \frac{\phi_j}{dh}(a_k)\omega_k(p)\right) = 0.$$

Hence, $\partial_q \omega(p,q)$ has only null periods and so $q \to \omega(p,q)$ is a single-valued function. Summing up these observations, we get the following, which is the main objective of this section.

Theorem. There exists a differential $\omega(p,q)$ on a compact bordered Riemann surface $\bar{R}$ such that

(B1) for every q in $\bar{R}$, $p \to \omega(p,q)$ is a meromorphic differential on $\bar{R}$ with only one pole--a simple pole--at q with residue $+1$;

(B2) for every p in $\bar{R}$, we choose a local variable z in a neighborhood of p and set $\omega(p,q) = f(z,q)dz$ with $z_0 = z(p)$; then $q \to f(z_0,q)$ is a meromorphic function on $\bar{R}$ having a simple pole at p with residue $+1$.

11C. We now verify the equation (14). Without loss of generality, we restrict ourselves to the case $j = 1$. We first move B_1 slightly within the region $R' \setminus \gamma'$ to obtain a simple closed curve, say B, satisfying the conditions: (i) B intersects with A_1 at a single point, b, distinct from any vertex of the polygon R_0', and (ii) $B \setminus \{b\}$ lies entirely in R_0'. Then the left-hand side of (14) is equal to $\int_B \phi_{p',p}$ and therefore we have only to show

$$\int_B \phi_{p',p} = -2\pi i \int_{p'}^{p} \phi_1 .$$

To proceed further, let V be a parametric disk centered at b such that its closure $Cl(V)$ does not meet γ' and also all the cycles A_j, B_j except A_1. We then take small arcs A' and B' of A_1 and B, respectively, containing the point b and contained in V. Let b' and b'' be the end points of B' so that $\partial B' = b' - b''$. Thus, setting $B'' = B \setminus B'$, we have $\partial B'' = b'' - b'$. From the fact $B = B' + B''$ follows that

$$\int_B \phi_{p',p} = \int_{B'} \phi_{p',p} + \int_{B''} \phi_{p',p}.$$

We now compute the two integrals on the right-hand member separately. First we note that the arc B'' lies in the polygon R_0' and is disjoint from γ'. So Th. 11A implies that

$$\int_{B''} \phi_{p',p} = \int_{b'}^{b''} \phi_{p',p} = \int_{p'}^{p} \phi_{b',b''}.$$

In order to deal with the integral over B', on the other hand, we use a modified polygon in place of R_0'. In fact, by means of a continuous deformation within V, the arc A' is changed to another arc A'' with the same end points such that A'' is disjoint from B' and $A' - A''$ is a closed curve of winding number $+1$ with respect to the point b''. Setting $A_1' = (A_1 \setminus A') + A''$, we construct our new polygon, R_1', out of

R_0' just by replacing A_1 by A_1'. Since V does not intersect with the arc γ', we have $\int_{A_1'} \phi_{p',p} = \int_{A_1} \phi_{p',p} = 0$. Let $\phi_{b'',b'}'$ be the analytic differential on R' having simple poles at b' and b'' with residue $+1$ and -1, respectively, and having null periods along all the cycles $A_1', A_2, \ldots, A_{g'}$. Thus, by Th. 11A applied to the polygon R_1', we have

$$\int_{B'} \phi_{p',p} = \int_{\gamma'} \phi_{b'',b'}'$$

and hence

$$\int_B \phi_{p',p} = \left(\int_{B'} + \int_{B''}\right) \phi_{p',p} = \int_{\gamma'} (\phi_{b',b''} + \phi_{b'',b'}').$$

Our construction of $\phi_{b'',b'}'$ shows that $\phi_{b',b''} + \phi_{b'',b'}'$ has no singularities and has null period along $A_2, \ldots A_{g'}$. As for the period along A_1, we have

$$\int_{A_1} (\phi_{b',b''} + \phi_{b'',b'}') = \int_{A_1} \phi_{b'',b'}' = \int_{A_1} \phi_{b'',b'}' - \int_{A_1'} \phi_{b'',b'}'$$

$$= \int_{A'-A''} \phi_{b'',b'}' = -2\pi i.$$

Since the differentials in $\Gamma_a(R')$ are determined by their A-periods, we conclude that $\phi_{b',b''} + \phi_{b'',b'}' = -2\pi i \phi_1$, as was to be proved. □

NOTES

Most results come from Chapters I, II, V of Ahlfors and Sario [AS] and Chapters 1, 3, 4 of Constantinescu and Cornea [CC], so that more information can be obtained from sources indicated respectively. The results in 10E, 10F and in Subsection 11 (Cauchy kernels) are taken from Kusunoki [42] (see also Hurwitz and Courant [35] for this matter).

CHAPTER II. MULTIPLICATIVE ANALYTIC FUNCTIONS

In our later discussion of Hardy classes an important role will be played by a certain family of multiple-valued analytic functions called multiplicative analytic functions. These are defined as multiple-valued analytic functions having single-valued modulus. The first objective of this chapter is to give a precise definition to such functions. In fact, we will define them in terms of two equivalent notions: line bundles and characters of the fundamental group. We shall then show that, on a compact bordered Riemann surface, every line bundle admits nonvanishing bounded holomorphic sections. The second purpose is to observe the order structure in the space of harmonic functions, leading to the so-called inner-outer factorization of multiplicative analytic functions.

§1. MULTIPLICATIVE ANALYTIC FUNCTIONS

1. The First Cohomology Group

1A. Let R be a Riemann surface. For every open covering $\mathcal{V} = \{V_i : i \in I\}$ of R, we denote by $Z^1(\mathcal{V};\mathbb{T})$ the totality of <u>1-cocycles</u> $\{\xi_{ij}\}$ over $\mathcal{V}$ with values in the group $\mathbb{T}$ of complex numbers of modulus one, i.e. ξ_{ij} in $\mathbb{T}$ is assigned to every pair i, j in I with $V_i \cap V_j \neq \emptyset$ in such a way that

$$\xi_{ij}\xi_{jk} = \xi_{ik} \tag{1}$$

if $V_i \cap V_j \cap V_k \neq \emptyset$. Two 1-cocycles $\{\xi_{ij}\}$ and $\{\xi'_{ij}\}$ in $Z^1(\mathcal{V};\mathbb{T})$ are said to be <u>equivalent</u> if there exists an element θ_i in $\mathbb{T}$ for each i in I such that $\xi'_{ij} = \theta_i^{-1}\xi_{ij}\theta_j$ if $V_i \cap V_j \neq \emptyset$. The set $H^1(\mathcal{V};\mathbb{T})$ is then defined to be the set of equivalence classes in $Z^1(\mathcal{V};\mathbb{T})$, which forms a group with respect to the pointwise multiplication, i.e. $\{\xi_{ij}\}\{\eta_{ij}\} = \{\xi_{ij}\eta_{ij}\}$.

Next, take any two open coverings $\mathcal{V} = \{V_i : i \in I\}$ and $\mathcal{V}' = \{V'_{i'} : i' \in I'\}$ of R such that $\mathcal{V}'$ is a refinement of $\mathcal{V}$ with refining map μ. We may regard the refining map μ as a map from I'

into I so that $\mu(V'_{i'}) = V_{\mu(i')}$. Let $\{\xi_{ij}\}$ be any element of $Z^1(\mathcal{V};\mathbb{T})$. If $V'_{i'} \cap V'_{j'} \neq \emptyset$ for some i', j' in I', then we have $V_{\mu(i')} \cap V_{\mu(j')} \supseteq V'_{i'} \cap V'_{j'} \neq \emptyset$ and therefore $\xi_{\mu(i'),\mu(j')}$ is defined. We set $\xi'_{i'j'} = \xi_{\mu(i'),\mu(j')}$. Then $\{\xi'_{i'j'}\}$ forms a 1-cocycle over the covering $\mathcal{V}'$. Since the correspondence $\{\xi_{ij}\} \to \{\xi'_{i'j'}\}$ is seen to preserve the equivalence relation, we get a map $\iota(\mathcal{V}',\mathcal{V})$: $H^1(\mathcal{V};\mathbb{T}) \to H^1(\mathcal{V}';\mathbb{T})$. We also see that the map $\iota(\mathcal{V}',\mathcal{V})$ does not depend on the choice of refining maps and is a group homomorphism. Moreover, if $\mathcal{V}''$ is a refinement of $\mathcal{V}'$, then we have $\iota(\mathcal{V}'',\mathcal{V}) = \iota(\mathcal{V}'',\mathcal{V}') \circ \iota(\mathcal{V}',\mathcal{V})$. This means that $\{H^1(\mathcal{V};\mathbb{T});\ \iota(\mathcal{V}',\mathcal{V})\}$ forms a direct system of groups when $\mathcal{V}$ varies over all open coverings of R. The direct limit of this system is denoted by $H^1(R;\mathbb{T})$ and is called the <u>first cohomology group</u> of R with coefficients in $\mathbb{T}$.

1B. <u>Theorem</u>. The group $H^1(R;\mathbb{T})$ is canonically isomorphic with the character group $F_0(R)^*$ of the fundamental group $F_0(R)$ of R.

<u>Proof</u>. The proof is divided into several steps. We begin with some definitions. An open covering $\mathcal{V}$ of R is called <u>faithful</u> if (i) $\mathcal{V}$ consists of simply connected open sets and (ii) for each triple V_1, V_2, V_3 in $\mathcal{V}$ the union $V_1 \cup V_2 \cup V_3$ is included in a simply connected subset of R whenever it is connected. Since R is separable, it is not hard to see that R admits a faithful covering and that every open covering has a refinement which is faithful. So we restrict ourselves to faithful coverings in the proof. By a <u>chain</u> of a given covering, say $\mathcal{V}$, we mean a finite sequence $\gamma = \{V_{j(1)},\ldots,V_{i(m)}\}$ of elements in $\mathcal{V}$ such that $V_{i(\nu)} \cap V_{i(\nu+1)} \neq \emptyset$ for every $\nu = 1,\ldots, m-1$. A chain of $\mathcal{V}$ is called a subdivision of a chain $\gamma = \{V_{i(1)},\ldots,V_{i(m)}\}$ if it is obtained from γ by repeating some of the elements without changing the order; e.g. $\{V_1,V_1,V_2,V_3,V_3,V_3\}$ is a subdivision of $\{V_1,V_2,V_3\}$. On the other hand, by a <u>partition</u> of an arc c on R we mean a partition of c into a finite number of consecutive arcs, say $\sigma_1,\ldots, \sigma_m$, so that $c = \sigma_1\sigma_2\cdots\sigma_m$. A subpartition of a partition of c has an obvious meaning. A chain $\gamma = \{V_{i(1)},\ldots,V_{i(m)}\}$ is said to <u>cover</u> an arc c if there exists a partition $c = \sigma_1\cdots\sigma_m$ of c with $\sigma_\nu \subseteq V_{i(\nu)}$ for every ν. In this case, we say that $c = \sigma_1\cdots\sigma_m$ is <u>adapted</u> to γ. When $c = \sigma_1\cdots\sigma_m$ is adapted to γ, each subpartition of $\sigma_1\cdots\sigma_m$ is adapted to some subdivision of γ.

(a) First we define a homomorphism of $H^1(R;\mathbb{T})$ into $F_0(R)^*$. Let $\xi \in H^1(R;\mathbb{T})$ and $\bar{c} \in F_0(R)$. Choose any representatives $\{\xi_{ij}\}$ and c of ξ and $\bar{c}$, respectively. Namely, $\{\xi_{ij}\}$ is a 1-cocycle over a covering $\mathcal{V} = \{V_i : i \in I\}$ and c is a closed curve issuing from the origin O. Let $\gamma = \{V_{i(1)},\ldots,V_{i(m)}\}$ be a chain of $\mathcal{V}$, which covers c, and set

$$(2) \qquad F(c;\gamma;\{\xi_{ij}\}) = \prod_{\nu=1}^{m} \xi_{i(\nu)i(\nu+1)}$$

with $i(m+1) = i(1)$. We have to show that the product depends only on ξ and $\bar{c}$.

(b) $F(c;\gamma;\{\xi_{ij}\})$ does not depend on the choice of γ, fixing the representative $\{\xi_{ij}\}$. To see this, let γ_1 be another chain of $\mathcal{V}$ covering c. If γ_1 is a subdivision of γ, then, using the property $\xi_{ii} = 1$, we see that $F(c;\gamma_1;\{\xi_{ij}\}) = F(c;\gamma;\{\xi_{ij}\})$. Next, if $\gamma_1 = \{V_{j(1)},\ldots,V_{j(m)}\}$ and if a partition $c = \sigma_1\cdots\sigma_m$ of c is adapted to both γ and γ_1, then $\sigma_\nu \subseteq V_{i(\nu)} \cap V_{j(\nu)}$ for $\nu = 1,\ldots, m$ and therefore

$$V_{i(\nu)} \cap V_{i(\nu+1)} \cap V_{j(\nu)} \cap V_{j(\nu+1)} \neq \emptyset$$

for $\nu = 1,\ldots, m$ with $i(m+1) = i(1)$ and $j(m+1) = j(1)$. Thus, in view of (1), we get

$$\begin{aligned} F(c;\gamma_1;\{\xi_{ij}\}) &= \prod_{\nu=1}^{m} \xi_{j(\nu)j(\nu+1)} \\ &= \prod_{\nu=1}^{m} \left(\xi_{j(\nu)i(\nu)}\xi_{i(\nu)i(\nu+1)}\xi_{i(\nu+1)j(\nu+1)}\right) \\ &= \prod_{\nu=1}^{m} \xi_{i(\nu)i(\nu+1)} = F(c;\gamma;\{\xi_{ij}\}). \end{aligned}$$

Finally, let γ_1 be any chain of $\mathcal{V}$ covering c. Then we can find subdivisions γ' and γ_1' of γ and γ_1, respectively, to which a partition $c = \sigma_1\cdots\sigma_\ell$ of c is adapted simultaneously. Thus we get

$$F(c;\gamma_1;\{\xi_{ij}\}) = F(c;\gamma_1';\{\xi_{ij}\}) = F(c;\gamma';\{\xi_{ij}\}) = F(c;\gamma;\{\xi_{ij}\}),$$

as was to be proved. So we may write $F(c;\{\xi_{ij}\})$ in place of $F(c;\gamma;\{\xi_{ij}\})$.

(c) Let $\{\xi'_{i'j'}\}$ be another representative of ξ associated with a covering $\mathcal{V}' = \{V'_{i'} : i' \in I'\}$. Since $\{\xi_{ij}\}$ and $\{\xi'_{i'j'}\}$ are equivalent, there exist a covering $\mathcal{V}'' = \{V''_{i''} : i'' \in I''\}$, which is a common refinement of $\mathcal{V}$ and $\mathcal{V}'$ with refining maps μ and μ', respectively, and a system $\{\theta_{i''} : i'' \in I''\}$ with values in $\mathbb{T}$ such that

$$\xi'_{\mu'(i'')\mu'(j'')} = \theta_{i''}\xi_{\mu(i'')\mu(j'')}\theta_{j''}^{-1}$$

for each pair i'', j'' in I'' with $V''_{i''} \cap V''_{j''} \neq \emptyset$. Now let $\gamma'' = \{V''_{i''(1)},\ldots,V''_{i''(m)}\}$ be a chain of $\mathcal{V}''$ covering the curve c. Write, for simplicity, $i(\nu) = \mu(i''(\nu))$, $i'(\nu) = \mu'(i''(\nu))$ for $\nu = 1,\ldots, m$. Then, $\mu(\gamma'') = \{V_{i(1)},\ldots,V_{i(m)}\}$ (resp. $\mu'(\gamma'') = \{V_{i'(1)},\ldots,V_{i'(m)}\}$) is a chain of $\mathcal{V}$ (resp. $\mathcal{V}'$) covering the curve. Then, using the convention $i(m+1) = i(1)$ etc. as before, we have

$$
\begin{aligned}
F(c;\{\xi'_{i'j'}\}) = F(c;\mu'(\gamma'');\{\xi'_{i'j'}\}) &= \prod_{\nu=1}^{m} \xi'_{i'(\nu)i'(\nu+1)} \\
&= \prod_{\nu=1}^{m} \left(\theta_{i''(\nu)}\xi_{i(\nu)i(\nu+1)}\theta_{i''(\nu+1)}^{-1}\right) \\
&= \prod_{\nu=1}^{m} \xi_{i(\nu)i(\nu+1)} \\
&= F(c;\mu(\gamma'');\{\xi_{ij}\}) = F(c;\{\xi_{ij}\}).
\end{aligned}
$$

This means that $F(c;\{\xi_{ij}\})$ depends only on the cohomology class of $\{\xi_{ij}\}$. So we may write $F(c;\xi)$ instead of $F(c;\{\xi_{ij}\})$.

(d) Let c_1 and c_2 be two representatives of $\bar{c}$ in $F_0(R)$. If the deformation between c_1 and c_2 is very small, then they are covered by the same chain, say γ, of $\mathcal{V}$ so that

$$F(c_1;\xi) = F(c_1;\gamma;\{\xi_{ij}\}) = F(c_2;\gamma;\{\xi_{ij}\}) = F(c_2;\xi).$$

In the general case, we can find a finite sequence of closed paths c'_1, ..., c'_k issuing from O such that $c'_1 = c_1$, $c'_k = c_2$ and such that, for every $\nu = 1,\ldots, k-1$, c'_ν and $c'_{\nu+1}$ are covered by the same chain. It follows that $F(c_1;\xi) = F(c'_1;\xi) = F(c'_2;\xi) = \cdots = F(c'_k;\xi) = F(c_2;\xi)$. Hence, $F(c;\xi)$ depends only on the homotopy class $\bar{c}$ of c. Namely, the expression in (2) is determined by ξ and $\bar{c}$. This quantity is denoted by $F(\bar{c};\xi)$.

(e) As is easily seen, $F(\bar{c}_1\bar{c}_2;\xi) = F(\bar{c}_1;\xi)F(\bar{c}_2;\xi)$ for any $\bar{c}_1$ and $\bar{c}_2$ in $F_0(R)$. Thus $F(\cdot;\xi)$ is a character of $F_0(R)$ for every fixed ξ in $H^1(R;\mathbb{T})$. This character is denoted by F_ξ.

(f) Given any ξ, η from $H^1(R;\mathbb{T})$ we find representatives $\{\xi_{ij}\}$, $\{\eta_{ij}\}$ associated with the same covering $\mathcal{V} = \{V_i\colon i \in I\}$. So, using the same notation as in (a), we have

$$
\begin{aligned}
F(\bar{c};\xi\eta) &= F(c;\gamma;\{\xi_{ij}\eta_{ij}\}) \\
&= \prod_{\nu=1}^{m} \xi_{i(\nu)i(\nu+1)}\eta_{i(\nu)i(\nu+1)} \\
&= F(\bar{c};\xi)F(\bar{c};\eta).
\end{aligned}
$$

It follows that the map $\xi \to F_\xi$ is a homomorphism of $H^1(R;\mathbb{T})$ into $F_0(R)^*$.

(g) We now define a homomorphism of $F_0(R)^*$ into $H^1(R;\mathbb{T})$, which will turn out to be the inverse of the map $\xi \to F_\xi$.

Let F be any character of $F_0(R)$. Given any covering $V = \{V_i: i \in I\}$ of R, we choose, for every $i \in I$ once for all, a point z_i from V_i and a curve c_i joining the origin 0 to the point z_i. If $V_i \cap V_j \neq \emptyset$, then we take a curve, c_{ij}, joining z_i to z_j within the union $V_i \cup V_j$. The curve c_{ij} is unique up to homotopy equivalence, for $V_i \cup V_j$ is included in a simply connected subset of R. We set

$$\xi_{ij} = F(c_i c_{ij} c_j^{-1})$$

for every pair i, j in I with $V_i \cap V_j \neq \emptyset$. Then $\{\xi_{ij}\}$ is a 1-cocycle over V. In fact, if $V_i \cap V_j \cap V_k \neq \emptyset$, then $V_i \cup V_j \cup V_k$ is included in a simply connected subset of R and so c_{ik} is homotopic to $c_{ij}c_{jk}$. It follows that $c_i c_{ik} c_k^{-1}$ is homotopic to the composite $(c_i c_{ij} c_j^{-1})\cdot(c_j c_{jk} c_k^{-1})$ and therefore that

$$\xi_{ik} = F(c_i c_{ik} c_k^{-1}) = F(c_i c_{ij} c_j^{-1}) F(c_j c_{jk} c_k^{-1}) = \xi_{ij}\xi_{jk},$$

as was to be proved. It is a routine matter to verify that distinct choices of V, z_i or c_i only result in equivalent cocycles and consequently that the above definition gives a unique element, say ξ_F, in $H^1(R;\mathbb{T})$. Since the multiplication in $F_0(R)^*$ is defined to be pointwise, we see that the map $F \to \xi_F$ is a homomorphism of $F_0(R)^*$ into $H^1(R;\mathbb{T})$.

(h) Take any F in $F_0(R)^*$ and set $\xi = \xi_F$. To show that $F = F_\xi$, we choose, as in (g), $V = \{V_i: i \in I\}$, z_i, c_i and c_{ij}, and then set $\xi_{ij} = F(c_i c_{ij} c_j^{-1})$. Let c be any closed curve issuing from 0 and let $\gamma = \{V_{i(1)},\ldots,V_{i(m)}\}$ be a chain of V covering the curve c, so that there exists a partition $c = \sigma_1\cdots\sigma_m$ of c with $\sigma_\nu \subseteq V_{i(\nu)}$ for $\nu = 1,\ldots, m$. Denote the initial point of σ_ν by z_0^ν for $\nu = 1,\ldots, m$, and set $z_0^{m+1} = z_0^1$. Then, join z_0^ν to $z_{i(\nu)}$ by a curve c'_ν within $V_{i(\nu)}$. Since $V_{i(\nu)} \cup V_{i(\nu+1)}$ is included in a simply connected subset of R, we see that σ_ν is homotopic to $c'_\nu c_{i(\nu)i(\nu+1)} c'^{-1}_{\nu+1}$. It follows that the curve c is homotopic to

$$c'_1 c_{i(1)}^{-1}(c_{i(1)}c_{i(1)i(2)}c_{i(2)}^{-1})\cdots(c_{i(m)}c_{i(m)i(1)}c_{i(1)}^{-1})c_{i(1)}c'^{-1}_1.$$

Thus we have

$$F_\xi(c) = \xi_{i(1)i(2)}\cdots\xi_{i(m)i(1)} = \prod_{\nu=1}^{m} F(c_{i(\nu)}c_{i(\nu)i(\nu+1)}c_{i(\nu+1)}^{-1})$$

$$= F(c_{i(1)}c_1'^{-1}cc_1'c_{i(1)}^{-1}) = F(c_{i(1)}c_1'^{-1})F(c)F(c_1'c_{i(1)}^{-1})$$

$$= F(c),$$

as was to be proved.

(i) Finally, take any ξ in $H^1(R;\mathbb{T})$ and set $F = F_\xi$. We show that $\xi = \xi_F$. Let $\{\xi_{ij}\}$ be a representative 1-cocycle of ξ associated with a covering $\mathcal{V} = \{V_i: i \in I\}$. Again take z_i, c_i and c_{ij} as in (g). For every j in I, let us choose a fixed chain $\gamma_j = \{V_{i(1,j)},\ldots,V_{i(m_j,j)}\}$ which covers c_j. Here we assume that $i(1,j)$ is the same for all j, say $i(0)$. Take any pair i, j in I with $V_i \cap V_j \neq \emptyset$. Then

$$(\xi_F)_{jk} = F(c_jc_{jk}c_k^{-1})$$

$$= \xi_{i(1,j)i(2,j)}\cdots\xi_{i(m_j,j),j}\xi_{jk}\xi_{k,i(m_k,k)}\cdots\xi_{i(2,k)i(1,k)}.$$

Writing

$$\theta_j = \xi_{i(1,j)2(j,j)}\cdots\xi_{i(m_j,j),j},$$

we have $(\xi_F)_{jk} = \theta_j\xi_{jk}\theta_k^{-1}$ for j, k in I. Namely, $\{(\xi_F)_{jk}\}$ is equivalent to $\{\xi_{ij}\}$. Hence, $\xi_F = \xi$. This completes the proof. □

2. Line Bundles and Multiplicative Analytic Functions

2A. We now consider line bundles over R. We do not define them as topological spaces with certain structure but just say that every element in $H^1(R;\mathbb{T})$ determines a line bundle (precisely, a unitary flat complex line bundle) over R. Using representatives, we say:

(A1) Every 1-cocycle $(\{\xi_{ij}\}, \{V_i\})_{i,j\in I}$ in $Z^1(\{V_i\};\mathbb{T})$ defines a line bundle ξ;

(A2) Two 1-cocycles $(\{\xi_{ij}\}, \{V_i\})_{i,j\in I}$ and $(\{\xi'_{ij}\}, \{V'_i\})_{i,j\in I'}$ define the same line bundle, if there exists a common refinement $\{V''_\alpha\}_{\alpha\in I''}$ of $\{V_i\}$ and $\{V'_i\}$ with refining maps $\mu: I'' \to I$ and $\mu': I'' \to I'$, respectively, such that $\{\xi_{\mu(\alpha)\mu(\beta)}\}$ and $\{\xi'_{\mu'(\alpha)\mu'(\beta)}\}$ are equivalent 1-cocycles over $\{V''_\alpha\}_{\alpha\in I''}$; namely, there exists a system $\{\theta_\alpha\}_{\alpha\in I''}$ in $\mathbb{T}$ satisfying

$$\xi'_{\mu'(\alpha)\mu'(\beta)} = \theta_\alpha^{-1}\xi_{\mu(\alpha)\mu(\beta)}\theta_\beta \tag{3}$$

for every pair α, β in I'' with $V''_\alpha \cap V''_\beta \neq \emptyset$.

For a line bundle ξ over R, we define a section f of ξ over

R as follows: when a representative $(\{\xi_{ij}\}, \{V_i\})_{i,j\in I}$ is chosen for ξ, the corresponding representative of f is a collection $(\{f_i\}, \{V_i\})_{i\in I}$ of complex-valued functions f_i, f_i being defined on V_i, such that

(4) $$f_i(z) = \xi_{ij} f_j(z)$$

on $V_i \cap V_j$ whenever $V_i \cap V_j \neq \emptyset$. If $(\{\xi'_{ij}\}, \{V'_i\})_{i,j\in I'}$ is another representative of ξ and if $\{\xi_{ij}\}$ and $\{\xi'_{ij}\}$ are connected by the relation (3), then $(\{f'_i\}, \{V'_i\})_{i\in I'}$ with f'_i being a function on V'_i is defined to be the representative of f with respect to the system $(\{\xi'_{ij}\}, \{V'_i\})_{i,j\in I'}$ when we have for every $\alpha \in I''$

(5) $$f'_{\mu'(\alpha)} = \theta_\alpha^{-1} f_{\mu(\alpha)}$$

on V''. The relations (4) and (5) imply that, for every $z \in R$, the modulus $|f_i(z)|$ is the same for any representative $\{f_i\}$ of f and any i with $z \in V_i$. The common value is called the modulus of f and is denoted by $|f(z)|$.

A section f of ξ is called holomorphic (resp. meromorphic) if every f_i is holomorphic (resp. meromorphic) on V_i for a representative $(\{f_i\}, \{V_i\})_{i\in I}$. The definition is clearly independent of the choice of representatives. We denote by $\mathcal{H}(R,\xi)$ (resp. $M(R,\xi)$) the set of holomorphic (resp. meromorphic) sections of ξ over R.

2B. We next discuss analytic continuation of holomorphic or meromorphic sections of the given line bundle ξ. For this purpose, we take once for all a point $\mathcal{O}$, called the origin, and have the following

Theorem. Let ξ be a line bundle over R and $F = F_\xi$ the character of $\mathcal{F}_0(R)$ corresponding to ξ under the canonical isomorphism of $H^1(R;\mathbb{T})$ onto $\mathcal{F}_0(R)^*$. Let $f = (\{f_i\}, \{V_i\})$ be any meromorphic section of ξ and let $f_{i(0)}$ be any branch of f with $\mathcal{O} \in V_{i(0)}$. Then the analytic continuation of $f_{i(0)}$ is possible along each closed curve c issuing from $\mathcal{O}$ and the resulting function element at $\mathcal{O}$ is equal to $F_\xi(c) f_{i(0)}$.

Proof. Let $\gamma = \{V_{i(1)},\ldots,V_{i(m)}\}$ be a chain of $\{V_i\}$ covering the curve c with $i(1) = i(0)$. The formula (4) shows that $\xi_{ij} f_j$ is the direct analytic continuation of f_i to V_j. It follows that the function element to be obtained by continuing $f_{i(0)}$ analytically along c is exactly $\xi_{i(1)i(2)} \cdots \xi_{i(m-1)i(m)} \xi_{i(m)i(0)} f_{i(0)}$, which is equal to $F_\xi(c) f_{i(0)}$ in view of the definition (1). □

As far as analytic functions are concerned, a line bundle ξ in $H^1(R;\mathbb{T})$ and the corresponding character F_ξ in $\mathcal{F}_0(R)^*$ can be identified, so that ξ will be used to denote a character of $\mathcal{F}_0(R)$.

2C. We call a nonnegative extended real-valued function u on R a locally meromorphic modulus (abbreviated to "l.m.m.") if, for every point a in R, there exists a neighborhood V of a and a meromorphic function f on V such that $u = |f|$ on V. If, in this definition, f can always be chosen to be holomorphic, u is called a locally analytic modulus (abbreviated to "l.a.m."). If u is an l.m.m., then $\log u$ is harmonic wherever $0 < u < \infty$.

Theorem. Let u be a nonnegative extended real-valued function on R not identically vanishing. Then u is an l.m.m. (resp. l.a.m.) if and only if there exist a line bundle ξ and a meromorphic (resp. holomorphic) section f of ξ such that $u(z) = |f(z)|$ on R. The line bundle ξ is determined uniquely by u, which we call the bundle (or the character) of u.

The proof is omitted, for it is rather straightforward.

2D. A (multiple-valued) meromorphic function f on R is called multiplicative if its modulus $|f|$ is single-valued. Here, f is always assumed to be its maximal analytic extension. For any multiplicative meromorphic function f, its modulus $|f|$ is an l.m.m. on R and therefore, by Th. 2C, determines a unique line bundle over R, which is called the line bundle of f and is denoted by ξ_f.

Let us fix a point 0 in R, the origin of R. To every multiplicative meromorphic function f we associate, once and for all, a single-valued branch, say f_0, at the point 0, which we call the principal branch of f. Two multiplicative meromorphic functions are defined to be the same if they have the same principal branch at 0. As we saw in Th. 2B, f_0 can be continued analytically along any closed curve c issuing from 0 and the resulting function element at 0 is equal to $\xi_f(c)f_0$, regarding ξ_f as a character of $\mathcal{F}_0(R)$. In this sense, ξ_f is also called the character of f. Let us denote by $\mathcal{M}_0(R,\xi)$ the set of all multiplicative meromorphic functions of character ξ. $\mathcal{M}_0(R,\xi)$ forms a linear space if the sum of two functions f, g in $\mathcal{M}_0(R,\xi)$ is defined as the multiplicative meromorphic function whose principal branch is $f_0 + g_0$.

2E. So far, our sections of a line bundle ξ take only numerical values. Besides these, we can define a section ω of ξ whose values are first order differentials. The definition is entirely similar to the numerical ones: i.e. we have only to replace f_i by a differential ω_i on V_i in the definition in 2A. If every ω_i is a holomorphic (resp. meromorphic) differential, then ω is called a holomorphic (resp. meromorphic) differential section of ξ. The set of all holomorphic (resp. meromorphic) differential sections of ξ over R is denoted by $\Gamma\mathcal{H}(R,\xi)$ (resp. $\Gamma\mathcal{M}(R,\xi)$).

3. Existence of Holomorphic Sections

3A. Let $\overline{R}$ be a compact bordered Riemann surface with genus g and ℓ boundary components γ_k, $k = 1,\ldots,\ell$. Let A_j, B_j with $j = 1,\ldots,g$ and C_k with $k = 1,\ldots,\ell-1$ be a canonical basis of 1-cycles in R, where $\{A_j, B_j : j = 1,\ldots,g\}$ is a canonical sequence in the sense of Ch. I, 2B and C_k is homologous to γ_k for $k = 1,\ldots,\ell-1$. Here we may assume that, among these cycles, every A_j (resp. B_j) intersects only with B_j (resp. A_j) and every C_k is disjoint from all others. Let $\omega(p,q)$ be the Cauchy differential on $\overline{R}$ defined by Th. 11B, Ch. I. For any arc γ on R we set

$$\Omega_\gamma(q) = \int_\gamma \omega(p,q).$$

We shall look into the properties of Ω_γ for $\gamma = A_j$, B_j and C_k.

(i) Ω_{A_j} is single-valued and holomorphic on $R \setminus A_j$, so that the differential $d\Omega_{A_j}$ has no periods along any cycles disjoint from A_j. Moreover, we have

$$\int_{B_j} d\Omega_{A_j} = 2\pi i$$

for $j = 1,\ldots,g$. In fact, let $\{V, z\}$ be a parametric disk centered at the intersection p_j of A_j and B_j and take small arcs A'_j and B'_j of A_j and B_j, respectively, containing p_j and contained in V. Then we have

$$\Omega_{A_j}(q) = \int_{A'_j} \omega(p,q) + \int_{A_j \setminus A'_j} \omega(p,q).$$

The second term of the right-hand side is single-valued and holomorphic on $\overline{R}$ off $A_j \setminus A'_j$, so that

$$\int_{B_j} d\Omega_{A_j} = \int_{B_j} \partial_q\Big(\int_{A_j'} \omega(p,q)\Big) = \Big(\int_{B_j'} + \int_{B_j\setminus B_j'}\Big)\partial_q\Big(\int_{A_j'} \omega(p,q)\Big)$$

$$= \int_{B_j'-B_j''} \partial_q\Big(\int_{A_j'} \omega(p,q)\Big),$$

where B_j'' is a continuous deformation of B_j', within V, such that B_j'' has the same end points as B_j' and is disjoint from A_j' and such that $B_j' - B_j''$ is a closed curve of winding number $+1$ with respect to the terminal point of A_j'. In terms of local variables z and ζ representing p and q, respectively, in the parametric disk V,

$$\omega(z,\zeta) = [(z-\zeta)^{-1} + R(z,\zeta)]dz,$$

where $R(z,\zeta)$ is holomorphic in two variables. Denoting by z_1 and z_2 the initial and the terminal points of A_j', we thus have

$$\int_{A_j'} \omega(p,\zeta) = \log[(z_2-\zeta)/(z_1-\zeta)] + R_1(\zeta),$$

with some holomorphic function $R_1(\zeta)$. So we get

$$\int_{B_j} d\Omega_{A_j} = \int_{B_j'-B_j''} \{(\frac{1}{\zeta-z_2} - \frac{1}{\zeta-z_1})d\zeta + dR_1\} = 2\pi i,$$

as desired.

(ii) Similarly, the only nonzero period of $d\Omega_{B_j}$ is seen to be the one along A_j:

$$\int_{A_j} d\Omega_{B_j} = -2\pi i.$$

(iii) Finally, let γ_k' be an arc from a point on γ_ℓ to a point on γ_k, $k = 1,\ldots,\ell-1$, intersecting with C_k at a single point and not with other C_j and γ_j, $j \neq k$ and $j \leqq \ell-1$. We set $\Omega_{C_k}'(q) = \Omega_{\gamma_k'}(q)$. Supposing that γ_k is positively oriented with respect to R, we see that the only nonzero period of $d\Omega_{C_k}'$ occurs along C_k:

$$\int_{C_k} d\Omega_{C_k}' = -2\pi i$$

for $k = 1,\ldots,\ell-1$.

These observations prove the following

<u>Theorem</u>. There exists a function holomorphic on $\overline{R}$ and having any prescribed periods along cycles in the canonical basis.

3B. We are now able to construct multiplicative holomorphic functions with any prescribed line bundle or character.

Theorem. Every line bundle on a compact bordered Riemann surface $\overline{R}$ has nonvanishing (or invertible) bounded holomorphic sections, i.e. holomorphic sections f such that f^{-1} are also bounded.

Proof. Let A_j, $j = 1,\ldots, 2g+\ell-1$, be a canonical basis of 1-cycles in R, where we assume as above that R has genus g and ℓ boundary contours. Let ξ be a line bundle on R. By Th. 1B, ξ can be identified with a character of the fundamental group $F_0(R)$. Choose, once for all for every $j = 1,\ldots, 2g+\ell-1$, an arc A_j' joining the origin 0 to a point on A_j and set $c_j = A_j'A_jA_j'^{-1}$. In an obvious sense, c_j is considered as a closed curve issuing from 0. We then define a real number ξ_j, for every j, by $\xi(c_j) = \exp(2\pi i\xi_j)$. By the preceding theorem there exists a holomorphic function h on $\overline{R}$ such that the period of h along A_j is equal to ξ_j for every $j = 1,\ldots, 2g+\ell-1$. If we set $f = \exp(2\pi ih)$, then it is an invertible bounded holomorphic section of the line bundle ξ, as was to be proved. □

§2. LATTICE STRUCTURE OF HARMONIC FUNCTIONS

4. Basic Structure

4A. Let $C(R,\mathbb{R})$ be the linear space of all real-valued continuous functions on a Riemann surface R. We give $C(R,\mathbb{R})$ the topology τ_c of uniform convergence on compact subsets of R. Since R has a countable dense subset, the space $C(R,\mathbb{R})$ is seen to be metrizable. Let $HP = HP(R)$ denote the set of all nonnegative harmonic functions on R. HP is a cone in the sense that (i) $\alpha u \in HP$ for every u in HP and every nonnegative real number α and (ii) $u+v \in HP$ for any u and v in HP.

Theorem. The cone HP is locally compact with respect to the topology τ_c.

Proof. By use of the Harnack inequality it is seen that, for every compact subset E of R and every point a in R, there exists a constant $C > 0$ depending only on E and a and satisfying

$$\sup\{u(z): z \in E\} \leq Cu(a)$$

for any u in HP. So, for every fixed a in R and every fixed real numbers $0 \leqq \alpha \leqq \beta$, the set of functions $\{u \in HP: \alpha \leqq u(a) \leqq \beta\}$ is equicontinuous on every compact subset of R and therefore is a compact subset of HP with respect to the topology τ_c. From this the theorem follows at once. □

<u>Corollary</u>. For every a in R the set

$$HP(a) = HP(R;a) = \{u \in HP: u(a) = 1\}$$

is convex, compact and metrizable.

4B. $HP' = HP'(R)$ denotes the real linear span of HP in the space $C(R,\mathbb{R})$. It is the space of functions on R which can be represented as the difference of two nonnegative harmonic functions on R. The cone HP defines an ordering relation $\leqq$ in HP' in the usual way. Namely, for u, v in HP', we define $u \leqq v$ if and only if $v - u \in HP$. This ordering relation is compatible with the linear space structure of HP'. In fact, HP' becomes a vector lattice. Thus, every pair u, v of functions in HP' has the least upper bound $u \vee v$ and the greatest lower bound $u \wedge v$ in HP', which are exactly the ones we defined in Ch. I, 4C. By use of local compactness of HP or otherwise we see that HP' is order complete, i.e. every subset of HP' which is bounded from above (resp. below) by some element in HP' has the least upper bound (resp. the greatest lower bound) in the space HP'. In particular, the cone HP itself forms a lattice with respect to the order defined by HP.

4C. Let E be a locally convex topological linear space and take a convex cone C in E with vertex at the origin 0 of E. Suppose that all the rays (issuing from 0) in C intersect a closed hyperplane H not passing through 0. The intersection $C \cap H$ is denoted by B. The set B is then convex and is called a <u>base</u> of C. We can define an ordering relation in E by setting $x \leqq y$ whenever $y - x \in C$. We also suppose that B is compact. Then it is seen that a point x in B is an extreme point of the convex set B if and only if it has the following property: $y \in C$ and $y \leqq x$ imply $y = \alpha x$ for some $\alpha \geqq 0$. A point x in C having this property is sometimes called <u>minimal</u>. The fundamental result in this respect is the following theorem due to Choquet:

<u>Theorem</u>. Suppose that the base B of a convex cone C is compact and metrizable. Then the following hold:

(a) The set $\mathcal{E}$ of extreme points of B is a nonvoid G_δ-subset of B and every x in B is the barycenter of a probability measure μ on B supported by $\mathcal{E}$, i.e. there exists a probability Borel measure μ on B such that $\mu(\mathcal{E}) = 1$ and $x = \int_{\mathcal{E}} y\, d\mu(y)$, where the integral is defined in the sense of the weak topology of E, meaning that for every continuous linear functional ℓ on E we have $\ell(x) = \int_{\mathcal{E}} \ell(y)\, d\mu(y)$.

(b) If C is a lattice with respect to the ordering relation defined by C, then every x in B is the barycenter of an at most one probability measure μ supported by $\mathcal{E}$.

Our observation in 4A and 4B shows that the cone HP satisfies all the requirements in the Choquet theorem and so we have the following

Corollary. Let a be in R and denote by $\mathcal{E}(a)$ the set of extreme points of the convex set $HP(a) = \{u \in HP: u(a) = 1\}$. Then $\mathcal{E}(a)$ is a nonvoid G_δ-subset of $HP(a)$ and every u in $HP(a)$ is the barycenter of a unique probability measure μ on $HP(a)$ with $\mu(\mathcal{E}(a)) = 1$, i.e. $u = \int_{\mathcal{E}(a)} v\, d\mu(v)$ in the metric topology of $HP(a)$. In particular, we have $u(z) = \int_{\mathcal{E}(a)} v(z)\, d\mu(v)$ for every z in R.

4D. We consider harmonic functions with singularities. Let $\mathcal{S}$ be the disjoint union of the space $HP'(R \setminus Z)$, where Z runs over all discrete subsets of R. For every pair of elements u_1, u_2 in $\mathcal{S}$ we define $u_1 \sim u_2$ if $u_1(z) = u_2(z)$ outside some discrete subset Z of R. The relation $\sim$ is an equivalence relation in $\mathcal{S}$ and the quotient space $\mathcal{S}/\sim$ is denoted by $SP' = SP'(R)$. Clearly two elements in the same $HP'(R \setminus Z)$ are identical in SP' if and only if they are identical in $HP'(R \setminus Z)$. So every $HP'(R \setminus Z)$ is regarded as a subspace of SP'. The sum $u_1 + u_2$, u_i being in $HP'(R \setminus Z_i)$ with $i = 1, 2$, in SP' is defined as (the equivalence class of) the function $(u_1 + u_2)(z) = u_1(z) + u_2(z)$ for z in $R \setminus (Z_1 \cup Z_2)$. We say $u_1 \leqq u_2$ if $u_1(z) \leqq u_2(z)$ outside some discrete subset of R.

Theorem. SP' is a vector lattice with respect to the pointwise operations defined above. It induces the original vector lattice structure in every $HP'(R \setminus Z)$. SP' is order complete.

The proof is straightforward and is omitted. We note finally that, if u is in $HP'(R \setminus Z)$ with discrete Z, every point in Z is either a logarithmic singularity of u or a removable one.

5. Orthogonal Decomposition

5A. For u in SP' we set $|u| = u \vee (-u)$ and call it the absolute value (in SP') of u. Let u, v be in SP'. We say that u and v are orthogonal if $|u| \wedge |v| = 0$. If this is the case, then we write $u \perp v$. If u in SP' is expressed as $u = u_1 + u_2$ in SP' with $u_1 \perp u_2$, then the expression is called an orthogonal decomposition of u. Typical examples are as follows:

(a) For each u in SP' we set $u^+ = u \vee 0$ and $u^- = (-u) \vee 0$, which we call the positive and the negative parts of u, respectively. We then have $u = u^+ - u^-$, $|u| = u^+ + u^-$ and $u^+ \perp u^-$.

(b) For any subset Y of SP' we define $Y^{\perp}$ to be the set of all u in SP' which are orthogonal to every element in Y. Then $Y^{\perp}$ is a band of SP', i.e. $Y^{\perp}$ is a linear subspace of SP' and has the properties: (i) if $u \in SP'$ and $|u| \leq |v|$ for some v in $Y^{\perp}$, then $u \in Y^{\perp}$; and (ii) for any nonvoid subset Y' of $Y^{\perp}$, bounded above in SP', the least upper bound of Y' in the lattice SP' belongs to $Y^{\perp}$. And we have a direct sum decomposition of the form $SP' = Y^{\perp} + Y^{\perp\perp}$. Moreover, the projection onto each summand is positive. Namely, if $u = u_1 + u_2$ with $u_1 \in Y^{\perp}$ and $u_2 \in Y^{\perp\perp}$, then $u \geqq 0$ implies both $u_1 \geqq 0$ and $u_2 \geqq 0$. When Y consists of only the constant function 1 on R, then we write $I(R) = \{1\}^{\perp}$ and $Q(R) = I(R)^{\perp} = \{1\}^{\perp\perp}$. Functions in $I(R)$ (resp. $Q(R)$) are called inner (resp. quasibounded).

Theorem. Both $I(R)$ and $Q(R)$ are bands of SP' and SP' is the orthogonal direct sum of these bands. The projections pr_I and pr_Q of SP' onto $I(R)$ and $Q(R)$, respectively, are both positive.

This is an easy consequence of the general theory of vector lattices. For any u in SP', $pr_I(u)$ and $pr_Q(u)$ are called the inner and the quasibounded parts of u, respectively.

5B. Theorem. For every u in SP' its quasibounded part $pr_Q(u)$ has only removable singularities, so that the inner part $pr_I(u)$ of u has the same singularities as u.

Proof. We may assume $u \geqq 0$ without loss of generality. If we set $u_n = u \wedge n$, $n = 1, 2, \ldots$, then u_n are bounded above and so are harmonic on R. The sequence $\{u_n\}$ is monotonically increasing and is majorized by the superharmonic function u. So $\lim_{n\to\infty} u_n$ $(= v$, say) exists and is harmonic everywhere on R. We claim that $v = pr_Q(u)$. If $s \in I(R)$, then $|u_n| \wedge |s| = u_n \wedge |s| \leqq n \wedge |s| = 0$. So $u_n \in Q(R)$.

Since $Q(R)$ is complete, we see that $v \in Q(R)$. On the other hand, we have $(u-v) \wedge 1 = 0$. To see this, we set $w = (u-v) \wedge 1$. Then, $w \leqq u-v \leqq u-u_n$ for $n = 1, 2, \ldots$ and $w \leqq 1$. It follows that $w + u_n = w + (u \wedge n) \leqq n+1$ and therefore $w + u_n \leqq u \wedge (n+1) = u_{n+1}$. Thus we have $w \leqq u_{n+1} - u_n$. Since $\{u_n\}$ is a convergent sequence, we have $u_{n+1} - u_n \to 0$. This means that $w = 0$ on R and so $u - v \in I(R)$. Hence, $v = \mathrm{pr}_Q(u)$. The theorem is clear from these observations. □

5C. The above argument also shows the following

Theorem. For every u in SP' we have

$$\mathrm{pr}_Q(u) = \lim_{m\to\infty} \lim_{n\to\infty} [(-m) \vee (n \wedge u)].$$

5D. We apply the preceding results to l.m.m.'s (see 2C). An l.m.m. u on R is said to be of bounded characteristic if $\log u$ belongs to SP'. It is called inner (resp. outer) if $\log u \in I(R)$ (resp. $Q(R)$). Let u be any l.m.m. of bounded characteristic. Then $\mathrm{pr}_Q(\log u)$ is, by Th. 5B, everywhere harmonic and $\mathrm{pr}_I(\log u)$ is harmonic wherever $\log u$ is. We set

$$u_I = \exp(\mathrm{pr}_I(\log u)) \quad \text{and} \quad u_Q = \exp(\mathrm{pr}_Q(\log u)).$$

The functions u_I and u_Q are called the inner and the outer factors of u, respectively. Since $\mathrm{pr}_Q(\log u)$ has no singularities, the outer factor u_Q is always an l.a.m.

A multiplicative meromorphic function f on R is said to be of bounded characteristic if so is the corresponding l.m.m. $u = |f|$. Suppose that this occurs. Let f_I (resp. f_Q) be a multiplicative meromorphic (resp. holomorphic) function such that $u_I = |f_I|$, $u_Q = |f_Q|$ and $f_0 = (f_I)_0 (f_Q)_0$, where the suffix 0 denotes the principal branch at the point 0. These functions f_I and f_Q are determined up to a constant factor of modulus one and are called the inner and outer factors of f, respectively.

Finally we will speak of common inner factors. Let us consider bounded inner l.a.m.'s on R. We begin with the trivial remark that $u \leqq 1$ for any bounded inner l.a.m. u. For two bounded inner l.a.m.'s u_1 and u_2 we say that u_1 divides u_2 if $u_2/u_1 \leqq 1$. If this is the case, then u_2/u_1 is necessarily a bounded inner l.a.m., so that u_1 is called an inner factor of u_2. Let J be a set of l.a.m.'s of bounded characteristic on R such that the inner factor u_I of every u in J is bounded. Then, $A = \{\log u_I : u \in J\}$ forms a subset of

I(R), which is bounded above by the constant function 0. Since I(R) is a band of SP'(R), the set A has the least upper bound, say v_0, in SP'(R), which is a nonpositive element in I(R). Set $u_0 = \exp v_0$. Then u_0 is an inner l.a.m. on R such that (i) u_0 divides every u_I with u in J; and (ii) if a bounded inner l.a.m. u_1 divides all u_I with u in J, then u_1 divides u_0. It is seen that such a u_0 is determined uniquely by these two properties. This u_0 is called the greatest common inner factor of J. If a bounded inner l.a.m. u_0 satisfies (i) but not necessarily (ii), then it is called a common inner factor of J.

A parallel description is possible for multiplicative meromorphic functions on R. But the details look obvious and are thus omitted.

NOTES

Gunning [13] is our main source for cohomology groups and line bundles on Riemann surfaces. See also Weyl [68] for multiplicative functions. The discussion in 3A is adapted from Kusunoki [42].

Some basic facts on vector lattices of harmonic functions are in Chapter 2 of Constantinescu and Cornea [CC]. For Choquet's representation theorem we refer the reader to Phelps' lecture note [53]. Neville [47] contains a discussion on harmonic functions with singularities, together with a proof of Theorem 4D.

CHAPTER III. MARTIN COMPACTIFICATION

Compactification is often a powerful tool for attacking problems on noncompact spaces. As we shall see, this vague idea indeed suggests a right way in the case we are dealing with. Among several kinds of compactification fitted to Riemann surfaces, we are going to make use of Martin's. A naive reason for our choice is that it provides us with the most natural compactification of the open unit disk: the closed unit disk.

In this chapter we define Martin's compactification of open Riemann surfaces and discuss boundary behavior of harmonic functions related to this compactification. Most results in §§1 and 2 can be found in the book of Constantinescu and Cornea [CC], but we pick out and reorganize some ingredients of Chapters 6, 13 and 14 of [CC] for convenience of our later use. The results here are standard. In §3 we discuss the relation between Martin's compactification and covering maps. It is shown in particular that harmonic measures on Martin's boundaries are preserved under covering maps.

Throughout this chapter R will denote an open Riemann surface.

§1. COMPACTIFICATION

1. Definition

1A. Let Q be a set of continuous functions on R with values in the extended real line $[-\infty, +\infty]$ and $C_K = C_K(R,\mathbb{R})$ the set of real-valued continuous functions on R with compact support. For every $f \in Q \cup C_K$ let I_f be the line $[-\infty, +\infty]$. We then define I^Q as the cartesian product of all I_f with $f \in Q \cup C_K$, which is equipped with the usual weak topology. By the classical Tychonoff theorem the space I^Q is a compact Hausdorff space. Define a map ι of R into I^Q in such a way that the f-th coordinate of $\iota(z)$ with $z \in R$ is equal to $f(z)$ for any $f \in Q \cup C_K$. Since functions in C_K separate points in R in the sense that, for any two distinct points in R, we can find a function in C_K taking distinct values at these points, the map ι is an injection and, in fact, is a homeomorphism of R into I^Q. So we can identify R with the subspace $\iota(R)$ of the compact space I^Q under the map ι. Let R_Q^* be the closure of R in the space I^Q. As a closed subspace of the compact space I^Q, R_Q^* is a compact space and includes R as an open dense subset. We call R_Q^* the <u>Q-compactification</u> of R.

Let $f \in Q$ be fixed. For every $y \in I^Q$ we denote by $F_f(y) = y_f$ the f-th coordinate of the point y. The definition of the product topology implies that the function F_f is continuous on I^Q with values in $[-\infty, +\infty]$. Since $F_f(\iota(z)) = f(z)$ for $z \in R$ and since we have identified z with $\iota(z)$, F_f is a continuous extension of the function f to I^Q. Since R is dense in R_Q^*, f thus has a unique continuous extension to R_Q^*.

1B. Let, once and for all, R be a hyperbolic Riemann surface and take a point $0 \in R$, which is held fixed and is called the origin of R. Let $g_a(z) = g(a,z)$ be the Green function for R with pole at $a \in R$ (Ch. I, 6A). We note in passing that $g(a,z)$ is symmetric in both variables, i.e. $g(a,z) = g(z,a)$. Let Q_M be the family of functions $b \to g(b,a)/g(b,0) = k(b,a)$ on R, where the parameter a runs through R. We use the convention $k(b,0) = 1$. The Q-compactification of R with $Q = Q_M$ is called the <u>Martin compactification</u> of R and is denoted by R^*. The difference $\Delta = \Delta(R) = R^* \setminus R$ is called the <u>Martin boundary</u> of R. Since R contains a countable dense subset, it follows that R^* is a compact metric space.

By the remark in 1A every member $b \to k(b,a)$ in Q_M can be extended by continuity to the compact space R^*. The extended function is denoted by the same symbol $b \to k(b,a)$, where the variable b now ranges over R^*. For every fixed $b \in R^*$ the function $z \to k(b,z)$, $z \in R$, is called the <u>Martin function</u> with pole at b (and with origin 0), which is sometimes written as k_b. If $b \in \Delta$, then the function k_b is a positive harmonic function on R. We see also that the function $(b,z) \to k(b,z)$ is a continuous function on $R^* \times R$ with values in $[0, +\infty]$.

2. <u>Integral Representation</u>

2A. <u>Theorem</u>. For any $u \in HP(R)$ there exists a nonnegative measure μ, supported by Δ, such that

$$u(z) = \int_\Delta k(b,z)d\mu(b)$$

for every $z \in R$.

Proof. Let $\{R_n : n = 1, 2, \ldots\}$ be a regular exhaustion of R with $0 \in R_1$ (Theorem 1A, Ch. I). Then by Theorem 6E, Ch. I, the function $u^R_{E_n}$ with $E_n = Cl(R_n)$ is a potential on R and is equal to u on R_n. So there exists a nonnegative measure ν_n supported by ∂R_n such that

$$u(z) = u^R_{E_n}(z) = \int g(b,z)d\nu_n(b)$$

for every $z \in R_n$. We fix $z \in R$ and consider all large n for which $z \in R_n$. By setting $d\mu_n(b) = g(b,0)d\nu_n(b)$, we have

$$u(z) = \int g(b,z)d\nu_n(b) = \int k(b,z)d\mu_n(b).$$

Clearly, μ_n is nonnegative and $\int d\mu_n = \int g(b,0)d\nu_n(b) = u(0)$. The set $\{\mu_n\}$ is thus bounded in the space of measures on R^*, so that there exists a subsequence $\{\mu_{n(i)} : i = 1, 2, \ldots\}$ of $\{\mu_n\}$ converging vaguely to a nonnegative measure μ on R^*. It follows that

$$u(z) = \int k(b,z)d\mu(b)$$

for every $z \in R$. Since the boundary ∂R_n tends to Δ as $n \to \infty$, we see that the measure μ is carried by the Marting boundary Δ. □

2B. We combine our preceding observation with that of Ch. II, 4A-4C. Let $b \in \Delta$. Then the function k_b is a positive harmonic function with $k_b(0) = 1$, so that k_b belongs to $HP(0)$. Since both Δ and $HP(0)$ are compact metric spaces and since the map $(b,z) \to k(b,z)$ is continuous on $\Delta \times R$, we see that the map $j: b \to k_b$ of Δ into $HP(0)$ is continuous. Moreover, j is an injection. In fact, let b, b' be any two distinct points in Δ. The definition of the topology of R^* says that there exists an $f \in Q_M \cup C_K$ for which $f(b) \neq f(b')$. Since every function in C_K vanishes on the boundary Δ, the above f must belong to Q_M. Namely, there exists a point $a \in R$ such that $k(b,a) \neq k(b',a)$ and therefore $k_b \neq k_{b'}$, as was to be shown. So the map j is a homeomorphism of Δ into $HP(0)$. Since $k_b(0) = k_{b'}(0) = 1$, we see also that k_b and $k_{b'}$ with $b \neq b'$ cannot be proportional.

We set $\tilde{\Delta} = j(\Delta)$. Then $\tilde{\Delta}$ is a compact subset of $HP(0)$ and it follows from Theorem 2A that, for each $u \in HP(\mathcal{O})$, there exists a probability measure μ on $\tilde{\Delta}$ such that

$$u(z) = \int_{\tilde{\Delta}} v(z)d\mu(v)$$

for every $z \in R$. This means that $\tilde{\Delta}$ should contain all extreme points of $HP(\mathcal{O})$. We define $\Delta_1 = \Delta_1(R)$ as the inverse image under j of the set of extreme points of the convex set $HP(\mathcal{O})$. The set Δ_1 is exactly the totality of $b \in \Delta$ for which k_b is a minimal harmonic function on R. Points in Δ_1 are called the minimal points in Δ. Corollary 4C, Ch. II, thus shows the following

Theorem. (a) The set Δ_1 is a G_δ-subset of Δ (and also of R^*).

(b) For every $u \in HP(R)$ there exists a unique nonnegative measure μ supported by Δ_1 such that

$$u = \int_{\Delta_1} k_b \, d\mu(b),$$

where the integral converges with respect to the metric topology of $HP(R)$. In particular, we have

$$u(z) = \int_{\Delta_1} k_b(z)d\mu(b)$$

for every $z \in R$.

Let $u \in HP'(R)$. Writing $u = u^+ - u^-$ with $u^+, u^- \in HP(R)$ and using the above result, we see that

$$u = \int_{\Delta_1} k_b \, d\mu(b)$$

for some finite Borel measure μ on Δ_1. The uniqueness result in the statement (b) above shows that such a measure μ on Δ_1 is also determined uniquely by u. This measure is denoted by μ_u and is called the canonical measure for u.

2C. Let $M(\Delta_1)$ be the set of finite real Borel measures on Δ_1. It is a linear space over $\mathbb{R}$ and has a natural ordering relation, i.e. $\mu \leqq \nu$ for $\mu, \nu \in M(\Delta_1)$ if and only if $\mu(E) \leqq \nu(E)$ for every Borel subset E of Δ_1. In view of the remark at the end of 2B, we see that the correspondence $u \to \mu_u$ gives a linear isomorphism of $HP'(R)$ onto $M(\Delta_1)$, which is obviously order-preserving. So $u \perp v$ in $HP'(R)$ if and only if μ_u and μ_v are mutually singular.

Let $\chi = \chi_R$ be the measure in $M(\Delta_1)$ corresponding to the constant function identically equal to 1. We call χ the harmonic measure (on the boundary Δ_1) of R at the point 0. A harmonic function $u \in HP'(R)$ is called singular if $u \in I(R) \cap HP'(R)$, where $I(R)$ was defined in Ch. II, 5A. Our observation in Ch. II, 5A shows the following

Theorem. The correspondence $u \to \mu_u$ gives an order-preserving linear isomorphism of $HP'(R)$ onto $M(\Delta_1)$. u is quasibounded (resp. singular) if and only if μ_u is absolutely continuous (resp. singular) with respect to the measure χ.

Thus every $u \in Q(R)$ determines a unique χ-summable function, f^*, on Δ_1 such that $d\mu_u = f^* d\chi$ or, equivalently,

$$u = \int_{\Delta_1} f^*(b) k_b \, d\chi(b).$$

Indeed, $k_b(a)d\chi(b)$, $a \in R$, is seen to be the harmonic measure of R at the point a. We regard χ as basic among measures on the boundary Δ_1. So we say that a condition is satisfied a.e. on Δ_1 when it is satisfied a.e. with respect to the measure χ.

3. The Dirichlet Problem

3A. We will consider the Dirichlet problem for R^*. Let f be an extended real-valued function on $\Delta = \Delta(R)$. Let $\overline{S}(f)$ be the class of superharmonic functions s on R such that

(A1) s is bounded below,

(A2) $\liminf_{R \ni z \to b} s(z) \geqq f(b)$ for every $b \in \Delta$.

We set $\underline{S}(f) = -\overline{S}(-f)$. Suppose that both $\overline{S}(f)$ and $\underline{S}(f)$ are nonvoid and let $\overline{H}[f]$ (resp. $\underline{H}[f]$) be the pointwise infimum (resp. supremum) on R of functions in $\overline{S}(f)$ (resp. $\underline{S}(f)$). Then we have the following: (i) $s' \geqq s''$ for any $s' \in \overline{S}(f)$ and $s'' \in \underline{S}(f)$; (ii) $\overline{S}(f)$ and $\underline{S}(f)$ are Perron families; (iii) $\overline{H}[f]$ and $\underline{H}[f]$ are harmonic on R and $\underline{H}[f] \leqq \overline{H}[f]$. We call f resolutive, if both $\overline{S}(f)$ and $\underline{S}(f)$ are nonvoid and $\overline{H}[f] = \underline{H}[f]$. The common function is denoted by $H[f]$ and is called the solution of the Dirichlet problem for the Martin compactification R^* with the boundary function f.

3B. We reformulate Riesz's theorem (Theorem 6F, Ch. I) in the following form.

Theorem. Let u be a positive superharmonic function on R and E a closed subset of R. Suppose that $u(0) < +\infty$. Then there exists a positive measure μ supported by the closure E^* of E in R^* such that

$$u_E^R(z) = \int k(b,z)d\mu(b)$$

for every $z \in R$. Here, the restriction $\mu|R$ of μ to R is determined uniquely.

Proof. Let $\{E_n: n = 1, 2,...\}$ be a nondecreasing sequence of compact subsets of E such that $\cup_{n=1}^{\infty} E_n = E$. We have $u_{E_n}^R \leqq u_{E_{n+1}}^R \leqq u_E^R$. Set $s(z) = \lim_{n\to\infty} u_{E_n}^R(z)$ for $z \in R$. It is easy to see that $s(z)$ is a nonnegative superharmonic function and $s(z) \leqq u_E^R(z)$. Since $u_{E_n}^R \geqq u$ q.e. on E_n for every n by Theorem 6E-(b), Ch. I, we see that $s(z) \geqq u(z)$ q.e. on every E_n and therefore on E. It follows that $u_E^R \leqq s$ and hence

$$u_E^R = s = \lim_{n\to\infty} u_{E_n}^R . \tag{1}$$

By Theorem 6E-(c), Ch. I, all $u_{E_n}^R$ are potentials. So every E_n carries a positive measure ν_n such that

$$u_{E_n}^R(a) = \int g(a,z)d\nu_n(z).$$

Since we have

$$\int g(0,z)d\nu_n(z) = u_{E_n}^R(0) \leqq u(0) < +\infty, \tag{2}$$

ν_n has no mass at the point 0. Setting $d\mu_n(z) = g(0,z)d\nu_n(z)$, we get a finite measure on E_n satisfying

$$u_{E_n}^R(a) = \int k(z,a)d\mu_n(z).$$

From (2) follows that $\{\mu_n: n = 1, 2,...\}$ forms a bounded set of measures on the compact set E^*. Thus, passing to a subsequence if necessary, we may assume that $\{\mu_n\}$ converges vaguely to a positive measure μ on E^* We therefore have

$$\int k(z,a)d\mu(z) = \lim_{N\to\infty} \int \min\{k(z,a), N\}\, d\mu(z) \tag{3}$$

$$= \lim_{N\to\infty} \lim_{n\to\infty} \int \min\{k(z,a), N\}\, d\mu_n(z)$$

$$\leqq \lim_{n\to\infty} \int k(z,a) d\mu_n(z)$$

$$= \lim_{n\to\infty} u^R_{E_n}(a) = u^R_E(a).$$

On the other hand, let $a \in R$ be held fixed and then take $\alpha > 0$ so large that $K_\alpha = \{z \in R\colon g(a,z) \geqq \alpha\}$ is compact. Since $\min\{g_a,\alpha\}$ is a potential which is harmonic except at points in ∂K_α, there exists (Theorem 6B, Ch. I) a positive measure λ_α on ∂K_α such that

$$\min\{g(a,z), \alpha\} = \int g(z,b)\, d\lambda_\alpha(b).$$

Setting $z = a$, we see that $\int d\lambda_\alpha(b) = 1$. By Fubini's theorem we have

$$\text{(4)} \qquad \int u^R_{E_n}(b) d\lambda_\alpha(b) = \int \left[\int k(z,b) d\lambda_\alpha(b)\right] d\mu_n(z).$$

We note that

$$\int k(z,b) d\lambda_\alpha(b) = \frac{\min\{g(z,a), \alpha\}}{g(z,O)}$$

is continuous and finite throughout R^* as a function in z. So letting $n \to \infty$ in (4) and using (1), we get

$$\text{(5)} \qquad \int u^R_E(b) d\lambda_\alpha(b) = \int \left[\int k(z,b) d\lambda_\alpha(b)\right] d\mu(z)$$

$$= \int \frac{\min\{g(z,a), \alpha\}}{g(z,O)}\, d\mu(z).$$

Since u^R_E is lower semicontinuous, $\int d\lambda_\alpha = 1$ and the supports of λ_α are shrinking to the point a as $\alpha \to +\infty$, we see that

$$u^R_E(a) \leqq \liminf_{\alpha\to+\infty} \int u^R_E(b) d\lambda_\alpha(b).$$

The monotone convergence theorem shows that the last member of (5) tends to $\int k(z,a)\, d\mu(z)$ as $\alpha \to +\infty$. Combining these with (3), we get the desired integral expression. Finally, unicity of $\mu|R$ follows from Theorem 6B, Ch. I. □

Corollary. Let $b \in \Delta_1(R)$ and let E^* be a closed subset of R^* such that $b \notin E^*$. Set $E = E^* \cap R$. Then $(k_b)^R_E$ is a potential.

Proof. By the preceding theorem we have

$$(k_b)_E^R(a) = \int k(z,a)d\mu(z) \tag{6}$$

for some positive measure μ supported by E^*. We set $\mu' = \mu|\Delta$ and $v(a) = \int k(z,a)\,d\mu'(z)$ for $a \in R$. Then $0 \leqq v \leqq (k_b)_E^R \leqq k_b$. Since k_b is minimal, we have $v = \alpha k_b$ for some α with $0 \leqq \alpha \leqq 1$. If $\alpha \neq 0$, then we set $d\nu = \alpha^{-1}d\mu'$ and have a contradictory expression $k_b(a) = \int k(z,a)\,d\nu(z)$. In fact, the closed support of ν does not contain the point b so that the extreme point k_b cannot be written in that way. Thus $\alpha = 0$ and therefore $(k_b)_E^R$ is a potential. □

3C. Theorem. Let u be a positive harmonic function on R and write $u = \int_\Delta k_b\,d\mu(b)$, where μ is a positive measure on Δ. If E is a closed subset of R, then $b \to (k_b)_E^R(a)$ is integrable for every fixed $a \in R$ and

$$u_E^R(a) = \int_\Delta (k_b)_E^R(a)d\mu(b). \tag{7}$$ ([CC], pp. 44-45)

Proof. If a is an interior point of E or a regular boundary point of $R \setminus E$, then $u_E^R(a) = u(a)$ and $(k_b)_E^R(a) = k_b(a)$, so that (7) holds. If $a \in R \setminus E$ and if G is the connected component of $R \setminus E$ containing a, then, as we have seen in Theorem 6E-(a), Ch. I,

$$u_E^R(a) = \int u(z)d\omega_a^G(z) = \int_\Delta \left[\int k_b(z)d\omega_a^G(z)\right]d\mu(b)$$

$$= \int_\Delta (k_b)_E^R(a)d\mu(b).$$

Finally, let $a \in E$ be an irregular boundary point of $R \setminus E$. Then, by Theorem 5B, Ch. I, the singleton $\{a\}$ forms a connected component of E. So we can find a sequence $\{G_n: n = 1, 2, \ldots\}$ of Jordan domains in R such that $a \in G_n$, $\partial G_n \subseteqq R \setminus E$, $\mathrm{Cl}(G_{n+1}) \subseteqq G_n$ and $\bigcap_{n=1}^\infty \mathrm{Cl}(G_n) = \{a\}$. Let ω_n be the harmonic measure of G_n at the point a. Then we have

$$(k_b)_E^R(a) = \lim_{n\to\infty} \int (k_b)_E^R(z)d\omega_n(z)$$

and therefore the function $b \to (k_b)_E^R(a)$ is measurable. Hence,

$$u_E^R(a) = \lim_{n\to\infty} \int u_E^R(z)d\omega_n(z)$$

$$= \lim_{n\to\infty} \int_\Delta \left(\int (k_b)^R_E(z) d\omega_n(z) \right) d\mu(b)$$

$$= \int_\Delta \left(\lim_{n\to\infty} \int (k_b)^R_E(z) d\omega_n(z) \right) d\mu(b)$$

$$= \int_\Delta (k_b)^R_E(a) d\mu(b),$$

as was to be proved. □

3D. We are now in a position to prove the main result of this section.

<u>Theorem</u>. The Martin compactification R^* is a resolutive compactification in the sense that every real-valued continuous function on the boundary Δ is resolutive. The harmonic measure, with support in Δ_1, of R at the point a is given by $k_b(a)d\chi(b)$. ([CC], p. 140)

<u>Proof</u>. Let f be any real-valued continuous function on Δ. We extend it to a continuous function on R^*, which is denoted by the same letter f. We may assume that $0 \leqq f \leqq 1$ on R^*. For every $n = 1, 2, \ldots$ we set

$$A_i = \{b \in \Delta_1 : \tfrac{1}{n}(i - \tfrac{1}{2}) \leqq f(b) < \tfrac{1}{n}(i + \tfrac{1}{2})\},$$

$$E_i^* = \{a \in R^* : f(a) \leqq \frac{i-1}{n}\} \cup \{a \in R^* : f(a) \geqq \frac{i+1}{n}\},$$

$$E_i = E_i^* \cap R,$$

and

$$u_i = \int_{A_i} k_b \, d\chi(b)$$

for $i = 0, 1, \ldots, n$. For every fixed i we see $A_i \cap E_i^* = \emptyset$ and so by Corollary 3B $(k_b)^R_{E_i}$ is a potential for any $b \in A_i$. By use of Theorem 3C we see that

$$(u_i)^R_{E_i} = \int_{A_i} (k_b)^R_{E_i} \, d\chi(b)$$

is also a potential. Since $(u_i)^R_{E_i} = u_i$ q.e. on E_i and $(i-1)/n < f < (i+1)/n$ on $R \setminus E_i$, we have

$$\frac{i-1}{n}(u_i - (u_i)^R_{E_i}) \leq fu_i \leq \frac{i+1}{n}u_i + (u_i)^R_{E_i}$$

q.e. on R and therefore

$$\sum_{i=0}^{n} \frac{i-1}{n}(u_i - (u_i)^R_{E_i}) \leq f \leq \sum_{i=0}^{n} \frac{i+1}{n}u_i + \sum_{i=0}^{n} (u_i)^R_{E_i}$$

q.e. on R. So we find a positive superharmonic function s' on R such that

$$f \leq \sum_{i=0}^{n} \frac{i+1}{n}u_i + \sum_{i=0}^{n} (u_i)^R_{E_i} + \varepsilon s'$$

on R for any $\varepsilon > 0$. This shows that the right-hand member belongs to the class $\overline{S}(f)$ and therefore majorizes the function $\overline{H}[f]$ on R. As $\varepsilon > 0$ is arbitrary and $\sum_{i=0}^{n} (u_i)^R_{E_i}$ is a potential, we have

$$\overline{H}[f] \leq \sum_{i=0}^{n} \frac{i+1}{n}u_i.$$

It is a simple matter to see that

$$\sum_{i=0}^{n} \frac{i+1}{n}u_i = \sum_{i=0}^{n} \int_{A_i} \frac{i+1}{n}k_b \, d\chi(b) \to \int_{\Delta_1} k_b f(b)d\chi(b)$$

as $n \to \infty$ and hence

$$\overline{H}[f] \leq \int_{\Delta_1} k_b f(b)d\chi(b).$$

Similarly, we have

$$\underline{H}[f] \geq \int_{\Delta_1} k_b f(b)d\chi(b).$$

By combining these inequalities we get the desired result. □

<u>Corollary</u>. A function f on Δ_1 is resolutive if and only if f is χ-summable. If this is the case, we have

$$H[f](a) = \int_{\Delta_1} f(b)k_b(a)d\chi(b)$$

for every $a \in R$.

The proof is similar to that of the classical case and is thus omitted (cf. Ch. I, 5D).

§2. FINE LIMITS

4. Definition of Fine Limits

4A. Let $b \in \Delta_1$. A subset E of R is called thin at b if the balayaged function $(k_b)_E^R$ is different from k_b. In this case $(k_b)_E^R$ is a potential. In fact, by Riesz's theorem (Theorem 6F, Ch. I), the superharmonic function $(k_b)_E^R$ is the sum of a nonnegative harmonic function u and a potential U. Since $u \leqq k_b$, the minimality of k_b implies that $u = \alpha k_b$ with $0 \leqq \alpha \leqq 1$. If $\alpha > 0$, then $(k_b)_E^R = \alpha k_b + U \leqq \alpha(k_b)_E^R + U$ q.e. on E and therefore everywhere on R. But this leads to a contradiction $(k_b)_E^R = k_b$.

If E and E' are thin at b, then so is the union $E \cup E'$, for $(k_b)_{E \cup E'}^R$ is majorized by the potential $(k_b)_E^R + (k_b)_{E'}^R$. Let $\mathcal{G}(b)$ be the collection of sets of the form $R \setminus E$ with E thin at b. We have seen that $\mathcal{G}(b)$ is a filter, i.e. every member of $\mathcal{G}(b)$ is nonempty and $\mathcal{G}(b)$ is closed under finite intersection. Corollary 3B says that $V \cap R \in \mathcal{G}(b)$ for any neighborhood V of $b \in \Delta_1$ in the space R^*.

We see further that, if a closed set $E \subsetneqq R$ is thin at b, $R \setminus E$ has a unique connected component that belongs to $\mathcal{G}(b)$. In fact, let G_1 and G_2 be two components belonging to $\mathcal{G}(b)$. Then, for each $j = 1, 2$, $R \setminus G_j$ is thin at b so that $(k_b)_{R\setminus G_j}^R$ is a potential. So

$$k_b \leqq (k_b)_R^R \leqq (k_b)_{R\setminus G_1}^R + (k_b)_{R\setminus G_2}^R$$

and thus k_b should be a potential, a contradiction.

4B. Let f be a map from R into a compact space X. For any $b \in \Delta_1$ we set

$$f^{\wedge}(b) = \cap\{\mathrm{Cl}(f(D)) : D \in \mathcal{G}(b)\}.$$

Since $\mathcal{G}(b)$ is a filter, the compactness of X implies that $f^{\wedge}(b)$ is a nonvoid compact subset of X and depends only on the values of f taken outside any compact subset of R. So the definition of $f^{\wedge}(b)$ is effective if f is defined off some compact subset of R. Let $\mathcal{D}(f)$ be the set of $b \in \Delta_1$ for which $f^{\wedge}(b)$ is a singleton. For every $b \in \mathcal{D}(f)$ we denote by $\hat{f}(b)$ the point contained in $f^{\wedge}(b)$ and call $\hat{f}(b)$ the fine limit of f at b. The function $\hat{f}$ defined on $\mathcal{D}(f)$ is called the fine boundary function for f. If f is a numerical function, then we take as X the Riemann sphere or the extended real line. For any open set $G \subsetneqq R$ we denote by $\Delta_1(G)$ the set of $b \in \Delta_1$ with $G \in \mathcal{G}(b)$. $\Delta_1(G)$ is a Borel subset of Δ_1. To see this, we may assume that G is connected. Take any $a \in G$ which is held fixed. Then $G \in \mathcal{G}(b)$ if and only if $(k_b)^R_{R\setminus G}(a) < k_b(a)$. Since $(k_b)^R_{R\setminus G}(a) = \int k_b(z)\, d\omega^G_a(z)$ and since the map $(b,z) \to k_b(z)$ is continuous on $\Delta \times R$, we see that $b \to (k_b)^R_{R\setminus G}(a)$ is a Borel measurable function. So $\Delta_1(G)$ is a Borel measurable subset of Δ_1.

<u>Theorem</u>. Let X be a compact metric space and let f be a continuous function from R into X. Then the function $\hat{f}: \mathcal{D}(f) \to X$ is measurable and therefore $\mathcal{D}(f)$ is a Borel subset of Δ_1. ([CC], p. 147)

<u>Proof</u>. For each $x \in X$ and $r > 0$ let $B(x;r)$ denote the open ball of center x and radius r. Let A be a closed subset of X. Since X is separable, we can find a sequence $\{x_n: n = 1, 2, \ldots\}$ which is dense in A. We claim

$$\hat{f}^{-1}(A) = \bigcap_{k=1}^{\infty} \bigcup_{n=1}^{\infty} \Delta_1(f^{-1}(B(x_n;k^{-1}))), \tag{8}$$

which shows the desired result, for each $\Delta_1(f^{-1}(B(x_n;k^{-1})))$ is a Borel set. It is easy to see that the left-hand side of (8) is included in the right. So let b be any point in the right-hand side. For every $k = 1, 2, \ldots$ there exists an $n = n(k)$ such that b belongs to the set $\Delta_1(f^{-1}(B(x_n;k^{-1})))$. Since X is compact, $\{x_{n(k)}: k = 1, 2, \ldots\}$ has a convergent subsequence whose limit is denoted by x. Since A is closed, we have $x \in A$. Take any $\varepsilon > 0$. Then there exists an index k such that $B(x_{n(k)};k^{-1}) \subseteq B(x;\varepsilon)$. Since $f^{-1}(B(x_n;k^{-1})) \in \mathcal{G}(b)$, we have $f^{-1}(B(x;\varepsilon)) \in \mathcal{G}(b)$ and so $f^\wedge(b) \subseteq \mathrm{Cl}(B(x;\varepsilon))$. As ε is arbitrary, we conclude that $f^\wedge(b) = \{x\}$. Hence $b \in \mathcal{D}(f)$ and $\hat{f}(b) = x$, the latter being contained in A. This shows the reverse inclusion, as was to be shown. □

5. <u>Analysis of Boundary Behavior</u>

5A. Let f be an extended real-valued function on R. We denote by $\overline{\mathcal{W}}(f)$ (resp. $\underline{\mathcal{W}}(f)$) the set of superharmonic (resp. subharmonic) functions s on R such that $s \geq f$ (resp. $s \leq f$) outside a compact set (depending on s) in R. If both $\overline{\mathcal{W}}(f)$ and $\underline{\mathcal{W}}(f)$ are nonvoid, then we define $\overline{h}[f]$ (resp. $\underline{h}[f]$) as the pointwise infimum (resp. supremum) of functions in $\overline{\mathcal{W}}(f)$ (resp. $\underline{\mathcal{W}}(f)$). Since $\overline{\mathcal{W}}(f)$ and $\underline{\mathcal{W}}(f)$ are Perron families, both $\overline{h}[f]$ and $\underline{h}[f]$ are harmonic on R. We see also that $\underline{h}[f] \leq \overline{h}[f]$. We call f <u>harmonizable</u> if both $\overline{\mathcal{W}}(f)$ and $\underline{\mathcal{W}}(f)$ are nonvoid and $\overline{h}[f] = \underline{h}[f]$. If this is the case, we denote the common function by $h[f]$. We say that f is a Wiener function on R if it is harmonizable and $|f|$ has a superharmonic majorant on R. We denote by $W(R)$ the set of Wiener functions on R. Our definition of Wiener functions differs from the one in the book [CC] in that we do not assume the quasicontinuity of f.

Theorem. For any Wiener function f on R we have $h[f] \in HP'(R)$ and there exists a potential U on R such that $h[f] - U \leq f \leq h[f] + U$ on R and such that, for any $\varepsilon > 0$, $h[f] - \varepsilon U \leqq f \leqq h[f] + \varepsilon U$ outside a compact set $K_\varepsilon \subsetneqq R$. Conversely, if there exist a harmonic function $u \in HP'(R)$ and a potential U on R with $u - U \leq f \leqq u + U$ on R, then f is a Wiener function and $h[f] = u$. ([CC], p. 54)

Proof. (i) Let f be a Wiener function with $u = h[f]$. It is almost clear that $u \in HP'(R)$. By the definition of $\bar{h}[f]$, there exists a sequence $\{s_n: n = 1, 2, \ldots\}$ in $\overline{\mathcal{W}}(f)$ such that

$$\sum_{n=1}^{\infty} (s_n(a^*) - u(a^*)) < \infty,$$

where a^* is any prescribed point in R. So the sum $\sum_{n=1}^{\infty}(s_n - u)$ is a superharmonic function on R. For every n we have a compact set $K_n \subsetneqq R$ such that $s_n \geqq f$ outside K_n. We choose a regular exhaustion $\{R_n: n = 1, 2, \ldots\}$ of R such that $K_j \subsetneqq R_n$ for $j = 1, \ldots, 2n$ (see Theorem 1A, Ch. I). Set

$$U_1 = \sum_{n=1}^{\infty} (s_n - u)^R_{Cl(R_n)},$$

which is a potential by Theorem 6E-(c), Ch. I. Let m be any positive integer. Then, for any $a \in R_{m+2j} \setminus R_{m+j}$ with $j = 1, 2, \ldots$, we have

$$u(a) + m^{-1}U_1(a) \geqq u(a) + m^{-1} \sum_{n=m+2j}^{2m+2j} (s_n - u)^R_{Cl(R_n)}(a)$$

$$\geqq m^{-1} \sum_{n=m+2j}^{2m+2j} s_n(a) \geqq f(a).$$

Namely, we have $u + m^{-1}U_1 \geqq f$ outside the compact set $Cl(R_{m+1})$. Let s_0 be the superharmonic majorant of $|f|$ on R and define a potential U_2 by setting $U_2 = (s_0)^R_{Cl(R_3)}$. Since $Cl(R_2) \subsetneqq R_3$, we have $|f| \leqq s_0 = U_2$ on $Cl(R_2)$. If we set $U' = U_1 + U_2$, then U' is a potential such that $u + U' \geqq f$ on R and, for any $\varepsilon > 0$, $u + \varepsilon U' \geqq f$ outside some compact set K'_ε. Similarly, we can find a potential U'' such that $u - U'' \leqq f$ on R and, for any $\varepsilon > 0$, $u - \varepsilon U'' \leqq f$ outside a compact set K''_ε. So $U = U' + U''$ is a potential, which meets the requirement of the first half of the theorem.

(ii) Next suppose that f is a potential U, i.e. f is a positive superharmonic function whose greatest harmonic minorant vanishes identically. We see at once that $f \geqq \overline{h}[f] \geqq \underline{h}[f] \geqq 0$. We set $s_n = f^R_{R\setminus R_n}$ for $n = 1, 2, \ldots$. Then, by Theorem 6E, Ch. I, s_n is superharmonic on R, such that s_n is harmonic on R_n, $s_n \leqq f$ on R and $s_n = f$ on $R \setminus R_n$. We have clearly $s_n \geqq s_{n+1} \geqq 0$ for every n. Set $u = \lim_{n\to\infty} s_n$. Then, since $s_n \in \overline{W}(f)$, $u \geqq \overline{h}[f]$. On the other hand, since u is harmonic on R and $u \leqq f$, we should have $u \leqq 0$. Hence, $\overline{h}[f] = \underline{h}[f] = 0$. Namely, f is a Wiener function with $h[f] = 0$.

(iii) Finally let f be a function on R such that $u - U \leqq f \leqq u + U$ with $u \in HP'(R)$ and a potential U. Then (ii) implies that $f \in W(R)$ and $h[f] = u$. This finishes the proof. □

5B. The following properties are easy consequences of Theorem 5A.

<u>Theorem</u>. (a) The correspondence $f \to h[f]$ is a linear map of $W(R)$ onto $HP'(R)$.

(b) The map $f \to h[f]$ is a lattice homomorphism, i.e. if f and f' belong to $W(R)$, then both $\max\{f,f'\}$ and $\min\{f,f'\}$ belong to $W(R)$ and

$$h[\max\{f,f'\}] = h[f] \vee h[f'], \quad h[\min\{f,f'\}] = h[f] \wedge h[f'].$$

(c) Let $f \in W(R)$ be a superharmonic function. Let $f = u + U$ be the decomposition of f by Riesz's theorem (Theorem 6F, Ch. I), where u is harmonic and U is a potential. Then $u = h[f]$. We have $h[f] = 0$ if and only if f is a potential.

(d) Let $f \in W(R)$ and set $L_\alpha = \{z \in R: f(z) = \alpha\}$ for $\alpha \in \mathbb{R}$. Then the balayaged function $1^R_{L_\alpha}$ is a potential except for an at most countable number of α. ([CC], p. 64)

<u>Proof</u>. We only state the proof of (d), for the others follow from 5A rather easily. Take any two real numbers $\alpha' < \alpha''$ and set

$$\alpha = \tfrac{1}{2}(\alpha' + \alpha''), \quad f' = \max\{\frac{\alpha - f}{\alpha - \alpha'}, 0\} \quad \text{and} \quad f'' = \max\{\frac{f - \alpha}{\alpha'' - \alpha}, 0\}.$$

Since f' and f'' are Wiener functions, Theorem 5A shows that there exist potentials U' and U'' such that $f' \leqq h[f'] + U'$ and $f'' \leqq h[f''] + U''$ on R. Since $f' = 1$ on $L_{\alpha'}$ (resp. $f'' = 1$ on $L_{\alpha''}$), we have

$$1^R_{L_{\alpha'}} \leqq h[f'] + U' \quad (\text{resp. } 1^R_{L_{\alpha''}} \leqq h[f''] + U'')$$

and therefore

(9) $$\min\{1^R_{L_{\alpha'}}, 1^R_{L_{\alpha''}}\} \leqq \min\{h[f'], h[f'']\} + U' + U''.$$

As we have, in view of (b),

$$h[\min\{h[f'],h[f'']\}] = h[f'] \wedge h[f''] = h[\min\{f',f''\}]$$

$$= h[0] = 0,$$

so the left-hand side of (9) is a potential. It follows that

$$h[1^R_{L_{\alpha'}}] \wedge h[1^R_{L_{\alpha''}}] = h[\min\{1^R_{L_{\alpha'}}, 1^R_{L_{\alpha''}}\}] = 0.$$

Since α', α'' are arbitrary, we have shown that the set of all $h[1^R_{L_\alpha}]$ with $\alpha \in \mathbb{R}$ forms a family of mutually orthogonal nonnegative harmonic functions. So we get

$$\sum_{\alpha\in\mathbb{R}} h[1^R_{L_\alpha}] = \bigvee_{\alpha\in\mathbb{R}} h[1^R_{L_\alpha}] \leqq 1$$

and consequently $h[1^R_{L_\alpha}] = 0$ except for countably many α. □

5C. Lemma. Let G be an open set in R and set $F = R \setminus G$. Let f be an extended real-valued continuous function on R such that the restriction of f to each connected component of G is a Wiener function. If 1^R_F is a potential, then $\hat{f}$ exists a.e. on $\Delta_1(G)$. This is in particular the case when F is a compact subset of R. ([CC], p. 150)

Proof. Let $L_\alpha = \{z \in R: f(z) = \alpha\}$ with $\alpha \in \mathbb{R}$. By Theorem 5B-(d) applied to each connected component, G', of G in place of R, we find a countable dense subset S of $\mathbb{R}$ such that $1^{G'}_{G'\cap L_\alpha}$ is a potential on G' for every $\alpha \in S$ and every G'.

If we set $F' = \{a \in R: 1^R_F(a) \leqq 1/2\}$, then F' is a nonvoid closed set and, by Theorems 6D and 6E-(a), Ch. I, $F \cap F'$ is a polar set. Let $\alpha \in S$ and define f_α on R by setting $f_\alpha = 1^{G'}_{F'\cap G'\cap L_\alpha}$ on G' and $= 0$ on F. Then we have the following: (a) f_α is continuous on R except at points belonging to a polar set; (b) $0 \leqq f_\alpha \leqq 1$; and (c) $f_\alpha | G' \leqq 1^{G'}_{G'\cap L_\alpha}$, which is a potential.

Let U be a potential which is unbounded on the set of irregular boundary points of G and set $s = 1^R_{L_\alpha}$. Moreover we set

$$s_\alpha = \begin{cases} U + s & \text{on } F \\ U + s^R_F + \min\{s - s^R_F, 1^{G'}_{G' \cap L_\alpha}\} & \text{on } G'. \end{cases}$$

Then s_α is superharmonic on R, for s_α is lower semi-continuous and is not larger than $U+s$ on G. Furthermore, s_α majorizes f_α on R, so that, in the notation of 5A, we have $\bar{h}[f_\alpha] \leqq s_\alpha$ and therefore

$$\bar{h}[f_\alpha] \leqq U + s^R_F + 1^{G'}_{G' \cap L_\alpha}$$

on G'. Since $1^{G'}_{G' \cap L_\alpha}$ is a potential on G', we have $\bar{h}[f_\alpha] \leqq U + s^R_F$ on G'. Since $\bar{h}[f_\alpha] \leqq s_\alpha = U + s$ on F and since U is unbounded at irregular boundary points of G, $\bar{h}[f_\alpha] \leqq U + s^R_F$ on F. Consequently, $\bar{h}[f_\alpha] \leqq U + s^R_F \leqq U + 1^R_F$ on R. We thus conclude that $\bar{h}[f_\alpha] = 0$, for U and 1^R_F are potentials on R. This means that f_α is a Wiener function on R with $h[f_\alpha] = 0$. Hence, by Theorem 5A, f_α is majorized by a potential.

We then claim that

$$1^R_{L_\alpha} \leqq 2(1^R_F) + f_\alpha \tag{10}$$

q.e. on R. This is clear on $F' \cap L_\alpha$ and outside of F', in view of the definition of F' and the fact that $F \cap F'$ is a polar set. Let G'' be a connected component of $G \setminus L_\alpha$. Then we have on G''

$$1^R_{L_\alpha} = H[1^R_{L_\alpha}; G''] \leqq H[2(1^R_F) + f_\alpha; G''] \leqq 2(1^R_F) + f_\alpha.$$

This shows (10), which in turn means that $1^R_{L_\alpha}$ is a potential on R.

Next we set

$$A_\alpha = \{b \in \Delta_1(G) \colon (k_b)^R_{L_\alpha} = k_b\},$$

which is a Borel subset of Δ_1 as noted in 4B. Then we have $\chi(A_\alpha) = 0$ for any $\alpha \in S$. To see this, we consider the harmonic function $u = \int_{A_\alpha} k_b d\chi(b)$. Noting that $0 \leqq u \leqq 1$ and using Theorem 3C, we have

$$0 \leqq u = \int_{A_\alpha} k_b \, d\chi(b) = \int_{A_\alpha} (k_b)^R_{L_\alpha} d\chi(b) = u^R_{L_\alpha} \leqq 1^R_{L_\alpha}.$$

As shown above, $1^R_{L_\alpha}$ is a potential and therefore $u = 0$. Hence, $\chi(A_\alpha) = 0$. We finally set $A = \cup\{A_\alpha \colon \alpha \in S\}$. As S is countable, it follows that $\chi(A) = 0$.

To conclude the proof, let $b \in \Delta_1(G) \setminus A$. Take any $\alpha \in S$. Then $(k_b)^R_{L_\alpha} < k_b$ and so $R \setminus L_\alpha \in \mathcal{G}(b)$. Let G_α be the unique connected component of $R \setminus L_\alpha$ such that $(k_b)^R_{R \setminus G_\alpha} < k_b$ (see 4A). Thus G_α belongs to $\mathcal{G}(b)$ and the range $f(G_\alpha)$, which is connected, is contained either in $(\alpha, +\infty]$ or in $[-\infty, \alpha)$. So, $f^\wedge(b)$ is contained either in $[\alpha, +\infty]$ or in $[-\infty, \alpha]$. This shows that $f^\wedge(b)$ is a singleton, i.e. $b \in \mathcal{D}(f)$. Hence, $\Delta_1(G) \setminus \mathcal{D}(f) \subseteqq A$ and consequently $\chi(\Delta_1(G) \setminus \mathcal{D}(f)) = 0$, as was to be proved. □

5D. The next result connects the fine limit $\hat{f}$ and the harmonic function $h[f]$ for a Wiener function f.

Theorem. (a) Let f be an extended real-valued continuous Wiener function on R. Then the quasibounded part of the harmonic function $h[f]$ is given by

$$\int_{\Delta_1} \hat{f}(b) k_b \, d\chi(b).$$

(b) If f is a bounded continuous function on R and if $\hat{f}$ is defined a.e. on Δ_1, then fu is a Wiener function for any quasibounded positive harmonic function u on R. ([CC], p. 151)

Proof. Let f be a continuous function on R such that $0 \leqq f \leqq 1$ and $\chi(\Delta_1 \setminus \mathcal{D}(f)) = 0$. Take any natural number n and set

$$A_i = \{b \in \mathcal{D}(f) : \tfrac{1}{n}(i - \tfrac{1}{2}) \leqq \hat{f}(b) < \tfrac{1}{n}(i + \tfrac{1}{2})\},$$

$$E_i = \{a \in R : f(a) \leqq \frac{i-1}{n}\} \cup \{a \in R ; f(a) \geqq \frac{i+1}{n}\},$$

and

$$u_i = \int_{A_i} k_b \hat{u}(b) d\chi(b)$$

for $i = 0, 1, \ldots, n$. If $b \in A_i$, then $R \setminus E_i \in \mathcal{G}(b)$ so that $(k_b)^R_{E_i}$ is a potential. Using Theorem 3C, we see that

$$(u_i)^R_{E_i} = \int_{A_i} (k_b)^R_{E_i} \hat{u}(b) d\chi(b)$$

is also a potential. As in 3D we have

$$\sum_{i=0}^{n} \frac{i-1}{n}(u_i - (u_i)^R_{E_i}) \leqq fu \leqq \sum_{i=0}^{n} \frac{i+1}{n} u_i + \sum_{i=0}^{n} (u_i)^R_{E_i}$$

q.e. on R. So there exists a positive superharmonic function s' on R such that

$$fu \leqq \sum_{i=0}^{n} \frac{i+1}{n} u_i + \sum_{i=0}^{n} (u_i)_{E_i}^{R} + \varepsilon s'$$

holds on R for any $\varepsilon > 0$. The right-hand member thus belongs to the class $\overline{W}(f)$ and so

$$\overline{h}[fu] \leqq \sum_{i=0}^{n} \frac{i+1}{n} u_i,$$

for $\sum_{i=0}^{n} (u_i)_{E_i}^{R}$ is a potential. Similarly,

$$\sum_{i=0}^{n} \frac{i-1}{n} u_i \leqq \underline{h}[fu].$$

Combining these inequalities, we see that fu is harmonizable and

$$h[fu] = \lim_{n\to\infty} \sum_{i=0}^{n} \frac{i+1}{n} u_i = \lim_{n\to\infty} \sum_{i=0}^{n} \int_{A_i} k_b \frac{i+1}{n} \hat{u}(b) d\chi(b)$$

$$= \int k_b \hat{f}(b)\hat{u}(b) d\chi(b).$$

This shows (b) and also (a) as well for a bounded f.

In order to show the statement (a) for a general f, we may assume that $f \geqq 0$. We set $f_n = \min\{f,n\}$ for $n = 1, 2, \ldots$. The quasi-bounded part of $h[f]$ is then given by

$$\lim_{n\to\infty} (h[f] \wedge n) = \lim_{n\to\infty} h[f_n] = \lim_{n\to\infty} \int k_b \hat{f}_n(b) d\chi(b)$$

$$= \int k_b \hat{f}(b) d\chi(b). \quad \Box$$

5E. As a special case of the preceding theorem, we get the following result.

Theorem. (a) If $u \in HP'(R)$, then $\hat{u}$ exists a.e. on Δ_1 and the quasibounded part $pr_Q(u)$ of u is given by

$$pr_Q(u) = \int_{\Delta_1} k_b \hat{u}(b) d\chi(b).$$

In particular, if u is a singular harmonic function on R, then $\hat{u} = 0$ a.e. on Δ_1.

(b) If u^* is a χ-summable function on Δ_1, then

$$u = H[u^*] = \int_{\Delta_1} k_b u^*(b) d\chi(b)$$

is a quasibounded harmonic function and $\hat{u} = u^*$ a.e. on Δ_1.

Proof. (a) This is immediate from the preceding theorem, for u is evidently a Wiener function and $u = h[u]$.

(b) That u is a quasibounded harmonic function was already indicated in 2C. So, by (a), $\hat{u}$ exists a.e. and $u = \int_{\Delta_1} k_b \hat{u}(b)\, d\chi(b)$. It follows again from Theorem 2C that $u^* = \hat{u}$ a.e. □

§3. COVERING MAPS

6. Correspondence of Harmonic Functions

6A. Let $(R, \mathcal{O})$ be a pair consisting of a hyperbolic Riemann surface R and a point $\mathcal{O} \in R$, called the origin of R. A map ϕ from another pair $(R', \mathcal{O}')$ onto $(R, \mathcal{O})$ is called a covering map if

(A1) ϕ is a holomorphic function from R' onto R,

(A2) every point $a \in R$ has an open neighborhood V such that every connected component of $\phi^{-1}(V)$ is conformally isomorphic with V under ϕ,

(A3) $\phi(\mathcal{O}') = \mathcal{O}$.

The triple $(R', \mathcal{O}', \phi)$ having these properties is called a covering triple of $(R, \mathcal{O})$. There exists a covering triple $(\tilde{R}, \tilde{\mathcal{O}}, \tilde{\phi})$ of $(R, \mathcal{O})$ which is the strongest in the sense that for each covering triple $(R', \mathcal{O}', \phi)$ of $(R, \mathcal{O})$ one can find a covering map ϕ' of $(\tilde{R}, \tilde{\mathcal{O}})$ onto $(R', \mathcal{O}')$ satisfying $\tilde{\phi} = \phi \circ \phi'$. Such a surface $(\tilde{R}, \tilde{\mathcal{O}})$ is determined uniquely up to a conformal isomorphism and is called the universal covering surface of $(R, \mathcal{O})$ ([AS], Ch. I, 18C). The uniformization theorem says that $(\tilde{R}, \tilde{\mathcal{O}})$ is conformally isomorphic with the pair $(\mathbb{D}, 0)$ of the open unit disk $\mathbb{D}$ and the origin 0 ([AS], Ch. III, 11G).

6B. Let $(R, \mathcal{O})$ and $(R', \mathcal{O}')$ be hyperbolic Riemann surfaces with origin and ϕ a covering map of $(R', \mathcal{O}')$ onto $(R, \mathcal{O})$. Let $(\mathbb{D}, 0, \phi_R)$ (resp. $(\mathbb{D}, 0, \phi_{R'})$) be the universal covering triple of $(R, \mathcal{O})$ (resp. $(R', \mathcal{O}')$) such that $\phi_R = \phi \circ \phi_{R'}$. Since ϕ is locally a conformal isomorphism, a function f on R is holomorphic (resp. harmonic, super- or sub-harmonic) if $f \circ \phi$ is holomorphic (resp. harmonic, super- or sub-harmonic) on R'.

Let T_R be the group of cover transformations for ϕ_R, i.e. the multiplicative group of fractional linear transformations τ of $\mathbb{D}$ onto itself such that $\phi_R \circ \tau = \phi_R$. If f is a function on R, then $F = f \circ \phi_R$ is a function on $\mathbb{D}$, which is T_R-invariant in the sense that $F \circ \tau = F$ for any $\tau \in T_R$. Conversely, every T_R-invariant function F on $\mathbb{D}$ is expressed in the form $f \circ \phi_R$ with a function f on R.

Theorem. (a) The map $u \to u \circ \phi$ is a lattice isomorphism of $HP'(R)$ into $HP'(R')$. So, $u \in HP'(R)$ is quasibounded (resp. singular) if and only if $u \circ \phi$ is quasibounded (resp. singular) on R'.

(b) Let s be a positive superharmonic function on R. Then s is a potential on R if and only if $s \circ \phi$ is a potential on R'.

(c) If a function f on R is a Wiener function, then $f \circ \phi$ is a Wiener function on R' and $h[f] \circ \phi = h[f \circ \phi]$.

Proof. (i) We first consider the case when $(R', \mathcal{O}', \phi)$ is the universal covering triple $(\mathbb{D}, 0, \phi_R)$ of $(R, \mathcal{O})$.

Let $u_i \in HP'(R)$, $i = 1, 2$, and we set $v = (u_1 \circ \phi_R) \vee (u_2 \circ \phi_R)$. Clearly, $v \leqq (u_1 \vee u_2) \circ \phi_R$. Since $v \geqq u_i \circ \phi_R$ for $i = 1, 2$, we have for any $\tau \in T_R$

$$v \circ \tau \geqq u_i \circ \phi_R \circ \tau = u_i \circ \phi_R$$

for $i = 1, 2$ and therefore $v \circ \tau \geqq v$. Since T_R is a group, we have $v \circ \tau = v$ for any $\tau \in T_R$. Thus there exists a harmonic function u on R such that $v = u \circ \phi_R$. It is easy to see that $u \geqq u_1 \vee u_2$ and so

$$(u_1 \vee u_2) \circ \phi_R \leqq u \circ \phi_R = v \leqq (u_1 \vee u_2) \circ \phi_R.$$

Consequently, we have $(u_1 \vee u_2) \circ \phi_R = (u_1 \circ \phi_R) \vee (u_2 \circ \phi_R)$, which shows that the correspondence $u \to u \circ \phi_R$ is a lattice isomorphism of $HP'(R)$ onto the set of T_R-invariant elements in $HP'(\mathbb{D})$. Since $1 \circ \phi_R = 1$, we see that a harmonic function $u \in HP'(R)$ is quasibounded (resp. singular) if and only if $u \circ \phi_R$ is. Thus (a) is proved.

Next let s be a positive superharmonic function on R and let v be the greatest harmonic minorant of $s \circ \phi_R$ on $\mathbb{D}$. By using the same argument as above we see that v is T_R-invariant. So there exists a nonnegative harmonic function u on R with the property $v = u \circ \phi_R$. It is easy to show that u is the greatest harmonic minorant of s. By Riesz's theorem (Theorem 6F, Ch. I) a positive superharmonic function is a potential if and only if its greatest harmonic minorant is identically zero. So s is a potential if and only if $s \circ \phi_R$ is. This proves (b).

Finally let f be a Wiener function on R. By Theorem 5A there exists a potential U on R such that $h[f] - U \leqq f \leqq h[f] + U$ on R. So we have $h[f] \circ \phi_R - U \circ \phi_R \leqq f \circ \phi_R \leqq h[f] \circ \phi_R + U \circ \phi_R$, where $h[f] \circ \phi_R$ belongs to $HP'(\mathbb{D})$ and $U \circ \phi_R$ is a potential by what we have seen above. By Theorem 5A $f \circ \phi_R$ is a Wiener function on $\mathbb{D}$ and $h[f \circ \phi_R] = h[f] \circ \phi_R$, which proves (c).

(ii) We consider the general case $\phi\colon (R', \mathcal{O}') \to (R, \mathcal{O})$. Let us take $u_i \in HP'(R)$, $i = 1, 2$, and set $v = (u_1 \circ \phi) \vee (u_2 \circ \phi)$. Then as before $v \leqq (u_1 \vee u_2) \circ \phi$. By using the fact $\phi_R = \phi \circ \phi_{R'}$ and our observations in (i), we have

$$\begin{aligned}(u_1 \vee u_2) \circ \phi_R &= (u_1 \vee u_2) \circ \phi \circ \phi_{R'} \geqq v \circ \phi_{R'} = \{(u_1 \circ \phi) \vee (u_2 \circ \phi)\} \circ \phi_{R'} \\ &= (u_1 \circ \phi \circ \phi_{R'}) \vee (u_2 \circ \phi \circ \phi_{R'}) = (u_1 \circ \phi_R) \vee (u_2 \circ \phi_R) \\ &= (u_1 \vee u_2) \circ \phi_R.\end{aligned}$$

Thus we have

$$(u_1 \vee u_2) \circ \phi = v = (u_1 \circ \phi) \vee (u_2 \circ \phi),$$

from which (a) follows. By the fact shown in (i), U is a potential on R if and only if $U \circ \phi_R$ is a potential on $\mathbb{D}$. And since $\phi_R = \phi \circ \phi_{R'}$, $U \circ \phi_R$ is a potential if and only if $U \circ \phi$ is a potential on R'. So (b) follows. (c) is a consequence of (a) and (b). □

7. Preservation of Harmonic Measures

7A. We again use our notations in 6B. Since R^* is a compact metric space, there exists a countable family C of real-valued continuous functions on R^* which separates the points of R^*. Let $f \in C$. By Theorem 5D-(b) the restriction $f|R$ is a continuous Wiener function on R and so by Theorem 6B-(c) $f \circ \phi$ is also a continuous Wiener function on R'. It follows from Lemma 5C $\widehat{f \circ \phi}$ is defined a.e. on $\Delta_1(R')$. We set

$$\mathcal{D}_0 = \cap\{\mathcal{D}(f \circ \phi)\colon f \in C\}.$$

Since C is countable and each $\mathcal{D}(f \circ \phi)$ has negligible complement in $\Delta_1(R')$, $\mathcal{D}_0$ has negligible complement in $\Delta_1(R')$. We claim that $\mathcal{D}_0 \subseteqq \mathcal{D}(\phi)$. In fact, let $b' \in \mathcal{D}_0$. Since every $f \in C$ is continuous on R^*, we have

$$\phi\hat{}(b') \subseteqq \bigcap_{f\in C} \{b \in R^*: f(b) = \widehat{f\circ\phi}(b')\}.$$

Since C separates the points in R^*, the right-hand side can contain only one point. So $b' \in \mathcal{D}(\phi)$ and

$$f(\hat{\phi}(b')) = \widehat{f\circ\phi}(b') \tag{11}$$

for any $f \in C$. Next we consider the function

$$h(z') = g_R(\mathcal{O},\phi(z'))$$

for $z' \in R'$ with the Green function g_R for R. By Theorem 6B-(b) the function h is a potential on R'. It is also continuous with values in $[0, +\infty]$ and so, by Theorem 5D-(a), $\hat{h}$ exists and vanishes a.e. on $\Delta_1(R')$. We define

$$\mathcal{D}_0(\phi) = \mathcal{D}(\phi) \cap \{b' \in \mathcal{D}(h): h(b') = 0\}.$$

Theorem. $\mathcal{D}_0(\phi)$ is a Borel subset of $\Delta_1(R')$, $\chi_{R'}(\mathcal{D}_0(\phi)) = 1$, and $\hat{\phi}$ maps $\mathcal{D}_0(\phi)$ into $\Delta(R)$.

Proof. Theorem 4B shows that $\mathcal{D}(\phi)$ and $\mathcal{D}(h)$ are Borel sets and that $\hat{h}$ is Borel measurable. So $\mathcal{D}_0(\phi)$ is a Borel set in $\Delta_1(R')$. We saw above that $\chi_{R'}(\mathcal{D}_0(\phi)) = 1$. Take any $b' \in \mathcal{D}_0(\phi)$. If $\hat{\phi}(b') \in R$, then we would find a constant $\alpha > 0$ and an open neighborhood V of $\hat{\phi}(b')$ in R such that $g_R(\mathcal{O},z) \geqq \alpha$ for $z \in V$. So $h(z') \geqq \alpha$ for any z' belonging to $\phi^{-1}(V)$. Since $\phi^{-1}(V)$ is easily seen to be a member in $G(b')$, we thus see $h(b') \geqq \alpha > 0$, which is a contradiction. Hence, $\hat{\phi}(b') \in \Delta(R)$. □

7B. The main result in this part is the following

Theorem. (a) $\hat{\phi}$ is a measure-preserving map of $\mathcal{D}_0(\phi)$ into $\Delta(R)$, i.e. $\chi_{R'}(\hat{\phi}^{-1}(A)) = \chi_R(A)$ for any χ_R-measurable subset A of $\Delta(R)$.

(b) If $f^* \in L^1(d\chi_R)$, then $f^*\circ\hat{\phi} \in L^1(d\chi_{R'})$, $\|f^*\|_1 = \|f^*\circ\hat{\phi}\|_1$, and

$$H[f^*]\circ\phi = H[f^*\circ\hat{\phi}], \tag{12}$$

where $\|\cdot\|_1$ denotes the L^1-norm and $H[\cdot]$ denotes the operation defining the solution of the Dirichlet problem as given in 3A.

Proof. (i) Let f^* be any real-valued continuous function on $\Delta(R)$ and extend it to a continuous function f on R^*. By adjoining f to the class C used above we see that $\mathcal{D}(f\circ\phi)\cap\mathcal{D}_0(\phi)$ has a negligible complement in $\Delta_1(R')$ and $f^*(\hat{\phi}(b')) = \widehat{f\circ\phi}(b')$ for any b' belonging to $\mathcal{D}_0(\phi)\cap\mathcal{D}(f\circ\phi)$. Since $f\circ\phi$ is a bounded Wiener function, $h[f\circ\phi]$ is quasibounded with boundary values $\widehat{f\circ\phi}$ and therefore $h[f\circ\phi] = H[\widehat{f\circ\phi}]$. So we have by Theorem 6B

$$(13) \qquad H[f^*]\circ\phi = h[f]\circ\phi = h[f\circ\phi] = H[\widehat{f\circ\phi}] = H[f^*\circ\hat{\phi}].$$

Hence (12) holds for any continuous function f^* on $\Delta(R)$.

(ii) Suppose next that f^* on $\Delta(R)$ is a lower semicontinuous function which is bounded below and χ_R-summable. We take a nondecreasing sequence $\{f_n^*: n = 1, 2,\ldots\}$ of continuous functions on $\Delta(R)$ converging pointwise to f^*. Since $H[\cdot]$ is a lattice isomorphism of the space $L^1(d\chi_R;\mathbb{R})$ (resp. $L^1(d\chi_{R'};\mathbb{R})$) of real-valued χ_R- (resp. $\chi_{R'}$-) summable functions onto the space $Q(R)$ (resp. $Q(R')$) of quasibounded harmonic functions on R (resp. R'), we have

$$H[f^*] = H[\sup_n f_n^*] = \bigvee_n H[f_n^*]$$

and, since $f^*\circ\hat{\phi} = \sup_n (f_n^*\circ\hat{\phi})$,

$$H[f^*\circ\hat{\phi}] = H[\sup_n (f_n^*\circ\hat{\phi})] = \bigvee_n H[f_n^*\circ\hat{\phi}].$$

By use of (13) and Theorem 6B we get

$$H[f^*]\circ\phi = (\bigvee_n H[f_n^*])\circ\phi = \bigvee_n (H[f_n^*]\circ\phi) = \bigvee_n H[f_n^*\circ\hat{\phi}]$$

$$= H[\sup_n (f_n^*\circ\hat{\phi})] = H[f^*\circ\hat{\phi}].$$

So (12) holds for our f^*. Similarly, (12) holds for any upper semicontinuous function which is bounded above and χ_R-summable.

(iii) Let f^* be any everywhere defined χ_R-summable function on $\Delta(R)$. Then there exist a nondecreasing sequence $\{g_n^*: n = 1, 2,\ldots\}$ of upper semicontinuous functions and a nonincreasing sequence $\{h_n^*: n = 1, 2,\ldots\}$ of lower semicontinuous functions on $\Delta(R)$ such that every g_n^* (resp. h_n^*) is bounded above (resp. below), $g_n^* \leq f^* \leq h_n^*$ and

$$\lim_{n\to\infty}\int g_n^*\, d\chi_R = \int f^*\, d\chi_R = \lim_{n\to\infty}\int h_n^*\, d\chi_R.$$

It follows that $\{H[g_n^*]\}$ is nondecreasing, $\{H[h_n^*]\}$ is nonincreasing and $H[g_n^*] \leq H[f^*] \leq H[h_n^*]$ for $n = 1, 2,\ldots$. Moreover

$$H[h_n^*](O) - H[g_n^*](O) = \int (h_n^* - g_n^*)\, d\chi_R \to 0.$$

So we have

$$\bigvee_n H[g_n^*] = H[f^*] = \bigwedge_n H[h_n^*].$$

Applying ϕ, which is a lattice isomorphism, and using (12) for g_n^* and h_n^*, we get

$$\bigvee_n H[g_n^* \circ \hat{\phi}] = \bigvee_n (H[g_n^*] \circ \phi) = H[f^*] \circ \phi = \bigwedge_n (H[h_n^*] \circ \phi)$$
$$= \bigwedge_n (H[h_n^* \circ \hat{\phi}]).$$

Furthermore, since $H[g_n^* \circ \hat{\phi}] \leqq H[f^* \circ \hat{\phi}] \leqq H[h_n^* \circ \hat{\phi}]$, $n = 1, 2, \ldots$, we see finally $H[f^*] \circ \phi = H[f^* \circ \hat{\phi}]$. Hence, (12) holds for every χ_R-summable function f^* on $\Delta(R)$, provided f^* is everywhere defined.

(iv) Let $A \subseteqq \Delta(R)$ be any measurable subset and let f^* be the characteristic function of A. Then, $f^* \circ \hat{\phi}$ is the characteristic function of the set $\hat{\phi}^{-1}(A)$. We see by (12), which is valid for this f^*, that $H[f^*] \circ \phi = H[f^* \circ \hat{\phi}]$. From this follows that

$$\chi_R(A) = \int f^*(b)\, d\chi_R(b) = H[f^*](O) = (H[f^*] \circ \phi)(O')$$
$$= H[f^* \circ \hat{\phi}](O') = \int (f^* \circ \hat{\phi})(b')\, d\chi_{R'}(b')$$
$$= \chi_{R'}(\hat{\phi}^{-1}(A)).$$

Hence, (a) holds. This implies in turn that the correspondence $f^* \to f^* \circ \hat{\phi}$ maps $L^1(d\chi_R; \mathbb{R})$ isometrically into $L^1(d\chi_{R'}; \mathbb{R})$ and that (12) holds when f^* is regarded as an equivalence class modulo null functions. This finishes the proof of the theorem. □

7C. We now look at the case of universal covering maps. First we identify the group T_R of cover transformations for a universal covering map $\phi_R: \mathbb{D} \to R$ with $\phi_R(0) = O$. Given $\tau \in T_R$, we take an arc γ' joining the origin 0 with $\tau(0)$ within $\mathbb{D}$. Then $\phi_R(\gamma')$ is a 1-cycle issuing from O. Since different choices for γ' result in homotopic 1-cycles in R, $\phi_R(\gamma')$ determines a unique element, say γ_τ, of $F_O(R)$.

<u>Theorem</u>. The map $\tau \to \gamma_\tau$ gives an isomorphism of the group T_R onto the fundamental group $F_O(R)$. ([AS], Ch. I, 19A)

7D. Using the notations in 7A with $R' = \mathbb{D}$ and $\phi = \phi_R$, we see that $\mathcal{D}_0(\phi_R)$ is T_R-invariant. Besides the properties mentioned in the theorems in 7A and 7B, we have the following

Theorem. Let $0 < p \leqq \infty$. Then the correspondence $f^* \to f^* \circ \hat{\phi}_R$ maps $L^p(d\chi_R)$ isometrically onto the subspace $L^p(d\sigma)_T$ of $L^p(d\sigma)$ consisting of all T_R-invariant elements, where $d\sigma$ denotes the normalized Lebesgue measure on the unit circle. Moreover $f^* \in H^p(d\chi_R)$ if and only if $f^* \circ \hat{\phi}_R \in H^p(d\sigma)$.

Proof. We first note that the Martin compactification of the open unit disk $\mathbb{D}$ is the closed unit disk, its Martin boundary is the unit circumference $\mathbb{T}$ and the corresponding harmonic measure on $\mathbb{T}$ at the origin is equal to the normalized Lebesgue measure $d\sigma(t) = d\sigma(e^{it}) = dt/2\pi$.

(i) Suppose that $p = 1$. If $f^* \in L^1(d\chi_R)$, then, by (12), we have $H[f^*] \circ \phi_R = H[f^* \circ \hat{\phi}_R]$. Since $f^* \circ \hat{\phi}_R$ is the fine boundary function for the T_R-invariant harmonic function $H[f^*] \circ \phi_R$, it is (equivalent to) a T_R-invariant function on $\mathbb{T}$. Conversely, let $u^* \in L^1(d\sigma)$ be a T_R-invariant function. Then the harmonic function $H[u^*]$ is a T_R-invariant harmonic function on $\mathbb{D}$. In view of Theorem 7C, we find a harmonic function v on R such that $v \circ \phi_R = H[u^*]$. Since $H[u^*]$ is quasibounded, Theorem 6B-(a) implies that v is quasibounded. So, by Theorem 5E, $v = H[\hat{v}]$ for the fine boundary function $\hat{v}$ of v. Again by (12) $v \circ \phi_R = H[\hat{v}] \circ \phi_R = H[\hat{v} \circ \hat{\phi}_R]$ and hence $u^* = \hat{v} \circ \hat{\phi}_R$ a.e. That the map $f^* \to f^* \circ \hat{\phi}_R$ is an isometry was shown in Theorem 7B.

(ii) Consider the general case $0 < p < \infty$. Let $f^* \in L^p(d\chi_R)$. Since $|f^* \circ \hat{\phi}_R|^p = |f^*|^p \circ \hat{\phi}_R$, $|f^*|^p \in L^1(d\chi_R)$ and $\| |f^*|^p \circ \hat{\phi}_R \|_1 = \| |f^*|^p \|_1$ by (i), we have $f^* \circ \hat{\phi}_R \in L^p(d\sigma)$ and $\|f^* \circ \hat{\phi}_R\|_p = \|f^*\|_p$. Conversely, let $u^* \in L^p(d\sigma)_T$. If $u^* \geqq 0$, then we find, by (i), an $h^* \in L^1(d\chi_R)$ with $h^* \circ \hat{\phi}_R = u^{*p}$ a.e. Letting $f^* = (h^*)^{1/p}$, we have $f^* \in L^p(d\chi_R)$ and $f^* \circ \hat{\phi}_R = u^*$ a.e. The result for an arbitrary u^* is then deduced by writing u^* as a linear combination of four positive elements in $L^p(d\sigma)_T$. The rest is now obvious. □

NOTES

The results on Martin compactification are adapted from Chapters 6, 13, 14 of Constantinescu and Cornea [CC]. The results in §3 form an improved version of those given in Hasumi [17].

CHAPTER IV. HARDY CLASSES

Some basic results are recalled here concerning Hardy classes H^p on the unit disk and on general Riemann surfaces. We mainly look into boundary behavior of functions in such classes. Results on Riemann surfaces are deduced from those for the unit disk by using results on covering maps given in Ch. III, §3. Finally, we determine the dual of h^∞--the harmonic version of H^∞--with respect to the β-topology.

§1. HARDY CLASSES ON THE UNIT DISK

1. Basic Definitions

1A. Let $\mathbb{D}$ be the open unit disk. The Green function $g_{\mathbb{D}}$ on $\mathbb{D}$ is given by $g_{\mathbb{D}}(a,z) = \log\{|1-\bar{a}z|/|z-a|\}$ and so the Martin functions with $\mathcal{O}$ being the origin 0 of the complex plane are given by $k_{\mathbb{D}}(b,z) = g_{\mathbb{D}}(b,z)/g_{\mathbb{D}}(b,0)$. For any fixed $z \in \mathbb{D}$ the function $b \to k_{\mathbb{D}}(b,z)$ can be extended continuously to the closed unit disk $\mathrm{Cl}(\mathbb{D})$, so that the Martin compactification of $\mathbb{D}$ is identified with the closed unit disk $\mathrm{Cl}(\mathbb{D})$ with the usual Euclidean topology. We see also that both $\Delta(\mathbb{D})$ and $\Delta_1(\mathbb{D})$ are equal to the usual unit circumference $\mathbb{T}$ (or $\partial\mathbb{D}$). For each point $e^{it} \in \mathbb{T}$ the Martin function $z \to k_{\mathbb{D}}(e^{it},z)$ with pole at e^{it} is exactly the Poisson kernel

$$P(r,\theta - t) = \frac{1-r^2}{1-2r\cos(\theta-t)+r^2}$$

with $z = re^{i\theta}$. The harmonic measure $d\chi_{\mathbb{D}}(e^{it})$ (on $\mathbb{T}$) of $\mathbb{D}$ at the origin 0 is seen to be the normalized Lebesgue measure $dt/2\pi$, which is denoted, in the following, by $d\sigma(e^{it})$ or $d\sigma(t)$.

1B. For $0 < p < \infty$ let $H^p(\mathbb{D})$ denote the space of analytic functions $f(z)$ on $\mathbb{D}$ such that $|f(z)|^p$ has a harmonic majorant. $H^\infty(\mathbb{D})$ denotes the space of bounded analytic functions on $\mathbb{D}$. Further, let $N(\mathbb{D})$ be the Nevanlinna class on $\mathbb{D}$, i.e. the space of analytic functions $f(z)$ on $\mathbb{D}$ such that $\log^+|f(z)|$ has a harmonic majorant or, equivalently, $\log|f| \in SP'(\mathbb{D})$ (Ch. II, 5D). Then

$$N(\mathbb{D}) \supseteqq H^p(\mathbb{D}) \supseteqq H^q(\mathbb{D}) \supseteqq H^\infty(\mathbb{D})$$

for $0 < p < q < \infty$. For $f \in H^p(\mathbb{D})$, $0 < p < \infty$, we set

$$\|f\|_p = ((\mathrm{LHM}(|f|^p)(0))^{1/p},$$

where LHM stands for the least harmonic majorant. For $f \in H^\infty(\mathbb{D})$ we set $\|f\|_\infty = \sup\{|f(z)|: z \in \mathbb{D}\}$. We note in passing that $|f(z)|^p$ is subharmonic for any analytic function f on $\mathbb{D}$ and any $0 < p < \infty$ and therefore that $|f(z)|^p$ has the least harmonic majorant provided it has a superharmonic majorant.

1C. Along with the Hardy classes $H^p(\mathbb{D})$ we consider analogous classes consisting of harmonic functions. Namely, let $h^p(\mathbb{D})$, $1 \leqq p < \infty$, be the space of complex-valued harmonic functions $f(z)$ on $\mathbb{D}$ such that $|f(z)|^p$ has a harmonic majorant. $h^\infty(\mathbb{D})$ denotes the complex-valued bounded harmonic functions on $\mathbb{D}$.

Lemma. If $f(z)$ is harmonic and if $\Phi(x,y)$ is a convex function on $\mathbb{R}^2$, then $\Phi \circ f$ is subharmonic. In particular, $|f(z)|^p$ is subharmonic for any harmonic f and any $1 \leqq p < \infty$.

Proof. We set $u = \mathrm{Re}(f)$ and $v = \mathrm{Im}(f)$, so that u and v are real harmonic functions. Let $|z - z_0| \leqq r_0$ be any closed disk in the domain of $f(z)$. Since u and v are harmonic,

$$(u(z_0),v(z_0)) = \int_{\mathbb{T}} (u(z_0+r_0e^{it}),v(z_0+r_0e^{it}))d\sigma(t).$$

Since Φ is convex,

$$\begin{aligned}\Phi \circ f(z_0) &= \Phi(u(z_0),v(z_0)) \\ &\leqq \int_{\mathbb{T}} \Phi(u(z_0+r_0e^{it}),v(z_0+r_0e^{it}))d\sigma(t) \\ &= \int_{\mathbb{T}} (\Phi \circ f)(z_0+r_0e^{it})d\sigma(t).\end{aligned}$$

Hence $\Phi \circ f$ is seen to be subharmonic. The latter fact follows from the fact that $(x^2 + y^2)^{p/2}$ is a convex function if $1 \leqq p < \infty$. □

The lemma then implies that, in case $1 \leqq p < \infty$, $|f|^p$ has the least harmonic majorant whenever it has a harmonic majorant. We define as before $\|f\|_p = ((\mathrm{LHM}(|f|^p))(0))^{1/p}$ for $1 \leqq p < \infty$ and $\|f\|_\infty = \sup\{|f(z)| : z \in \mathbb{D}\}$. A complex-valued harmonic function f is called

<u>quasibounded</u> if both Re(f) and Im(f) are quasibounded. We denote by $h^1_Q(\mathbb{D})$ the set of quasibounded elements belonging to $h^1(\mathbb{D})$.

2. Some Classical Results

2A. We show that every function in $H^p(\mathbb{D})$ has radial limits along almost all radii.

<u>Theorem</u>. (a) Let $f \in H^p(\mathbb{D})$ with $0 < p \leqq \infty$. Then f admits radial boundary values $f^*(e^{i\theta}) = \lim_{r \to 1-0} f(re^{i\theta})$ for almost all $e^{i\theta} \in \mathbb{T}$ and $f^* \in L^p(d\sigma)$. For $0 < p < \infty$ the net $\{f_r: 0 < r < 1\}$ with $f_r(e^{i\theta}) = f(re^{i\theta})$ converges to f^* in $L^p(d\sigma)$ as $r \to 1-0$ and the LHM of $|f(z)|^p$ is the Poisson integral of $|f^*|^p$, i.e.

$$(\mathrm{LHM}(|f|^p))(z) = \int_{\mathbb{T}} P(r, \theta - t)|f^*(e^{it})|^p d\sigma(t)$$

for $z = re^{i\theta} \in \mathbb{D}$, and therefore it is quasibounded. If $f \not\equiv 0$, then $\log|f^*|$ is summable on $\mathbb{T}$.

(b) The map $f \to f^*$ is an injection of $H^p(\mathbb{D})$ into $L^p(d\sigma)$ satisfying $\|f\|_p = \|f^*\|_p$, where $\|f^*\|_p$ denotes the usual norm of the space $L^p(d\sigma)$. If $1 \leqq p \leqq \infty$, then the functional $\|\cdot\|_p$ defined in 1B is a norm, which makes $H^p(\mathbb{D})$ into a Banach space.

We denote by $H^p(d\sigma)$ the set of f^* with $f \in H^p(\mathbb{D})$, so that $H^p(\mathbb{D})$ is isometrically isomorphic with $H^p(d\sigma)$ under the map $f \to f^*$. The proof will be stated in 2B-2E, where we only deal with the case $0 < p < \infty$, for the case $p = \infty$ is easily covered by the discussion to follow.

2B. First consider the special case $p = 2$. Let $f \in H^2(\mathbb{D})$ and let u_f be the least harmonic majorant of $|f|^2$ on $\mathbb{D}$. Let $0 < r_1 < r_2 < \cdots < 1$ be a fixed sequence tending to 1 and set $f_n(e^{it}) = f(r_n e^{it})$ for $n = 1, 2, \ldots$. Then,

$$\int_{\mathbb{T}} |f_n(e^{it})|^2 d\sigma(t) \leqq \int_{\mathbb{T}} u_f(r_n e^{it}) d\sigma(t) = u_f(0)$$

for $n = 1, 2, \ldots$. So $\{f_n\}$ is a bounded sequence in $L^2(d\sigma)$. Since $L^2(d\sigma)$ is a separable Hilbert space, every closed ball is metrizable and compact with respect to the weak topology and therefore the sequence $\{f_n\}$ has a convergent subsequence. Namely, there exist a subsequence $\{f_{n(j)}: j = 1, 2, \ldots\}$ and a function $F \in L^2(d\sigma)$ such that

$$\int_{\mathbb{T}} f_{n(j)} k \, d\sigma \to \int_{\mathbb{T}} Fk \, d\sigma$$

for any $k \in L^2(d\sigma)$ as $j \to \infty$. Setting $k(e^{it}) = P(r,\theta - t)$ with $z = re^{i\theta} \in \mathbb{D}$, we have

$$f(r_{n(j)}z) = \int_{\mathbb{T}} f_{n(j)}(e^{it})P(r,\theta - t)d\sigma(t)$$

$$\to \int_{\mathbb{T}} F(e^{it})P(r,\theta - t)d\sigma(t)$$

and therefore

$$f(z) = \int_{\mathbb{T}} F(e^{it})P(r,\theta - t)d\sigma(t).$$

Fatou's theorem (Theorem A1.2, Appendix) then implies that, for almost every $e^{i\theta}$, $f(re^{i\theta})$ tends to $F(e^{i\theta})$ as r tends increasingly to 1. For any such θ, define $f^*(e^{i\theta}) = F(e^{i\theta})$, so that $f^* \in L^2(d\sigma)$ and $f^*(e^{i\theta}) = \lim_{r\to 1-0} f(re^{i\theta})$ a.e. We thus have

$$f(z) = \int_{\mathbb{T}} f^*(e^{it})P(r,\theta - t)d\sigma(t)$$

for $z = re^{i\theta} \in \mathbb{D}$. By use of the Schwarz inequality

(1) $$|f(z)|^2 \leqq \int_{\mathbb{T}} |f^*(e^{it})|^2 P(r,\theta - t)d\sigma(t).$$

The right-hand side, say u, gives a harmonic majorant of $|f|^2$ and so

(2) $$|f(z)|^2 \leqq u_f(z) \leqq u(z)$$

on $\mathbb{D}$. Since u is seen to be quasibounded, so is u_f. Thus u_f is equal to the Poisson integral of its radial boundary function. Taking radial limits in (2), we have $|f^*(e^{it})|^2 \leqq u_f^*(e^{it}) \leqq u^*(e^{it}) = |f^*(e^{it})|^2$ a.e. and therefore $u = u_f$. It follows that

$$\|f\|_2^2 = (\mathrm{LHM}(|f|^2))(0) = \int_{\mathbb{T}} |f^*|^2 d\sigma = \|f^*\|_2^2.$$

Namely, $f \to f^*$ gives an isometry of $H^2(\mathbb{D})$ into $L^2(d\sigma)$. It follows from (1) that the L^2-convergence of f^*'s implies the almost uniform convergence of $f(z)$'s on $\mathbb{D}$. Hence, $H^2(\mathbb{D})$ is a Hilbert space, which is canonically isometrically isomorphic with a subspace of $L^2(d\sigma)$.

2C. Suppose that $f \in H^2(\mathbb{D})$ with $f \not\equiv 0$. If f has a zero of order $m \geqq 0$ at the origin, then we set $f_0(z) = f(z)/z^m$. Then $f_0(0) \neq 0$, $f_0 \in H^2(\mathbb{D})$ and indeed $|f_0(z)|^2 \leqq u_f(z)$. Let $a_1, a_2, \ldots$ be the zeros of f_0, repeated according to multiplicity and ordered so that $0 < |a_1| \leqq |a_2| \leqq \cdots$. By Jensen's formula (cf. Rudin [59], p. 330)

$$\sum_{|a_k|<r} \log \frac{r}{|a_k|} = -\log|f_0(0)| + \int_{\mathbb{T}} \log|f_0(re^{it})|\,d\sigma(t)$$

$$\leqq -\log|f_0(0)| + \int_{\mathbb{T}} |f_0(re^{it})|^2 d\sigma(t)$$

$$\leqq -\log|f_0(0)| + \|f_0\|_2^2.$$

Letting $r \to 1-0$, we get $\sum_{k=1}^{\infty} \log(1/|a_k|) < \infty$, which is equivalent to:

(3) $$\prod_{k=1}^{\infty} |a_k| \quad \text{is convergent.}$$

This in turn implies that the Blaschke product with zeros $\{a_k\}$

$$B(z) = \prod_{k=1}^{\infty} \frac{|a_k|}{a_k} \frac{a_k - z}{1 - \bar{a}_k z}$$

converges almost uniformly on $\mathbb{D}$ and is an inner function, i.e. $|B(z)| \leqq 1$ on $\mathbb{D}$ and $|B^*(e^{it})| = 1$ a.e. on $\mathbb{T}$.

To see this, let $B_n(z)$ denote the n-th partial product of $B(z)$. Then, every B_n is analytic on the closed unit disk and $|B_n^*(e^{it})| \equiv 1$ on $\mathbb{T}$. Moreover, for $m > n$,

$$\|B_m - B_n\|_2^2 = \int_{\mathbb{T}} |B_m^*(e^{it}) - B_n^*(e^{it})|^2 d\sigma(t)$$

$$= 2\int_{\mathbb{T}} (1 - \mathrm{Re}(B_m^*(e^{it})/B_n^*(e^{it}))d\sigma(t)$$

$$= 2(1 - \mathrm{Re}[\int_{\mathbb{T}} (B_m^*(e^{it})/B_n^*(e^{it}))d\sigma(t)])$$

$$= 2(1 - \mathrm{Re}[B_m(0)/B_n(0)])$$

$$= 2(1 - \prod_{k=n+1}^{m} |a_k|) \to 0$$

as $n \to \infty$, because of (3). The sequence $\{B_n(z)\}$ is thus a Cauchy sequence in $H^2(\mathbb{D})$. By our observation in 2B, there exists an element $F \in H^2(\mathbb{D})$ such that $B_n \to F$ in $H^2(\mathbb{D})$ and $B_n^* \to F^*$ in $L^2(d\sigma)$.

Since B_n converge almost uniformly to F, we conclude that $F = B$. So, there exists a subsequence of B_n^* which converges a.e. to B^*. Hence, $|B^*| = 1$ a.e. on $\mathbb{T}$.

2D. Continuing the discussion of 2C, we set $h_n(z) = f_0(z)/B_n(z)$ for $n = 1, 2, \ldots$ and $h(z) = f_0(z)/B(z)$. Since $B_n(z)$ converge to $B(z)$ almost uniformly on $\mathbb{D}$, $h_n(z)$ converge almost uniformly to $h(z)$ on $\mathbb{D}$, so that h is analytic on $\mathbb{D}$.

Let n be fixed. Then, for any $0 < \varepsilon < 1$, there exists a $\delta > 0$ such that $1 - \varepsilon \leqq |B_n(z)| \leqq 1$ for $1 - \delta \leqq |z| \leqq 1$ and so

$$|h_n(z)|^2 \leqq |f_0(z)|^2/(1-\varepsilon)^2 \leqq u_f(z)/(1-\varepsilon)^2$$

for $1 - \delta \leqq |z| \leqq 1$. Since $|h_n(z)|^2$ is subharmonic, $|h_n(z)|^2 \leqq u_f(z)/(1-\varepsilon)^2$ everywhere on $\mathbb{D}$. As ε is arbitrary, $|h_n(z)|^2$ is majorized on $\mathbb{D}$ by the same $u_f(z)$ for every n. From this follows that $|h(z)|^2 \leqq u_f(z)$. Hence, $h \in H^2(\mathbb{D})$ with

$$(4) \qquad |h^*(e^{it})| = |f_0^*(e^{it})|/|B^*(e^{it})| = |f_0^*(e^{it})| = |f^*(e^{it})| \quad \text{a.e.}$$

We thus have

$$(5) \qquad f(z) = z^m B(z) h(z),$$

where $h(z)$ has no zeros in $\mathbb{D}$.

Finally, we show that $\log|f^*|$ is integrable. Since $\log|h(z)| \leqq |h(z)|^2/2 \leqq u_f(z)/2$ and since $h(z)$ has no zeros in $\mathbb{D}$,

$$v(z) = -\log|h(z)| + u_f(z)/2$$

is a nonnegative harmonic function on $\mathbb{D}$. By Theorem 2A, Ch. III (or directly), there exists a nonnegative measure $d\mu$ on $\mathbb{T}$ such that

$$v(z) = \int_{\mathbb{T}} P(r, \theta - t)\, d\mu(e^{it})$$

for $z = re^{i\theta} \in \mathbb{D}$. Fatou's theorem then says that $v^*(e^{it})d\sigma(t)$ is the absolutely continuous part of the measure $d\mu(e^{it})$ and therefore v^* is summable. Since $v^*(e^{it}) = -\log|h^*(e^{it})| + u_f^*(e^{it})/2 = -\log|h^*(e^{it})| + |f^*(e^{it})|^2$ and since $f^* \in L^2(d\sigma)$, $\log|h^*(e^{it})|$ must be summable. Hence, by (4), $\log|f^*(e^{it})|$ is summable.

2E. Turning to the general case, let us take any $f \in H^p(\mathbb{D})$,

$f \not\equiv 0$, with $0 < p < \infty$. Let u be the LHM of $|f|^p$ on $\mathbb{D}$. Let $\tilde{u}$ be the harmonic conjugate of u on $\mathbb{D}$ and set

$$f_1(z) = \exp(-\frac{1}{p}(u(z) + i\tilde{u}(z))).$$

Then f_1 is a nonvanishing analytic function with $\|f_1\|_\infty \leqq 1$ and

$$|f| \leqq u^{1/p} \leqq \exp(u/p) = |f_1|^{-1}.$$

Setting $f_2 = ff_1$, we have $f_2 \in H^\infty(\mathbb{D})$, $f_2 \not\equiv 0$, $\|f_2\|_\infty \leqq 1$ and $f = f_2/f_1$. Since $H^\infty(\mathbb{D}) \subseteqq H^2(\mathbb{D})$, our observation in 2D shows that f_1^* and f_2^* exist a.e. and that both $\log|f_1^*|$ and $\log|f_2^*|$ are summable. It follows that $\log|f^*| = \log|f_2^*| - \log|f_1^*|$ is also summable. We then apply the factorization formula (5) to f_2 and get $f_2(z) = z^m B(z)\ell(z)$, where B is the Blaschke product formed of the zeros $\neq 0$ of f_2, and ℓ has no zeros in $\mathbb{D}$ with $\|\ell\|_\infty \leqq 1$. So we have $f(z) = z^m B(z)h(z)$, where $h(z) = \ell(z)/f_1(z)$ has no zeros in $\mathbb{D}$.

Set, as before, $f_0(z) = f(z)/z^m$ and $h_n(z) = f_0(z)/B_n(z)$, $n = 1, 2, \ldots$, where B_n is the finite Blaschke product formed of the first n zeros $a_1, \ldots, a_n$. Our argument in 2D then shows that $|h_n(z)|^p \leqq u(z)$ for $n = 1, 2, \ldots$ and thus $|h(z)|^p \leqq u(z)$. Hence $h \in H^p(\mathbb{D})$ and

$$|h^*(e^{it})| = \lim_{r\to 1-0} |f(re^{it})|/r^m|B(re^{it})| = |f^*(e^{it})| \quad \text{a.e.}$$

on $\mathbb{T}$. Since $|f(z)|^p$ is subharmonic on $\mathbb{T}$, we see that the integrals $\int_{\mathbb{T}} |f(re^{it})|^p \, d\sigma(t)$ increase as $r \to 1-0$. By Fatou's lemma we have

$$\begin{aligned}\int_{\mathbb{T}} |f^*(e^{it})|^p d\sigma(t) &= \int_{\mathbb{T}} (\lim_{n\to\infty} |f(r_n e^{it})|^p) d\sigma(t) \\ &\leqq \liminf_{n\to\infty} \int_{\mathbb{T}} |f(r_n e^{it})|^p d\sigma(t) \\ &\leqq u(0) < \infty.\end{aligned}$$

Hence, $f^* \in L^p(d\sigma)$ and so $h^* \in L^p(d\sigma)$.

Since h has no zeros in $\mathbb{D}$, there exists an analytic function $h_0 \in H^2(\mathbb{D})$ such that $h = h_0^{2/p}$. Let u_0 be the LHM of $|h_0|^2$. As shown in 2B, u_0 is quasibounded and $v_0 = -\log|h_0| + u_0/2$ is a nonnegative harmonic function on $\mathbb{D}$. Since the projections pr_I and pr_Q in the space $SP'(\mathbb{D})$ are positive, both $pr_I(v_0)$ and $pr_Q(v_0)$ are nonnegative. Moreover, since u_0 is quasibounded, $pr_I(\log|h_0|) = -pr_I(v_0) \leqq 0$ and thus, in view of Fatou's theorem (Appendix A.1.3),

$$\begin{aligned} |h_0(z)| &= \exp(\log|h_0(z)|) \\ &= \exp[\mathrm{pr}_I(\log|h_0(z)|) + \mathrm{pr}_Q(\log|h_0(z)|)] \\ &\leqq \exp(\mathrm{pr}_Q(\log|h_0(z)|)) \\ &= \exp[\int_{\mathbb{T}} P(r,\theta - t)\log|h_0^*(e^{it})|\, d\sigma(t)] \end{aligned}$$

for $z = re^{i\theta}$. For the function f we get

$$\begin{aligned} |f(z)|^p &\leqq |h(z)|^p = |h_0(z)|^2 \\ &\leqq \exp[\int_{\mathbb{T}} P(r,\theta - t)\log|h_0^*(e^{it})|^2\, d\sigma(t)] \\ &\leqq \int_{\mathbb{T}} |h_0^*(e^{it})|^2 P(r,\theta - t)d\sigma(t) \\ &= \int_{\mathbb{T}} |f^*(e^{it})|^p P(r,\theta - t)d\sigma(t) \end{aligned}$$

for $z = re^{i\theta} \in \mathbb{D}$. Just as in 2B, we infer that

$$(\mathrm{LHM}(|f|^p))(z) = \int_{\mathbb{T}} |f^*(e^{it})|^p P(r,\theta - t)d\sigma(t),$$

for $z = re^{it} \in \mathbb{D}$, which is quasibounded, and therefore $\|f\|_p = \|f^*\|_p$. This means finally that, for $1 \leqq p < \infty$, $\|\cdot\|_p$ is a norm, which makes $H^p(\mathbb{D})$ into a Banach space. This completes the proof of Theorem 2A. □

2F. As for functions in $h^p(\mathbb{D})$ we have the following

<u>Theorem</u>. Let $f \in h^p(\mathbb{D})$ with $1 \leqq p \leqq \infty$. Then the following hold:

(a) f has radial boundary values $f^*(e^{it})$ a.e. on $\mathbb{T}$ and the function f^* belongs to $L^p(d\sigma)$.

(b) If $1 < p \leqq \infty$, then f is quasibounded and is the Poisson integral of f^* with $\|f\|_p = \|f^*\|_p$.

(c) If $p = 1$, then the Poisson integral of f^* is equal to the quasibounded part, $\mathrm{pr}_Q(f)$, of f and $\|\mathrm{pr}_Q(f)\|_1 = \|f^*\|_1$.

The map $f \to f^*$ is an isometric isomorphism of $h^p(\mathbb{D})$ onto $L^p(d\sigma)$, if $1 < p \leqq \infty$, and of $h^1_Q(\mathbb{D})$ onto $L^1(d\sigma)$.

<u>Proof</u>. Since the case $p = \infty$ is rather obvious, we only consider the case $1 \leqq p < \infty$. Let $0 < r_1 < r_2 < \cdots < 1$ be a fixed sequence tending to 1 and set $f_n(e^{it}) = f(r_n e^{it})$. Then

$$\int_{\mathbb{T}} |f_n(e^{it})|^p d\sigma(t) \leqq \int_{\mathbb{T}} u(r_n e^{it}) d\sigma(t) = u(0),$$

where u is a harmonic majorant of $|f|^p$. So $\{f_n\}$ is a bounded sequence in $L^p(d\sigma)$. If $1 < p < \infty$, then $L^p(d\sigma)$ is a reflexive Banach space; so, by passing to a subsequence if necessary, we may assume that $\{f_n\}$ converges weakly in $L^p(d\sigma)$, i.e. there exists an element, f^*, in $L^p(d\sigma)$ such that

$$\int_{\mathbb{T}} f_n(e^{it}) k(e^{it}) d\sigma(t) \to \int_{\mathbb{T}} f^*(e^{it}) k(e^{it}) d\sigma(t)$$

for any $k \in L^{p'}(d\sigma)$ with $p^{-1} + p'^{-1} = 1$. Setting $k(e^{it}) = P(r, \theta - t)$ with any fixed $z = re^{i\theta} \in \mathbb{D}$, we get

$$f(r_n z) = \int_{\mathbb{T}} f_n(e^{it}) P(r, \theta - t) d\sigma(t) \to \int_{\mathbb{T}} f^*(e^{it}) P(r, \theta - t) d\sigma(t)$$

and therefore

$$f(z) = \int_{\mathbb{T}} f^*(e^{it}) P(r, \theta - t) d\sigma(t).$$

By Fatou's theorem, $\lim_{r \to 1-0} f(re^{i\theta}) = f^*(e^{i\theta})$ a.e. This also shows that f is quasibounded. By Hölder's inequality

$$|f(z)|^p \leqq \int_{\mathbb{T}} |f^*(e^{it})|^p P(r, \theta - t) d\sigma(t)$$

and the argument in 2B shows that the right-hand side gives the least harmonic majorant of $|f|^p$, which implies $\|f\|_p = \|f^*\|_p$.

Finally let $p = 1$. Then the sequence $\{f_n d\sigma: n = 1, 2, \ldots\}$ is bounded in the space $M(\mathbb{T})$ of finite Borel measures on $\mathbb{T}$. Since the closed ball of $M(\mathbb{T})$ is compact and metrizable with respect to the weak topology $w(M(\mathbb{T}), C(\mathbb{T}))$, the sequence contains a convergent subsequence. Namely, there exist a subsequence $\{f_{n(j)} d\sigma: j = 1, 2, \ldots\}$ and a finite Borel measure $d\nu$ on $\mathbb{T}$ such that

$$\int_{\mathbb{T}} k(e^{it}) f_{n(j)}(e^{it}) d\sigma(t) \to \int_{\mathbb{T}} k(e^{it}) d\nu(e^{it})$$

for any $k \in C(\mathbb{T})$. By setting $k(e^{it}) = P(r, \theta - t)$, $z = re^{i\theta} \in \mathbb{D}$, we get as above

$$f(z) = \int_{\mathbb{T}} P(r, \theta - t) d\nu(e^{it}).$$

By Fatou's theorem again, $f^* d\sigma$ is exactly the absolutely continuous part of $d\nu$, so that

$$\mathrm{pr}_Q(f)(z) = \int_{\mathbb{T}} f^*(e^{it})P(r,\theta - t)d\sigma(t).$$

This implies at once that $\|\mathrm{pr}_Q(f)\|_1 = \|f^*\|_1$. □

§2. HARDY CLASSES ON HYPERBOLIC RIEMANN SURFACES

3. Boundary Behavior of H^p and h^p Functions

3A. We look at Hardy classes on a hyperbolic Riemann surface R with origin $\mathcal{O}$. The basic definitions are the same as in the case of the unit disk but are repeated for the sake of completeness. For $0 < p < \infty$, $H^p(R)$ denotes the set of analytic functions f on R such that $|f|^p$ has a harmonic majorant. $H^\infty(R)$ denotes the set of bounded analytic functions on R. For $f \in H^p(R)$ with $0 < p < \infty$, we put

$$\|f\|_p = (\mathrm{LHM}(|f|^p)(\mathcal{O}))^{1/p},$$

where LHM stands for the least harmonic majorant on the surface R. For $f \in H^\infty(R)$ we set $\|f\|_\infty = \sup\{|f(z)|: z \in R\}$. It is clear that $H^p(R)$, $0 < p \leqq \infty$, are complex linear spaces and that $H^\infty(R)$ is an algebra over the complex field. We also define $h^p(R)$, $1 \leqq p < \infty$, to be the space of complex-valued harmonic functions f on R such that $|f|^p$ has a harmonic majorant, and $h^\infty(R)$ the space of complex-valued bounded harmonic functions on R. A complex-valued harmonic function f is called quasibounded if both $\mathrm{Re}(f)$ and $\mathrm{Im}(f)$ are quasibounded. And $h^1_Q(R)$ denotes the set of quasibounded elements in $h^1(R)$. We define the functionals $\|\cdot\|_p$ even on $h^p(R)$, $1 \leqq p \leqq \infty$, by means of the above formulas. It is trivial to see that $H^p(R) \subseteqq h^p(R)$ for $1 \leqq p \leqq \infty$. By using a universal covering map $\phi_R: \mathbb{D} \to R$ with $\phi_R(0) = \mathcal{O}$ and Theorems 6B and 7B, Ch. III, we easily deduce the following result from Theorem 2A.

<u>Theorem</u>. (a) Let $f \in H^p(R)$ with $0 < p < \infty$. Then f admits fine boundary values $\hat{f}(b)$ a.e. on Δ_1 and $\hat{f} \in L^p(d\chi)$. For $0 < p < \infty$ the LHM of $|f|^p$ is exactly the Poisson integral $H[|\hat{f}|^p]$ of $|\hat{f}|^p$ and therefore is quasibounded. If $f \not\equiv 0$, then $\log|\hat{f}|$ is summable on Δ_1.

(b) The map $f \to \hat{f}$ is an injection of $H^p(R)$ into $L^p(d\chi)$ satisfying $\|f\|_p = \|\hat{f}\|_p$. If $1 \leqq p \leqq \infty$, then $\|\cdot\|_p$ is a norm, which makes $H^p(R)$ into a Banach space.

We define a subspace $H^p(d\chi)$ of $L^p(d\chi)$ by setting

$$H^p(d\chi) = \{\hat{f} \in L^p(d\chi): f \in H^p(R)\},$$

which is isometrically isomorphic with $H^p(R)$ under the map $f \to \hat{f}$.

3B. Similarly, Theorem 2F implies the following

Theorem. Let $1 \leqq p \leqq \infty$. Then, for each $f \in h^p(R)$, the fine boundary function $\hat{f}$ is defined a.e. on Δ_1 and belongs to $L^p(d\chi)$. Put $S(f) = \hat{f}$. Then S is a linear map of $h^p(R)$ into $L^p(d\chi)$ such that

(a) S is isometric and surjective for $1 < p \leqq \infty$, in which case $f = H[\hat{f}]$ for any $f \in h^p(R)$, i.e. $H[\cdot]$ is the inverse of the map S;

(b) S is norm-decreasing for $p = 1$ and is isometric as well as surjective on the space $h^1_Q(R)$;

(c) S is isometric and injective on each $H^p(R)$ for $1 \leqq p \leqq \infty$.

4. Some Results on Multiplicative Analytic Functions

4A. As another application of the covering map $\phi_R: \mathbb{D} \to R$, we prove the following result on LHM's of multiplicative analytic functions.

Theorem. Let u be an l.a.m. (Cf. II, 3C) on R such that either u is bounded or u^p has a harmonic majorant for some $1 \leqq p < \infty$. Then we have the following:

(a) $\mathrm{LHM}(u^p)$ is quasibounded, if $p \neq \infty$.

(b) u is of bounded characteristic, i.e. $\log u \in SP'(R)$.

(c) $(\log u) \vee 0$ is quasibounded.

(d) The inner factor $u_I = \exp(\mathrm{pr}_I(\log u))$ is a bounded l.a.m. and $\|u_I\|_\infty = 1$.

Proof. In view of Theorem 6B, Ch. III, we have only to prove the statements in the case $R = \mathbb{D}$. We may also assume that $p < \infty$. The statements (a), (b) and (c) are almost clear from our discussion in 2A-2E. To see (d), we note that $-\log u$ is a superharmonic function having a quasibounded harmonic minorant, e.g. $h = -p^{-1}(\mathrm{LHM}(u^p))$. Thus the function $-\log u - h$ is nonnegative and therefore belongs to $SP'(\mathbb{D})$. By use of Theorem 5A, Ch. II, we decompose this into the inner and the quasibounded parts, i.e. $-\log u - h = v_1 + v_2$ with $v_1 \in I(R)$ and $v_2 \in Q(R)$. Since the projections are positive, we have $v_1 \geqq 0$ and $v_2 \geqq 0$. Since h is quasibounded, we have in particular $\mathrm{pr}_I(\log u) = -v_1 \leqq 0$. Moreover, we have $\inf_{z\in R} v_1(z) = 0$, for v_1 is inner.

Hence $u_I = \exp(-v_1) \leqq 1$ and $\sup_{z\in R} u_I(z) = 1$, as desired. □

4B. We also compute $\hat{u}$ for $u \in SP'(R)$.

Theorem. If $v \in SP'(R)$, then $\hat{v}$ exists a.e. on Δ_1 and $\hat{v} = \hat{v}_q$ a.e. on Δ_1, where $v_q = pr_Q(v)$ is the quasibounded part of v.

Proof. Since $v = v^+ - v^-$ in view of Ch. II, 5A, we may suppose that v is nonnegative. Now let us decompose v into its inner and quasibounded parts v_i and v_q, i.e. $v = v_i + v_q$. Since $v \geqq 0$, we have $v_i \geqq 0$ and $v_q \geqq 0$. By Theorem 5B, Ch. II, v_q is everywhere harmonic on R and the singularities of v_i, if any, are isolated. Let $a \in R$ be a singularity of v_i. Since v_i has only logarithmic singularities, we have $v_i(z) = c\log|z| + h(z)$ in a parametric disk centered at a, where $h(z)$ is a harmonic function. The function v_i being nonnegative, we see that $c \leqq 0$ and therefore v_i is superharmonic in a neighborhood of a. As a is arbitrary, v_i is superharmonic on R. Thus, by Riesz's theorem (Theorem 6F, Ch. I) we have $v_i = h_s + h_p$, where h_s is a nonnegative harmonic function and h_p is a potential. Since v_i is inner, so is h_s. Combined with Lemma 5C and Theorem 5D, Ch. III, these observations imply that $\hat{h}_s$ and $\hat{h}_p$ exist and vanish a.e. on Δ_1. So $\hat{v}_i$ exists and vanishes a.e. on Δ_1. Moreover, $\hat{v}_q$ exists a.e. again by Lemma 5C, Ch. III. Hence, $\hat{v}$ exists a.e. and is equal to $\hat{v}_q$ a.e. on Δ_1. □

Corollary. If u is a nonzero l.m.m. of bounded characteristic, then $\hat{u}$ exists a.e. on Δ_1 and

$$\hat{u}_I = 1 \quad \text{and} \quad \hat{u} = \hat{u}_Q \quad \text{a.e.}$$

on Δ_1, where $u_I = \exp(pr_I(\log u))$ and $u_Q = \exp(pr_Q(\log u))$.

Proof. Set $v = \log u$. Since $v \in SP'(R)$ by the assumption, the preceding theorem shows that $\hat{v}$ exists a.e. on Δ_1 and is equal to $\hat{v}_q$ a.e. with $v_q = pr_Q(v)$. Hence we have $\hat{u} = \exp\hat{v} = \exp\hat{v}_q = \hat{u}_Q$. It is now trivial that $\hat{u}_I = 1$ a.e. □

5. The β-Topology

5A. We will describe the β-topology for the space $h^\infty(R)$. For this purpose we first recall some basic facts in the theory of locally convex linear spaces. Let E be a locally convex linear space over the complex field. Precisely speaking, this is defined by the following conditions:

(A1) E is a linear space over the complex field;

(A2) E is a Hausdorff space such that addition and scalar multiplication are each continuous in both variables jointly and such that the origin 0 has a basis of neighborhoods consisting of convex subsets of E.

Let E' denote the dual of E, which is defined as the linear space of all continuous linear functionals on E. For each $x \in E$ and each $x' \in E'$ we write $\langle x,x'\rangle$ in place of $x'(x)$. Then the correspondence $(x,x') \to \langle x,x'\rangle$ defines a bilinear form on $E \times E'$, which is nondegenerate, meaning that $\{x \in E: \langle x,x'\rangle = 0$ for all $x' \in E'\} = \{0\}$ and $\{x' \in E': \langle x,x'\rangle = 0$ for all $x \in E\} = \{0\}$. For a subset A of E (resp. E') we set $A^\circ = \{x' \in E': |\langle x,x'\rangle| \leqq 1$ for all $x \in A\}$ (resp. $\{x \in E: |\langle x,x'\rangle| \leqq 1$ for all $x' \in A\}$) and call it the <u>polar</u> of A in E' (resp. in E). A subset A of E is called <u>bounded</u> if to each neighborhood U of 0 in E corresponds a positive number $\alpha > 0$ satisfying $A \subseteqq \alpha U$. On the other hand, a subset A of E (resp. E') is called weakly bounded (resp. weakly* bounded) if $\sup\{|\langle x,x'\rangle|: x \in A\} < \infty$ for every $x' \in E'$ (resp. $\sup\{|\langle x,x'\rangle|: x' \in A\} < \infty$ for every $x \in E$). We note in passing that a subset of E is bounded if and only if it is weakly bounded. The <u>strong topology</u> $s(E',E)$ is then defined as the topology of uniform convergence on bounded subsets of E. In other words, it is the locally convex topology for E' such that the totality of polars A° with A ranging over all bounded subsets of E forms a basis of neighborhoods of 0. The space E' with the topology $s(E',E)$ is called the <u>strong dual</u> of E and is denoted by E'_s. The dual of the topological linear space E'_s is called the <u>bidual</u> of E and is denoted by E''. Identification of each $x \in E$ with a linear functional $x' \to \langle x,x'\rangle$ on E' gives us a natural inclusion $E \subseteqq E''$. We call E <u>semireflexive</u> if $E = E''$.

Finally, the weak-(E,E') (resp. weak-(E',E)) topology, written as $w(E,E')$ (resp. $w(E',E)$), is defined as the weakest locally convex topology for E (resp. for E') that makes every functional $x' \in E'$ (resp. $x \in E$) continuous.

<u>Theorem</u>. A locally convex linear space E is semireflexive if and only if every bounded subset of E is relatively $w(E,E')$-compact. (cf. Kelley and Namioka [37], p. 190)

5B. We now look at the space $h^\infty(R)$ of complex-valued bounded harmonic functions on R. Let us define in it two kinds of topology. The first one is the usual norm topology, which is given by the sup-norm

(6) $$\|h\|_\infty = \sup\{|h(z)|: z \in R\}$$

for $h \in h^\infty(R)$. The second one is the β-topology (or the strict topology), which is the main objective of this section. In order to define this, let $C_0(R)$ be the space of all complex-valued continuous functions f on R that vanish at infinity, i.e. $\{z \in R: |f(z)| \geqq \varepsilon\}$ is compact for any $\varepsilon > 0$. Clearly, $C_0(R)$ forms a Banach space with respect to the usual addition and scalar multiplication of functions and the sup-norm of the form (6).

Now, for every $f \in C_0(R)$ we define a seminorm N_f in $h^\infty(R)$ by setting $N_f(h) = \|fh\|_\infty$. The totality $\{N_f: f \in C_0(R)\}$ of seminorms determines a locally convex topology for $h^\infty(R)$, which we call the $\underline{\beta\text{-topology}}$. In other words, the collection of sets of the form

$$V_f = \{h \in h^\infty(R): \|fh\|_\infty < 1\},$$

with f ranging over $C_0(R)$, makes up a basis of neighborhoods of 0 for this topology. The space $h^\infty(R)$ equipped with the β-topology is denoted by $h^\infty_\beta(R)$.

5C. We want to determine the dual of $h^\infty_\beta(R)$, which we call the β-dual of $h^\infty(R)$. Let $M_b(R)$ be the space of all complex-valued bounded Borel measures on R. This forms a Banach space with respect to the usual addition and scalar multiplication of measures and the total variation norm $\|\mu\| = \int_R |d\mu|$. Each $\mu \in M_b(R)$ determines a linear functional on $C_0(R)$ by means of the formula

(7) $$\langle h,\mu\rangle = \int_R h d\mu$$

for $h \in C_0(R)$. As we see easily by means of Riesz's representation theorem, the space $M_b(R)$ is identified via (7) with the dual $C_0(R)'$ of the Banach space $C_0(R)$. On the other hand, we set

$$N(R) = \{\mu \in M_b(R): \langle h,\mu\rangle = 0 \ \text{ for all } \ h \in h^\infty(R)\}$$

and denote by $M_b^*(R)$ the (algebraic) quotient space $M_b(R)/N(R)$. Then we have the following:

Theorem. The space $h^\infty_\beta(R)$ is semireflexive and its dual $h^\infty_\beta(R)'$ is (algebraically) identified with $M_b^*(R)$.

Proof. The proof is divided into three parts.

(a) First we show that the dual $h^\infty_\beta(R)'$ is algebraically equal to $M_b^*(R)$. Let $\mu \in M_b(R)$; then the linear functional $F_\mu: h \to \langle h,\mu\rangle$ is continuous on $h_\beta^\infty(R)$. To see this, we choose a sequence $\{K_n: n = 1, 2,\ldots\}$ of compact subsets of R such that $K_n \subseteqq \mathrm{Int}(K_{n+1})$, $R = \cup_{n=1}^\infty K_n$ and $|\mu|(R \setminus \mathrm{Int}(K_n)) \leqq 4^{-n}$ for $n = 1, 2,\ldots$. We then use Urysohn's lemma to define continuous functions f_n, $n = 1, 2,\ldots$, on R such that $f_n = 0$ on $R \setminus \mathrm{Int}(K_{n+1})$, $= 1$ on K_n, and $0 \leqq f_n \leqq 1$ everywhere on R. Then the function defined by

$$f = \sum_{k=1}^{\infty} 2^{-k} f_k$$

belongs to $C_0(R)$ and in fact $2^{-n-1} \leqq f \leqq 2^{-n+1}$ on $K_{n+1} \setminus K_n$ for $n = 0, 1,\ldots$, where K_0 denotes the empty subset. Finally, we define a measure μ' by setting $d\mu' = f^{-1}d\mu$. Then $\mu' \in M_b(R)$, because

$$|\mu'|(R) = \sum_{n=0}^{\infty} |\mu'|(K_{n+1} \setminus K_n) \leqq \sum_{n=0}^{\infty} 2^{n+1}|\mu|(K_{n+1} \setminus K_n)$$

$$\leqq 2|\mu|(K_1) + \sum_{n=1}^{\infty} 2^{n+1}\cdot 4^{-n} < \infty.$$

It follows that, for every $h \in h^\infty(R)$,

$$|F_\mu(h)| = |\langle h,\mu\rangle| = |\langle fh,\mu'\rangle| \leqq \|fh\|_\infty \|\mu'\|$$

$$= \|\mu'\| N_f(h).$$

Hence, F_μ is β-continuous. Since this is true of any $\mu \in M_b(R)$, we have shown that $M_b^*(R) \subseteqq h_\beta^\infty(R)'$.

In order to see the reverse inclusion, we take any β-continuous linear functional F on $h^\infty(R)$. Then there exists an $f \in C_0(R)$ such that $|F(h)| \leqq N_f(h)$ for every $h \in h^\infty(R)$. We then define a linear functional F_1 on the subspace $fh^\infty(R) = \{fh: h \in h^\infty(R)\}$ of $C_0(R)$ by setting $F_1(fh) = F(h)$. This is well-defined, for $fh = 0$ implies $|F(h)| \leqq N_f(h) = \|fh\|_\infty = 0$. Since we have $|F_1(fh)| \leqq \|fh\|_\infty$, F_1 is continuous on $fh^\infty(R)$ in the norm topology. By use of the Hahn-Banach theorem F_1 can be extended to a linear functional F_2 on the Banach space $C_0(R)$ without changing the bound. As we remarked above, the dual of $C_0(R)$ is identified with $M_b(R)$. Therefore, we can find a measure $\mu'' \in M_b(R)$ which represents the functional F_2. So we have

$$F(h) = F_1(fh) = F_2(fh) = \int_R fh\,d\mu''.$$

Setting $d\mu = f d\mu''$, we see that $\mu \in M_b(R)$ and $F(h) = \int_R h\,d\mu$ for all

$h \in h^{\infty}(R)$. This means that $h^{\infty}_{\beta}(R)' \subseteqq M^{*}_{b}(R)$, as was to be proved.

(b) Next we show that a subset of $h^{\infty}(R)$ is β-bounded--bounded in the β-topology--if and only it is bounded in the norm topology. Let A be any norm-bounded subset of $h^{\infty}(R)$. Then, for any $f \in C_0(R)$, we have

$$\sup\{N_f(h): h \in A\} = \sup\{\|fh\|_{\infty}: h \in A\}$$

$$\leqq \|f\|_{\infty}\sup\{\|h\|_{\infty}: h \in A\} < \infty$$

and therefore A is β-bounded.

Conversely, let A be any β-bounded subset of $h^{\infty}(R)$. Then A is bounded in the weak topology $w(h^{\infty}_{\beta}(R), h^{\infty}_{\beta}(R)')$. As we have seen in (a) that $h^{\infty}_{\beta}(R)' = M^{*}_{b}(R)$, it is an easy matter to verify that

(8) $$\sup\{|\langle h,\mu\rangle|: h \in A\} < \infty$$

for every fixed $\mu \in M_b(R)$. On the other hand, every $h \in h^{\infty}(R)$ can be viewed as a linear functional, say F_h, on the Banach space $M_b(R)$ via the formula $F_h(\mu) = \langle h,\mu\rangle$. Since $M_b(R)$ contains all point measures on R, we see that

$$\|F_h\| = \sup\{|\langle h,\mu\rangle|: \mu \in M_b(R),\ \|\mu\| \leqq 1\} = \|h\|_{\infty}$$

for every $h \in h^{\infty}(R)$ and therefore that the Banach space $h^{\infty}(R)$ is isometrically embedded in the dual $M_b(R)'$ of the Banach space $M_b(R)$. Our property (8) then says that A is weakly* bounded as a subset of $M_b(R)'$. By the principle of uniform boundedness (Dunford and Schwartz [8], p. 52), A is norm-bounded in the space $M_b(R)'$ and hence is norm-bounded in the space $h^{\infty}(R)$, as claimed.

(c) Finally we will show that $h^{\infty}_{\beta}(R)$ is semireflexive. By the fact mentioned in (b) as well as Theorem 5A, it is sufficient to prove that the ball $B = \{h \in h^{\infty}(R): \|h\|_{\infty} \leqq 1\}$ is $w(h^{\infty}(R), M_b(R))$-compact. Thus, let $\{h_{\lambda}: \lambda \in \Lambda\}$ be any net in B. Then, the net $\{\hat{h}_{\lambda}: \lambda \in \Lambda\}$ consisting of fine boundary functions for h_{λ}'s is contained in the closed unit ball, say B_1, of $L^{\infty}(d\chi)$ (Theorem 3B). Since $L^{\infty}(d\chi)$ is the dual of the Banach space $L^1(d\chi)$, B_1 is $w(L^{\infty}(d\chi), L^1(d\chi))$-compact by means of Alaoglu's theorem (Dunford and Schwartz [8], p. 424). So there exist a subnet $\{h_{\lambda'}: \lambda' \in \Lambda'\}$ of $\{h_{\lambda}\}$ and an element $h^{*} \in B_1$ such that $\{\hat{h}_{\lambda'}\}$ converges to h^{*} in the topology $w(L^{\infty}(d\chi), L^1(d\chi))$; namely, we have

$$\int_{\Delta_1} (\hat{h}_{\lambda'} - h^{*})f^{*}d\chi \to 0$$

for every $f^{*} \in L^1(d\chi)$. Since the map $h \to \hat{h}$ is a isometric linear map of the normed space $h^\infty(R)$ onto $L^\infty(d\chi)$ (Theorem 3B), there exists an $h \in h^\infty(R)$ such that $\|h\|_\infty \leqq 1$ and $\hat{h} = h^*$ a.e. Now we take any $\mu \in M_b(R)$ and set

$$f^*(b) = \int_R k_b(z)d\mu(z)$$

for any $b \in \Delta_1$. Then, f^* belongs to $L^1(d\chi)$; in fact

$$\int_{\Delta_1} |f^*(b)|d\chi(b) \leqq \int_{\Delta_1} [\int_R k_b(z)|d\mu(z)|]d\chi(b)$$

$$= \int_R [\int_{\Delta_1} k_b(z)d\chi(b)]|d\mu(z)|$$

$$= \int_R |d\mu| = \|\mu\| < \infty.$$

We thus have

$$\int_R (h_{\lambda'}(z) - h(z))d\mu(z) = \int_R [\int_{\Delta_1} (\hat{h}_{\lambda'}(b) - \hat{h}(b))k_b(z)d\chi(b)]d\mu(z)$$

$$= \int_{\Delta_1} (\hat{h}_{\lambda'}(b) - \hat{h}(b))f^*(b)d\chi(b)$$

$$= \int_{\Delta_1} (\hat{h}_{\lambda'}(b) - h^*(b))f^*(b)d\chi(b) \to 0.$$

Namely, the subnet $\{h_{\lambda'}\}$ converges to h with respect to the topology $w(h^\infty(R),M_b(R))$. This implies that B is $w(h^\infty(R),M_b(R))$-compact, as desired. □

5D. Here is another characterization of the dual of $h^\infty(R)$.

<u>Theorem</u>. A linear functional F on $h^\infty(R)$ is β-continuous if and only if there exists a function $f^* \in L^1(d\chi)$ such that

$$F(h) = \int_{\Delta_1} \hat{h}(b)f^*(b)d\chi(b) \tag{9}$$

for $h \in h^\infty(R)$. The correspondence $F \to f^*$ is a bijection from the β-dual $h^\infty_\beta(R)'$ of $h^\infty_\beta(R)$ onto the space $L^1(d\chi)$.

<u>Proof</u>. Let F be any β-continuous linear functional on $h^\infty(R)$. Then, by Theorem 5C, there exists a measure $\mu \in M_b(R)$ such that $F(h) = \int_R h(z)d\mu(z)$ for $h \in h^\infty(R)$. Since Theorem 3B shows that

$$h(z) = \int_{\Delta_1} \hat{h}(b)k_b(z)d\chi(b)$$

for $h \in h^\infty(R)$, we have by use of the Fubini theorem

$$F(h) = \int_{\Delta_1} \hat{h}(b)f^*(b)d\chi(b)$$

with $f^*(b) = \int_R k_b(z)d\mu(z) \in L^1(d\chi)$. Hence, (9) holds.

Conversely, let $f^* \in L^1(d\chi)$ and define a functional F on $h^\infty(R)$ by means of the formula (9). In order to prove that F is β-continuous, it suffices to show that the kernel $\mathrm{Ker}(F)$ of F, given by $\{h \in h^\infty(R): F(h) = 0\}$, is closed in the β-topology. Since the dual of $h^\infty_\beta(R)$ is equal to $M^*_b(R)$ by Theorem 5C, $\mathrm{Ker}(F)$ is β-closed if and only if it is $w(h^\infty(R),M^*_b(R))$-closed or, equivalently, $w(h^\infty(R),M_b(R))$-closed in $h^\infty(R)$. If we embed $h^\infty(R)$ in the dual $M_b(R)'$ of the Banach space $M_b(R)$ as in the part (b) of the proof of Theorem 5C, then it is sufficient to prove that $\mathrm{Ker}(F)$ is a $w(M_b(R)',M_b(R))$-closed subspace of $M_b(R)'$. In view of the Krein-Šmulian theorem (Dunford and Schwartz [8], p. 429), this happens if (and only if) the intersection of $\mathrm{Ker}(F)$ with the closed unit ball B' of $M_b(R)'$ is compact with respect to the topology $w(M_b(R)',M_b(R))$. If B denotes the closed unit ball of the normed space $h^\infty(R)$, then

$$\mathrm{Ker}(F) \cap B' = \{h \in \mathrm{Ker}(F): \|h\|_\infty \leqq 1\} = \mathrm{Ker}(F) \cap B.$$

So it is enough to show that $\mathrm{Ker}(F) \cap B$ is $w(h^\infty(R),M_b(R))$-compact. As we know already that B is $w(h^\infty(R),M_b(R))$-compact, so we have only to prove that $\mathrm{Ker}(F) \cap B$ is $w(h^\infty(R),M_b(R))$-closed. Thus, suppose that a net $\{h_\lambda: \lambda \in \Lambda\}$ in $\mathrm{Ker}(F) \cap B$ converge to some element $k \in B$ in the weak topology $w(h^\infty(R),M_b(R))$. We then take fine boundary functions to form a net $\{\hat{h}_\lambda: \lambda \in \Lambda\}$ in the closed unit ball B_1 of $L^\infty(d\chi)$. As before, there exist a subnet $\{h_{\lambda'}: \lambda' \in \Lambda'\}$ of $\{h_\lambda\}$ and an element $h^* \in B_1$ such that $\{\hat{h}_{\lambda'}\}$ converges to h^* in the weak* topology $w(L^\infty(d\chi),L^1(d\chi))$. In particular, we have

$$\int_{\Delta_1} (\hat{h}_{\lambda'} - h^*)f^*d\chi \to 0$$

for the function f^*. Since $h_{\lambda'} \in \mathrm{Ker}(F)$, $\int_{\Delta_1} \hat{h}_{\lambda'}f^*d\chi = F(h_{\lambda'}) = 0$ and therefore $\int_{\Delta_1} h^*f^*d\chi = 0$. If we determine $h_0 \in h^\infty(R)$ by the condition $\hat{h}_0 = h^*$ a.e., then the subnet $\{\hat{h}_{\lambda'}\}$ converges to $\hat{h}_0$ in the topology $w(L^\infty(d\chi),L^1(d\chi))$. So, as in the proof of Theorem 5C, we have

$h_{\lambda'} \to h_0$ in the topology $w(h^\infty(R), M_b(R))$. It then follows that $k = h_0$ and thus

$$F(k) = F(h_0) = \int_{\Delta_1} \hat{h}_0 f^* d\chi = \int_{\Delta_1} h^* f^* d\chi = 0;$$

namely, $k \in \mathrm{Ker}(F)$. Consequently, $\mathrm{Ker}(F) \cap B$ is $w(h^\infty(R), M_b(R))$-closed, as was to be shown.

Finally, the map $F \to f^*$ given by (9) is injective, for the set $\{\hat{h}: h \in h^\infty(R)\}$ coincides with $L^\infty(d\chi)$ as shown in Theorem 3B. □

As an easy consequence of the theorem we have the following

Corollary. A subspace of $h^\infty(R)$ is β-closed if and only if its image in $L^\infty(d\chi)$ under the map $h \to \hat{h}$ is closed in the weak* topology $w(L^\infty(d\chi), L^1(d\chi))$.

NOTES

There is a large literature on classical Hardy classes. Our source for §1 is Chapters 3, 4, 5 of Hoffman [34]. Results in Subsection 3 are taken from Hasumi [17] and Neville [47]. Results in Subsection 4 concerning harmonic majorants are adapted from Heins [31] and Parreau [51]. Results on the β-topology in Subsection 5 are taken, in a modified form, from Rubel and Shields [47].

CHAPTER V. RIEMANN SURFACES OF PARREAU-WIDOM TYPE

The main theme of the present notes begins with this chapter. Our first objective is to define Riemann surfaces of Parreau-Widom type and to prove H. Widom's fundamental theorem. The definition of surfaces of Parreau-Widom type is given according to Widom. Widom's theorem is stated in a form slightly weaker than the original one. The proof follows almost the same lines as that of Widom but we try to write down every detail with some minor modification. It is also shown that an arbitrary surface of Parreau-Widom type is obtained from a surface of the same type, which is regular in the sense of potential theory, by deleting a certain discrete subset. This fact indicates that the definition we are adopting is essentially the same as the one given earlier by M. Parreau. A few direct consequences of Widom's theorem are mentioned at the end of this chapter.

Unless stated otherwise, we denote by R a hyperbolic Riemann surface and by $g_a(z) = g(a,z)$ the Green function for R with pole at a point $a \in R$. A point $O \in R$ is fixed and is called the origin of R.

§1. DEFINITIONS AND FUNDAMENTAL PROPERTIES

1. Basic Definitions

1A. For every $\alpha > 0$ and every $a \in R$ we set

$$R(\alpha,a) = \{z \in R\colon g(a,z) > \alpha\}.$$

The property (A2) in Ch. I, 6A of the Green function and the maximum principle for harmonic functions show that every $R(\alpha,a)$ is a connected region in R and that $R \setminus R(\alpha,a)$ has no compact components. It then follows that for any $\alpha > 0$ a singular 1-chain in $R(\alpha,a)$ is homologous to zero in the surface R if and only if it is so in $R(\alpha,a)$. In case $\alpha > \alpha' > 0$ we have a natural inclusion relation $H_1(R(\alpha,a)) \subseteqq H_1(R(\alpha',a)) \subseteqq H_1(R)$ among singular homology groups (Theorem 2C, Ch. I). Let $B(\alpha,a)$ be the first Betti number of the region $R(\alpha,a)$, i.e. the number of generators of the group $H_1(R(\alpha,a))$. So, when a point $a \in R$

is held fixed, $B(\alpha,a)$ is a nonincreasing function in α and vanishes for all sufficiently large α, for the region $R(\alpha,a)$ is conformally isomorphic with an open disk for all large α.

Definition. A hyperbolic Riemann surface R is called a surface of Parreau-Widom type (abbreviated to a PW-surface or PWS) if

$$\int_0^\infty B(\alpha,a)d\alpha < \infty$$

for some $a \in R$.

1B. We first note that the above definition is independent of the choice of $a \in R$. In fact, take any two distinct points $a, a' \in R$. Let V and V' be parametric disks with centers a and a', respectively, having disjoint closures, and set $G = R \setminus (\mathrm{Cl}(V \cup V'))$. Since the boundary ∂G of G is compact and does not contain a and a', we can find a positive constant $A > 0$ such that $A^{-1}g(a',z) \leqq g(a,z) \leqq Ag(a',z)$ for any $z \in \partial G$. As a matter of fact, the same inequalities hold for every $z \in G$, for we know $g_a = H[g_a;G]$ and $g_{a'} = H[g_{a'};G]$ (Theorem 6A, Ch. I). If we denote by α_0 the minimum of $g(a,z)$ and $g(a',z)$ taken on the boundary ∂G, then for any $0 < \alpha < \alpha_0$ we have $R(\alpha,a) \subseteqq R(A^{-1}\alpha,a') \subseteqq R(A^{-2}\alpha,a)$ and therefore $B(\alpha,a) \leqq B(A^{-1}\alpha, a') \leqq B(A^{-2}\alpha,a)$. This means that

$$A^{-1}\int_0^{\alpha_0} B(\alpha,a)d\alpha \leqq \int_0^{\alpha_0/A} B(\alpha,a')d\alpha \leqq A\int_0^{\alpha_0/A^2} B(\alpha,a)d\alpha,$$

and therefore that two integrals $\int_0^\infty B(\alpha,a)\,d\alpha$ and $\int_0^\infty B(\alpha,a')\,d\alpha$ converge or diverge at the same time, as was to be proved.

1C. A Riemann surface R is called regular in the sense of potential theory if for some $a \in R$ the Green function $g_a(z)$ tends to zero as z tends to the point at infinity or, equivalently, the region $\{z \in R: g(a,z) \geqq \alpha\}$ is compact for any $\alpha > 0$. This property is again independent of the choice of a. A point $w \in R$ is a critical point of $g_a(z)$ if $\partial g_a/\partial z = \partial g_a/\partial \bar{z} = 0$ at $z = w$. Let $Z(a;R)$ be the set of all critical points of g_a, which we repeat according to multiplicity.

Theorem. If R is regular, then

$$\text{(1)} \qquad \int_0^\infty B(\alpha,a)d\alpha = \sum\{g(a,w): w \in Z(a;R)\}$$

for every $a \in R$. Such a surface is a PWS if and only if

(2) $$\sum\{g(a,w): w \in Z(a;R)\} < \infty$$

for some (and hence all) $a \in R$.

Proof. Since $Z(a;R)$ is discrete, we can find a sequence $\{\alpha_n: n = 1, 2,\ldots\}$ of positive numbers strictly decreasing to zero such that the level curves $\{z \in R: g_a(z) = \alpha_n\}$ do not contain any point in $Z(a;R)$. We set $\overline{R}_n = \mathrm{Cl}(R(\alpha_n,a))$. Since R is regular, every $\overline{R}_n$ is a compact bordered surface and its Green function $g_n(a,z)$ with pole a is equal to $g(a,z) - \alpha_n$. Let $\tilde{g}_n(a,z)$ denote the harmonic conjugate of $g_n(a,z)$. The function $\tilde{g}_n(a,z)$ is not necessarily single-valued but the differential $d(g_n(a,z) + i\tilde{g}_n(a,z))$ is single-valued and can be extended to a meromorphic differential τ on the double R_n' of R_n. So we can use the formula (9) in Ch. I, 10D, and have $\deg(\tau) = 2g_n' - 2$, where g_n' denotes the genus of R_n'. If $g_n(a,z)$ has N_n critical points in R_n, then τ should have $2N_n$ zeros and two poles, so that $\deg(\tau) = 2N_n - 2$. Since $g_n' = B(\alpha_n,a)$ is clear in view of Ch. I, 2B, we conclude that $B(\alpha_n,a) = g_n' = N_n$ for $n = 1, 2,\ldots$. It follows readily that

$$\int_{\alpha_n}^{\infty} B(\alpha,a)d\alpha = -\alpha_n B(\alpha_n,a) - \int_{\alpha_n}^{\infty} \alpha dB(\alpha,a)$$

$$= -\alpha_n B(\alpha_n,a) + \sum\{g_a(w): w \in Z(a;R),\ g_a(w) > \alpha_n\}.$$

So, if $\sum g(a,w) < \infty$, then $\int_0^\infty B(\alpha,a)d\alpha < \infty$. Conversely, if the integral $\int_0^\infty B(\alpha,a)d\alpha < \infty$, then

$$\alpha_n B(\alpha_n,a) \leq \int_0^{\alpha_n} B(\alpha,a)d\alpha \to 0$$

as $n \to \infty$, and thus $\sum g(a,w) < \infty$. Hence we get the equality (1). □

Parreau [52] considered regular Riemann surfaces R for which the inequality (2) holds. His aim was to discuss Dirichlet problems based on Green lines in such surfaces. As we shall see in 3B below, his definition is indeed general enough to cover in essence all surfaces of Parreau-Widom type. This is probably enough to legitimate our nomenclature for the surfaces we are dealing with.

2. Widom's Characterization

2A. A remarkable characteristic property due to Widom says that a hyperbolic Riemann surface is a PWS if and only if it has sufficiently many analytic functions. In order to give a precise statement of this fact, we need the notion of multiplicative analytic functions explained in Ch. II.

Given a line bundle $\xi \in H^1(R,\mathbb{T})$ over R, we consider the space $\mathcal{H}(R,\xi)$ of holomorphic sections of the bundle ξ. If $f \in \mathcal{H}(R,\xi)$, then Theorem 2C, Ch. II, says that $|f(z)|$ is an l.a.m. on R and therefore that $|f(z)|^p$ $(0 < p < \infty)$ is a subharmonic function on R. We set

$$\|f\|_p = \|f\|_{p,\mathcal{O}} = \{(\mathrm{LHM}(|f|^p))(\mathcal{O})\}^{1/p},$$

where LHM stands for the least harmonic majorant on the surface R. If $|f|^p$ has no harmonic majorant, then $\|f\|_p$ is defined to be $+\infty$. When $p = \infty$, we set

$$\|f\|_\infty = \sup\{|f(z)|: z \in R\}.$$

We then define for every $0 < p \leqq \infty$

$$\mathcal{H}^p(R,\xi) = \{f \in \mathcal{H}(R,\xi): \|f\|_p < \infty\}.$$

Obviously, $\mathcal{H}^p(R,\xi)$, $1 \leqq p \leqq \infty$, is a complex Banach space with respect to the norm $\|\cdot\|_p$, if addition and scalar multiplication are defined as in Ch. II, 2D.

2B. The fundamental result of Widom reads as follows:

Theorem. The following statements are equivalent:

(a) R is a PWS;

(b) $\mathcal{H}^\infty(R,\xi) \neq \{0\}$ for any line bundle $\xi \in H^1(R,\mathbb{T})$;

(c) $\mathcal{H}^1(R,\xi) \neq \{0\}$ for any line bundle $\xi \in H^1(R,\mathbb{T})$.

The original form of Widom's theorem asserts much more and in fact the statement (b) is valid not only for line bundles but also for any unitary flat vector bundle over R. The proof of the above theorem is rather long and is given in §§2 and 3.

3. Regularization of Surfaces of Parreau-Widom Type

3A. We shall show that every PWS is naturally embedded in some regular PWS. First we prove the following

Lemma. Let R be a hyperbolic Riemann surface for which the Betti numbers $B(\alpha,a)$ are finite for all $\alpha > 0$ and some (and hence all) a in R. Then there exist a regular hyperbolic Riemann surface $R^\dagger$ and a discrete subset Σ of $R^\dagger$ such that $R^\dagger$ has the same property concerning Betti numbers and R is conformally isomorphic with $R^\dagger \setminus \Sigma$. The regular surface $R^\dagger$ is determined uniquely by R up to a conformal isomorphism.

Proof. (a) First we assume that the first Betti number of R itself is finite. Then there exist a finite number of open subsets $V_1, \ldots, V_m$ of R satisfying the following:

(i) the closures $Cl(V_i)$ are mutually disjoint noncompact subsets of R;

(ii) for each i, the boundary ∂V_i of V_i is a simple closed analytic curve, say J_i, and there exists a conformal homeomorphism h_i of V_i onto the annulus $\{w \in \mathbb{C}: r_i < |w| < 1\}$ $(0 \leq r_i < 1)$ such that h_i is continuously extendable to a homeomorphism of $Cl(V_i)$ onto $\{w \in \mathbb{C}: r_i < |w| \leq 1\}$ which maps J_i onto the unit circle;

(iii) $R \setminus (\cup_{i=1}^m V_i)$ is a compact bordered Riemann surface.

If $r_i > 0$ for all i, then R itself is easily seen to be regular. So we may assume that $r_i = 0$ for some i. For the sake of simplicity, we assume that $r_i = 0$ for $i = 1, \ldots, s$ and $r_i > 0$ for $i = s+1, \ldots, m$. Take a set of s elements, say $B = \{b(1), \ldots, b(s)\}$, and form a formal union $R^\dagger = R \cup B$. To each point $a \in R$ we assign a parametric disk V_a, which is compatible with the given conformal structure of R. For each $i = 1, \ldots, s$, we put $V_{b(i)} = V_i \cup \{b(i)\}$ and

$$h_i^*(z) = \begin{cases} h_i(z) & \text{for } z \in V_i \\ 0 & \text{for } z = b(i). \end{cases}$$

We regard $(V_{b(i)}, h_i^*)$ as a parametric disk about the point $b(i)$. It follows from our construction that these parametric disks together define a conformal structure on $R^\dagger$ which induces on R the original structure of R. As is well known, every bounded harmonic function on the punctured unit disk $\{w \in \mathbb{C}: 0 < |w| < 1\}$ can be continuously extended to the full open unit disk so as to have a harmonic function. So the Green function $g(a,z)$ for R with pole $a \in R$ can always be extended by continuity to the Green function for $R^\dagger$. This function clearly vanishes on the ideal boundary of $R^\dagger$. We may thus conclude that the surface $R^\dagger$ is regular.

(b) We now consider the general case. Let us fix a point $a \in R$. Take any $\alpha > 0$ and put $R_\alpha = R(\alpha,a)$. Since the first Betti number of the surface R_α is equal to $B(\alpha,a)$ and so is finite. As we have shown in (a), R_α can be completed to a regular surface $R_\alpha^\dagger$ by adding a finite number of points to R_α. It is not difficult to show that, if $0 < \beta < \alpha$, the surface $R_\alpha^\dagger$ may be canonically identified with the subregion of the surface $R_\beta^\dagger$ defined by

$$\{z \in R_\beta^\dagger : g_\beta(a,z) > \alpha - \beta\},$$

where $g_\beta(a,z)$ denotes the Green function for $R_\beta^\dagger$ with pole a. With this identification for all α, β with $0 < \beta < \alpha$, we put

$$R^\dagger = \cup\{R_\alpha^\dagger : \alpha > 0\}.$$

If we give $R^\dagger$ the conformal structure induced from those of $R_\alpha^\dagger$, then $R^\dagger$ is a regular hyperbolic Riemann surface which satisfies the property of the lemma.

(c) It is a routine matter to show the uniqueness of the pair $(R^\dagger, \Sigma)$ up to a conformal isomorphism. □

3B. We are in a position to prove the following

<u>Theorem</u>. Let R be a hyperbolic Riemann surface for which the Betti numbers $B(\alpha,a)$ are finite for any $\alpha > 0$ and some (and hence all) a in R. Then the regular hyperbolic surface $R^\dagger$ obtained in Lemma 3A can be constructed as follows. Let Σ be the subset of the Martin boundary $\Delta(R)$ of R such that a point $b \in \Delta(R)$ belongs to Σ if and only if

$$\limsup_{R \ni z \to b} g(a,z) > 0.$$

Then the set Σ is at most countable and independent of the choice of a, and, for each $b \in \Sigma$, there exists a neighborhood V of b in the Martin compactification R^* of R with $V \cap \Delta(R) = \{b\}$. On the union $R^\dagger = R \cup \Sigma$ there exists a uniquely determined conformal structure which makes $R^\dagger$ into a hyperbolic Riemann surface, is compatible with the relative topology of $R^\dagger$ as a subspace of R^*, and satisfies the following properties:

(B1) $R^\dagger$ is a regular hyperbolic surface and the conformal structure of $R^\dagger$ induces on R the original conformal structure of R.

(B2) If V is a neighborhood of $b \in \Sigma$ in R^* such that the intersection $V \cap \Delta(R)$ reduces to the single point b, then every bounded harmonic function u on $V \cap R$ can be extended by continuity to the point b so as to get a harmonic function on V with respect to the conformal structure of $R^\dagger$.

(B3) The Green function $g^\dagger(a,z)$ for $R^\dagger$ with pole $a \in R$ is obtained by extending the Green function $g(a,z)$ for R to the points in Σ by continuity in the topology of R^*. Similarly, the Martin function $k^\dagger(b,z)$ for $R^\dagger$ with pole b and with respect to the origin $0 \in R$ is obtained by extending the Martin function $k(b,z)$ for R with respect to the same origin 0 to the points in Σ by continuity in the topology of R^*.

(B4) The Martin compactification of $R^\dagger$ can be identified with R^* and $\Delta_1(R) = \Delta_1(R^\dagger) \cup \Sigma$.

(B5) The harmonic measure $d\chi^\dagger$, supported on $\Delta_1(R^\dagger)$, of the surface $R^\dagger$ at the origin 0 is nothing but the restriction, to the set $\Delta_1(R^\dagger)$, of the harmonic measure $d\chi$, supported on $\Delta_1(R)$, of the surface R at the point 0.

(B6) If R is a PWS, then so is $R^\dagger$.

Proof. Let $(R^\dagger, \Sigma)$ be a pair given by Lemma 3A. Then the properties (B1) and (B2) are clearly satisfied. So we prove here that $R^\dagger$ has the remaining properties.

Property (B3): The property (B2) implies that, for each fixed z in R, the Green function $z' \to g(z',z)$ for R with pole $z \in R$ is extended by continuity to the Green function $z' \to g^\dagger(z',z)$ for $R^\dagger$. So the definition of the Martin functions shows that, for each fixed $z \in R$, the function $b \to k^\dagger(b,z)$ on $R^\dagger$ is obtained by extending, by continuity, the function $b \to k(b,z)$ on R to the points in Σ.

Property (B4): For each fixed $z \in R$, the function $b \to k^\dagger(b,z)$ on $R^\dagger$ can be extended to a continuous function on the Martin compactification, say $R^\#$, of $R^\dagger$ and therefore the function $b \to k(b,z)$ on R can also be extended to a continuous function on $R^\#$. Since the set of functions $b \to k^\dagger(b,z^\dagger)$ with $z^\dagger \in R^\dagger$ separates points of $R^\#$ and since R is dense in $R^\dagger$, the subfamily of functions $b \to k^\dagger(b,z)$ with $z \in R$ also separates points of $R^\#$. Hence, $R^\#$ can be identified with the Martin compactification of R, i.e. $R^* = R^\#$. In particular, we have $\Sigma = \Delta(R) \setminus \Delta(R^\dagger)$. Since $R^\dagger$ is regular, we conclude that a point $b \in \Delta(R)$ belongs to Σ if and only if

$$\lim_{R \ni z \to b} g(a,z) = g^\dagger(a,b) > 0.$$

If $b \in \Sigma$, then the function $z \to k(b,z)$ is equal to a constant multiple of $g^\dagger(b,z)$ and so is a minimal harmonic function on R. We thus have $\Sigma \subseteqq \Delta_1(R)$ and therefore $\Delta_1(R) = \Delta_1(R^\dagger) \cup \Sigma$.

Property (B5): This is clear from the above observation.

Property (B6): Let $B^\dagger(\alpha,a)$ be the first Betti number of the region $R^\dagger(\alpha,a) = \{z \in R^\dagger: g^\dagger(a,z) > \alpha\}$, where we assume $a \in R$. Then,

$B^\dagger(\alpha,a) \leq B(\alpha,a)$ for all $\alpha > 0$ and thus $\int_0^\infty B^\dagger(\alpha,a)d\alpha \leq \int_0^\infty B(\alpha,a)d\alpha$. So $R^\dagger$ is a PWS whenever R is. □

3C. We now compare the integrals of Betti numbers corresponding to R and $R^\dagger$.

Theorem. Let R be a PWS and let $a \in R$. Then the set $Z(a;R)$ of critical points of the function $z \to g(a,z)$ consists of those elements of the set $Z(a;R^\dagger)$ that lie in R and we have

$$\int_0^\infty B(\alpha,a)d\alpha = \int_0^\infty B^\dagger(\alpha,a)d\alpha + \sum\{g^\dagger(a,w): w \in \Sigma\}$$

$$= \sum\{g^\dagger(a,w): w \in Z(a;R^\dagger)\} + \sum\{g^\dagger(a,w): w \in \Sigma\}.$$

Proof. Let $\Sigma = \{w_1, w_2, \ldots\}$ and put $R_n = R^\dagger \setminus \{w_1,\ldots,w_n\}$ for $n = 1, 2, \ldots$. Denote by $B_n(\alpha,a)$ the first Betti number of the subregion $R_n(\alpha,a) = \{z \in R_n: g^\dagger(a,z) > \alpha\}$. Since $R_n = R_{n-1} \setminus \{w_n\}$, we have

$$B_n(\alpha,a) = \begin{cases} B_{n-1}(\alpha,a) & \text{for } \alpha > g^\dagger(a,w_n) \\ B_{n-1}(\alpha,a) + 1 & \text{for } \alpha < g^\dagger(a,w_n). \end{cases}$$

So we have

$$\int_0^\infty B_n(\alpha,a)d\alpha = \int_0^\infty B_{n-1}(\alpha,a)d\alpha + g^\dagger(a,w_n)$$

$$= \int_0^\infty B^\dagger(\alpha,a)d\alpha + \sum_{k=1}^{n} g^\dagger(a,w_k).$$

Since $B_n(\alpha,a)$ converges monotonically to $B(\alpha,a)$ as $n \to \infty$, we get the desired result. □

§2. PROOF OF WIDOM'S THEOREM (I)

In this section we prove the implication (a) ⇒ (b) in Widom's theorem stated in 2B.

4. Analysis on Regular Subregions

4A. We begin with the following classical theorem of Cauchy-Read. For an interesting proof of this crucial result we refer the reader to Heins [31], p. 75.

Theorem. Let G be any regular subregion of R in the sense of Ch. I, 1A and let $1 \leqq p \leqq \infty$. Then the following hold:

(a) Every $f \in H^p(G)$ has a nontangential boundary value $f^*(b)$ at almost every $b \in \partial G$ such that $f^* \in L^p(\partial G)$ with respect to the arc-length measure on ∂G and we have $\int_{\partial G} f^*\omega = 0$ for any holomorphic differential ω defined in a neighborhood of $\mathrm{Cl}(G)$.

(b) Let $u \in L^p(\partial G)$, $1 \leqq p \leqq \infty$, satisfy the equation $\int_{\partial G} u\omega = 0$ for any holomorphic differential ω defined in some neighborhood of $\mathrm{Cl}(G)$. Then there exists a unique $f \in H^p(G)$ such that u is equal a.e. on ∂G to the nontangential boundary function f^* of f.

4B. Let G be a regular region in R and let $\xi \in H^1(R,\mathbb{T})$ be a line bundle over R. We denote by $\Gamma M(\mathrm{Cl}(G),\xi)$ the set of meromorphic differential-sections of ξ defined in some neighborhood of $\mathrm{Cl}(G)$. Let C be a finite union of analytic curves such that $\mathrm{Cl}(G) \setminus C$ is simply connected. Since $G \setminus C$ is simply connected, every line bundle over $G \setminus C$ is trivial. If $\xi|G$ denotes the restriction of the bundle ξ to G, then every section in $\mathcal{H}(G,\xi|G)$ has a restriction to $G \setminus C$, which is represented by a single-valued holomorphic function. The set of such restrictions of sections in $\mathcal{H}^p(G,\xi|G)$ is denoted by the symbol $\tilde{\mathcal{H}}^p(G,\xi|G)$. Since the boundary ∂G of G consists of a finite number of analytic curves, every $\tilde{f} \in \tilde{\mathcal{H}}^p(G,\xi|G)$ has nontangential boundary values a.e. on ∂G, which we denote by the same symbol $\tilde{f}$.

Lemma. Assume that $\omega \in \Gamma M(\mathrm{Cl}(G),\xi)$ is nowhere vanishing and has only one pole, a simple pole at a point $a \in G \setminus C$ with residue of absolute value 1. Let $1 \leqq p < \infty$ and p' the conjugate exponent of p, i.e. $p' = p/(p-1)$. Then for any line bundle $\eta \in H^1(R;\mathbb{T})$ we have

$$(3)\qquad \inf\{\|\tilde{f}\|_{p',\omega} : \tilde{f} \in \tilde{\mathcal{H}}^{p'}(G,\eta|G),\ |\tilde{f}(a)| = 1\}$$
$$= \sup\{|\tilde{h}(a)| : \tilde{h} \in \tilde{\mathcal{H}}^p(G,\xi^{-1}\eta^{-1}|G),\ \|\tilde{h}\|_{p,\omega} = 1\},$$

where the exponent -1 denotes inverse element in the group $H^1(R;\mathbb{T})$ and

$$\|\tilde{f}\|_{q,\omega} = \begin{cases} \left(\frac{1}{2\pi}\int_{\partial G} |\tilde{f}|^q|\omega|\right)^{1/q} & \text{for } 1 \leqq q < \infty \\ \sup\{|f(z)| : z \in G\} & \text{for } q = \infty. \end{cases}$$

We note that the norm $\|\cdot\|_{q,\omega}$ is defined for any ω given only in some neighborhood of ∂G.

Proof. Take any $\tilde{h} \in \tilde{\mathcal{H}}^p(G,\xi^{-1}\eta^{-1}|G)$ and let $h \in \mathcal{H}^p(G,\xi^{-1}\eta^{-1}|G)$ be the corresponding section. Then $|h(z)|$ is a subharmonic function on G. So if we denote by ω_a^G the harmonic measure of G at the point a (Ch. I, 5D), then

$$(4) \qquad |\tilde{h}(a)| = |h(a)| \leqq \int_{\partial G} |h(\zeta)|\,d\omega_a^G(\zeta)$$

$$\leqq \max_{\zeta\in\partial G} (\omega_a^G(\zeta)/|\omega(\zeta)|) \int_{\partial G} |h||\omega|.$$

Since ω is non-vanishing along ∂G, $|\omega|$ is a strictly positive differential along ∂G and therefore $\max_{\zeta\in\partial G}(\omega_a^G(\zeta)/|\omega(\zeta)|) < \infty$. Thus by applying the Hölder inequality to the integral in the last member of (4) we see that the evaluation map $\varepsilon_a\colon \tilde{h} \to \tilde{h}(a)$ is a continuous linear functional on $\tilde{\mathcal{H}}^p(G,\xi^{-1}\eta^{-1}|G)$ with respect to the norm $\|\cdot\|_{p,\omega}$. The norm of the functional ε_a is then given by

$$\|\varepsilon_a\| = \sup\{|\tilde{h}(a)|\colon \tilde{h} \in \tilde{\mathcal{H}}^p(G,\xi^{-1}\eta^{-1}|G),\ \|\tilde{h}\|_{p,\omega} \leqq 1\},$$

which is exactly equal to the right-hand side of (3). We identify $\tilde{h}$ in $\tilde{\mathcal{H}}^p(G,\xi^{-1}\eta^{-1}|G)$ with its boundary values on ∂G, so that the space $\tilde{\mathcal{H}}^p(G,\xi^{-1}\eta^{-1}|G)$ can be viewed as a subspace of $L^p(\partial G,|\omega|/2\pi)$. By the Hahn-Banach theorem there exists a function $F \in L^{p'}(\partial G,|\omega|/2\pi)$ such that $\|F\|_{p'} = \|\varepsilon_a\|$ and

$$(5) \qquad \tilde{h}(a) = \frac{1}{2\pi i}\int_{\partial G} F\tilde{h}|\omega|$$

for every $\tilde{h} \in \tilde{\mathcal{H}}^p(G,\xi^{-1}\eta^{-1}|G)$.

By applying Theorem 3B, Ch. II, to a regular subregion containing Cl(G) we can find a holomorphic section $k \in \mathcal{H}^\infty(G,\eta^{-1}|G)$, which is non-vanishing on some neighborhood of Cl(G). Let ψ be any meromorphic differential in a neighborhood of Cl(G) with only one pole which is a simple pole at a with residue 1 (Theorem 11B, Ch. I). Finally let $\tilde{\omega}$ represent ω on $G \setminus C$. We may suppose that $\tilde{\omega}$ has residue 1 at the point a. Since ω is nowhere vanishing,

$$\tilde{k}\psi/\tilde{\omega} \in \tilde{\mathcal{H}}^\infty(G,\xi^{-1}\eta^{-1}|G) \subseteq \tilde{\mathcal{H}}^p(G,\xi^{-1}\eta^{-1}|G)$$

and so, by (5), we have

$$(6) \qquad \tilde{k}(a) = \frac{1}{2\pi i}\int_{\partial G} F\tilde{k}\psi|\omega|/\tilde{\omega}.$$

If α is any holomorphic differential on some neighborhood of Cl(G), then $\alpha + \psi$ has the same property as ψ, so that (6) remains to be valid if ψ is replaced by $\alpha + \psi$. It follows that $\int_{\partial G} F\tilde{k}\alpha|\omega|/\tilde{\omega} = 0$. Since this holds for all such differentials α, Theorem 4A-(b) implies that $F\tilde{k}|\omega|/\tilde{\omega}$ is equal to the boundary function f_0^* of some $f_0 \in H^{p'}(G)$. Moreover we have

$$f_0(a) = \frac{1}{2\pi i}\int_{\partial G} f_0^*\psi = \frac{1}{2\pi i}\int_{\partial G} F\tilde{k}|\omega|\psi/\tilde{\omega} = \tilde{k}(a).$$

We set $f = f_0/k$. Then $f \in \mathcal{H}^{p'}(G,\eta|G)$, $|f| = |F|$ on ∂G and $|f(a)| = 1$. Hence,

$$\inf\{\|\tilde{f}\|_{p',\omega}: \tilde{f} \in \tilde{\mathcal{H}}^{p'}(G,\eta|G),\ |f(a)| = 1\} \leqq \|F\|_{p'} = \|\varepsilon_a\|.$$

So in (3) the left-hand side is at most equal to the right-hand side.

Conversely, if $\tilde{f} \in \tilde{\mathcal{H}}^{p'}(G,\eta|G)$ with $|f(a)| = 1$ and $\tilde{h} \in \tilde{\mathcal{H}}^{p}(G,\xi^{-1}\eta^{-1}|G)$ with $\|\tilde{h}\|_{p,\omega} = 1$, then $\tilde{f}\tilde{h}\tilde{\omega}$ can be extended to a single-valued differential on G and so

$$|\tilde{h}(a)| = \left|\frac{1}{2\pi i}\int_{\partial G}\tilde{f}\tilde{h}\tilde{\omega}\right| \leqq \|\tilde{f}\|_{p',\omega}\|\tilde{h}\|_{p,\omega} = \|\tilde{f}\|_{p',\omega}.$$

This shows the reverse inequality, as desired. □

4C. The statement (b) in Theorem 2B is valid for any regular region G in R. More precisely, we show the following

<u>Theorem</u>. Let G be any regular region in R and take a point $a \in G$, which is held fixed. Then, for every p with $1 \leqq p \leqq \infty$,

$$\sup_{\xi}[\inf\{\|h\|_{p,a}: h \in \mathcal{H}^p(G,\xi),\ |h(a)| = 1\}]$$
$$= \exp[\textstyle\sum\{g(a,w): w \in Z(a;G)\}],$$

where $g(a,\cdot)$ denotes the Green function for G, ξ ranges over the group $H^1(G;\mathbb{T})$ and

$$\|f\|_{p,a} = \begin{cases} \left(-\frac{1}{2\pi}\int_{\partial G}|f(\zeta)|^p d\tilde{g}(\zeta,a)\right)^{1/p} & \text{for } 1 \leqq p < \infty \\ \sup\{|f(z)|: z \in G\} & \text{for } q = \infty. \end{cases}$$

The tilde here denotes conjugate harmonic function.

Before proceeding to the proof, we note that the above norm is identical with the one given in 2A in case $R = G$ and $O = a$. We also note that $Z(a,G)$ is a finite set.

Proof. Consider the meromorphic differential $dg(\cdot,a) + id\tilde{g}(\cdot,a)$, which has a simple pole at a with residue 1 and zeros at points belonging to $Z(a;G)$, the order of each zero being the same as the corresponding multiplicity in $Z(a;G)$. Let $\{w_1, w_2, \ldots, w_m\}$ be an enumeration of elements in $Z(a;G)$. For each j we set

$$A_j(z) = \exp[g(z,w_j) + i\int_a^z d\tilde{g}(\cdot,w_j)],$$

which is multiple-valued but has single-valued modulus, so that it defines a line bundle, say ξ_j, over G according to our observation in Ch. II, 2D. We then set

$$\omega' = \Big(\prod_{j=1}^{m} A_j\Big)[dg(\cdot,a) + id\tilde{g}(\cdot,a)].$$

Clearly, $\omega' \in \Gamma M(Cl(G),\xi)$ with $\xi = \prod_{j=1}^{m} \xi_j$. Moreover, it is nowhere vanishing on $Cl(G)$ and has only one pole--a simple pole--at the point a. Since the residue of ω' at a has absolute value

$$\Big|\prod_{j=1}^{m} A_j(a)\Big| = \exp[\prod_{j=1}^{m} g(a,w_j)] \quad (= r, \text{ say}),$$

$\omega = r^{-1}\omega'$ satisfies the hypothesis of Lemma 4B. Since $|\omega'| = -d\tilde{g}(\cdot,a)$ along ∂G, we have in case $q \neq \infty$

$$\|f\|_{q,a} = \|\tilde{f}\|_{q,\omega'} = r^{1/q}\|\tilde{f}\|_{q,\omega}.$$

Let $1 < p < \infty$ and set $p' = p/(p-1)$. Then Lemma 4B shows that

$$\begin{aligned}
&\inf\{\|f\|_{p',a}: f \in \mathcal{H}^{p'}(G,\eta),\ |f(a)| = 1\}\\
&= r^{1/p'}\inf\{\|\tilde{f}\|_{p',\omega}: \tilde{f} \in \tilde{\mathcal{H}}^{p'}(G,\eta),\ |\tilde{f}(a)| = 1\}\\
&= r^{1/p'}\sup\{|\tilde{h}(a)|: \tilde{h} \in \tilde{\mathcal{H}}^{p}(G,\xi^{-1}\eta^{-1}),\ \|\tilde{h}\|_{p,\omega} = 1\}\\
&= r^{1/p'}\sup\{|h(a)|: h \in \mathcal{H}^{p}(G,\xi^{-1}\eta^{-1}),\ \|h\|_{p,a} = r^{1/p}\}\\
&= \sup\{|h(a)|: h \in \mathcal{H}^{p}(G,\xi^{-1}\eta^{-1}),\ \|h\|_{p,a} = r\}
\end{aligned}$$

for any line bundle η over G. A similar computation is possible for $p = 1$, so that we have

$$\begin{aligned}
(7)\qquad &\inf\{\|f\|_{p',a}: f \in \mathcal{H}^{p'}(G,\eta),\ |f(a)| = 1\}\\
&= \sup\{|h(a)|: h \in \mathcal{H}^{p}(G,\xi^{-1}\eta^{-1}),\ \|h\|_{p,a} = r\}
\end{aligned}$$

for any line bundle η over G and for any p with $1 \leq p < \infty$. Since

$|h|^p$ is subharmonic, we have $|h(a)| \leqq \|h\|_{p,a}$ and therefore the value given by (7) is at most equal to r. Setting $\eta = \xi^{-1}$ in (7), we have $\xi^{-1}\eta^{-1} = \eta\eta^{-1}$ = the identity of $H^1(G;\mathbb{T})$. If h is identically equal to r, then $h(a) = r$ and $\|h\|_{p,a} = r$. This means that the right-hand side of (7) reaches r at this η. Hence

$$\sup_{\eta}[\inf\{\|f\|_{p',a}: f \in \mathcal{H}^{p'}(G,\eta),\ |f(a)| = 1\}] = r.$$

Writing p in place of p', we have seen that the theorem holds for all p with $1 < p \leqq \infty$.

Finally, we will dispose of the case $p = 1$. Letting $\xi^{-1}\eta^{-1} = \eta_1$, we have $\eta = \eta_1^{-1}\xi^{-1}$ so that, by (7),

$$\begin{aligned}
&\sup\{|h(a)|: h \in \mathcal{H}^1(G,\eta_1),\ \|h\|_{1,a} = 1\}\\
&\quad = r^{-1}\sup\{|h(a)|: h \in \mathcal{H}^1(G,\eta_1),\ \|h\|_{1,a} = r\}\\
&\quad = r^{-1}\inf\{\|f\|_{\infty}: f \in \mathcal{H}^{\infty}(G,\eta_1^{-1}\xi^{-1}),\ |f(a)| = 1\}\\
&\quad \geqq r^{-1}.
\end{aligned}$$

Here the value r^{-1} is reached at $\eta_1 = \xi^{-1}$ and $f \equiv 1$. Thus

$$\inf_{\eta}[\sup\{|h(a)|: h \in \mathcal{H}^1(G,\eta),\ \|h\|_{1,a} = 1\ ] = r^{-1},$$

which is easily seen to be equivalent to

$$\sup_{\eta}[\inf\{\|h\|_{1,a}: h \in \mathcal{H}^1(G,\eta),\ |h(a)| = 1\}] = r,$$

as was to be proved. □

5. Proof of Necessity

5A. We are going to prove (a) ⇒ (b) of Widom's theorem in the following form.

Theorem. Let R be any hyperbolic Riemann surface. Then, for every p with $1 \leqq p \leqq \infty$,

$$\sup_{\xi}[\inf\{\|f\|_{p,a}: h \in \mathcal{H}^p(R,\xi),\ |f(a)| = 1\}] = \exp\left(\int_0^{\infty} B(\alpha,a)d\alpha\right),$$

where the supremum is taken over all $\xi \in H^1(R,\mathbb{T})$, and $\|f\|_{p,a}$ is equal to $\{(LHM(|f|^p))(a)\}^{1/p}$ if $p < \infty$ and to $\sup\{|f(z)|: z \in R\}$ if $p = \infty$.

Proof. Let $\{R_n: n = 1, 2, \ldots\}$ be a regular exhaustion of R consisting of canonical subregions with $a \in R_1$ (Corollary 1C, Ch. I). We then denote by $g_n(a,z)$ the Green function for R_n with pole a, by $R_n(\alpha,a)$ the region $\{z \in R_n: g_n(a,z) > \alpha\}$ and by $B_n(\alpha,a)$ the first Betti number of $R_n(\alpha,a)$.

We first consider any finite p. Namely, we fix any p with $1 \leqq p < \infty$ and use the following notations:

$$M(\xi,a) = \inf\{\|f\|_{p,a}: f \in \mathcal{H}^p(R,\xi),\ |f(a)| = 1\} \quad \text{for} \quad \xi \in H^1(R;\mathbb{T}),$$

$$M_n(\xi,a) = \inf\{\|f\|_{p,a}: f \in \mathcal{H}^p(R_n,\xi),\ |f(a)| = 1\} \quad \text{for} \quad \xi \in H^1(R_n;\mathbb{T}),$$

$$M(a) = \sup\{M(\xi,a): \xi \in H^1(R;\mathbb{T})\},$$

$$M_n(a) = \sup\{M_n(\xi,a): \xi \in H^1(R_n;\mathbb{T})\}.$$

With this definition we claim

$$M(a) = \lim_{n\to\infty} M_n(a). \tag{8}$$

To see this, let $\xi \in H^1(R_{n+1};\mathbb{T})$; then $\xi|R_n \in H^1(R_n;\mathbb{T})$ so that

$$M_n(\xi|R_n,a) \leqq M_{n+1}(\xi,a) \leqq M_{n+1}(a).$$

Since R_n is canonical, we see by Theorem 3B, Ch. I, and Theorem 1B, Ch. II, that every line bundle over R_n is the restriction of a line bundle over R. In other words, the restriction map $\xi \to \xi|R_n$ is a surjection from $H^1(R;\mathbb{T})$ (resp. $H^1(R_{n+1};\mathbb{T})$) to $H^1(R_n;\mathbb{T})$. So, the above inequalities imply that $M_n(a) \leqq M_{n+1}(a) \leqq M(a)$ and consequently

$$\lim_{n\to\infty} M_n(a) \leqq M(a).$$

In order to get the reverse inequality, we may suppose that

$$\lim_{n\to\infty} M_n(a) < \infty.$$

Let us take any $\xi \in H^1(R;\mathbb{T})$. As explained in Ch. II, 1A, the bundle ξ is defined by means of an open covering $\{V'_\alpha\}$ in the form $(\{\xi_{\alpha\beta}\}, \{V'_\alpha\})$. Since R is separable, we may assume that $\{V'_\alpha\}$ is countable and that all V'_α as well as all nonempty $V'_\alpha \cap R_n$ are simply connected. Then we take a refinement $\{V_\alpha\}$ of the covering $\{V'_\alpha\}$ such that the closure $\mathrm{Cl}(V_\alpha)$ is a compact subset of V'_α for each α. The definition of $M_n(\xi|R_n,a)$ implies that, for each $n = 1, 2, \ldots$, there exists an element $f_n \in \mathcal{H}^p(R_n,\xi|R_n)$ satisfying $|f_n(a)| = 1$ and $\|f_n\|_{p,a} \leqq$

$M_n(\xi|R_n,a)+n^{-1}$. Let $\{f_{n\alpha}\}_\alpha$ be a representative of f_n with respect to the covering $\{V'_\alpha \cap R_n\}_\alpha$ of R_n and the representative $(\{\xi_{\alpha\beta}\}, \{V'_\alpha \cap R_n\}_\alpha)$ of $\xi|R_n$. Since each nonempty $V'_\alpha \cap R_n$ is simply connected, $f_{n\alpha}$ is defined as a single-valued holomorphic function on $V' \cap R_n$. Let u_n be the LHM of $|f_n|^p$ on R_n; then

$$\|f_n\|^p_{p,a} = -\frac{1}{2\pi}\int_{\partial R_n} |f_n(\zeta)|^p d\tilde{g}_n(\zeta,a) = u_n(a).$$

Take any compact set K included in some R_n. By use of Harnack's inequality, we find a constant $C > 0$, depending on K, such that we have $\sup_{z\in K} v(z) \leqq Cv(a)$ for any positive harmonic function v on R_n. Applying this fact to the functions u_j with $j \geqq n$, we have

$$\begin{aligned}|f_j(z)| &\leqq u_j(z)^{1/p} \leqq C^{1/p}u_j(a)^{1/p} = C^{1/p}\|f_j\|_{p,a} \\ &\leqq C^{1/p}(M_j(\xi|R_j,a)+\tfrac{1}{j}) \\ &\leqq C^{1/p}(\lim_{\ell\to\infty} M_\ell(a)+1) < \infty\end{aligned}$$

for every $z \in K$. Thus $\{|f_j(z)|: j = n, n+1,\ldots\}$ is uniformly bounded on any compact subset of R_n. Since every $\mathrm{Cl}(V_\alpha)$ is a compact subset of some R_n, we use a diagonal process to find a subsequence $\{f_{n(k)}: k = 1, 2,\ldots\}$ such that, for each α, $\{f_{n(k),\alpha}: k = 1, 2,\ldots\}$ converges uniformly on $\mathrm{Cl}(V_\alpha)$. Let f_α be the limit of the subsequence $\{f_{n(k),\alpha}: k = 1, 2,\ldots\}$; then $\{f_\alpha\}$ clearly represents a section f belonging to $\mathcal{H}^p(R,\xi)$ with $|f(a)| = 1$. Here, the only nontrivial thing to see is that $|f|^p$ has a harmonic majorant. First we note that $|f_{n(k)}(z)| \to |f(z)|$ uniformly on every compact subset of R. Next we choose any m, which is fixed for a moment. Then, for any $\varepsilon > 0$ there exists k_0 with $m < n(k_0)$ such that $||f(z)|^p - |f_{n(k)}(z)|^p| < \varepsilon$ on ∂R_m for every $k \geqq k_0$. It follows that

$$\begin{aligned}-\frac{1}{2\pi}\int_{\partial R_m} |f(\zeta)|^p d\tilde{g}_m(\zeta,a) &\leqq -\frac{1}{2\pi}\int_{\partial R_m} |f_{n(k)}(\zeta)|^p d\tilde{g}_m(\zeta,a) + \varepsilon \\ &\leqq -\frac{1}{2\pi}\int_{\partial R_{n(k)}} |f_{n(k)}(\zeta)|^p d\tilde{g}_{n(k)}(\zeta,a) + \varepsilon = \|f_{n(k)}\|^p_{p,a} + \varepsilon \\ &\leqq \left(M_{n(k)}(\xi|R_{n(k)},a) + \frac{1}{n(k)}\right)^p + \varepsilon \leqq \left(M_{n(k)}(a) + \frac{1}{n(k)}\right)^p + \varepsilon.\end{aligned}$$

As ε and $n(k)$ are arbitrary, we have, by letting $\varepsilon \to 0$ and $n(k) \to \infty$,

$$-\frac{1}{2\pi}\int_{\partial R_m} |f(\zeta)|^p d\tilde{g}_m(\zeta,a) \leqq \lim_{n\to\infty} M_n(a)^p.$$

We now move m and set

$$v_m(z) = -\frac{1}{2\pi}\int_{\partial R_m} |f(\zeta)|^p d\tilde{g}_m(\zeta,z)$$

on R_m with $m = 1, 2, \ldots$. Since $|f|^p$ is subharmonic, we easily see that $|f(z)|^p \leqq v_m(z) \leqq v_{m+1}(z)$ on R_m and $v_m(a) \leqq \lim_n M_n(a)^p < \infty$ for all m. Thus, by the Harnack theorem, the limit $v(z)$ of the sequence $\{v_m(z)\}$ is harmonic on R, $v(a) \leqq \lim_n M_n(a)^p$ and $|f(z)|^p$ is majorized by $v(z)$ everywhere on R. Hence, we have $f \in \mathcal{H}^p(R,\xi)$, $|f(a)| = 1$ and $\|f\|_{p,a} \leqq v(a)^{1/p} \leqq \lim_n M_n(a)$. This implies that $M(\xi,a) \leqq \lim_n M_n(a)$. As ξ is arbitrary, we conclude that $M(a) \leqq \lim_n M_n(a)$, which establishes the equation (8).

Now that we have (8), the desired conclusion can be drawn from Theorem 4C by a limiting process. Take any positive number $\alpha > 0$ distinct from critical values of any $g_n(a,\cdot)$. Thus each $R_n(\alpha,a)$ is a regular subregion of R. Since $g_n(a,z) < g_{n+1}(a,z)$ on $Cl(R_n)$, we have $Cl(R_n(\alpha,a)) \subsetneqq R_{n+1}(\alpha,a)$, and therefore have a natural map:

$$H_1(R_n(\alpha,a)) \to H_1(R_{n+1}(\alpha,a)),$$

which is injective because $R_n(\alpha,a)$ is seen to be a regular subregion of R_{n+1}. This means that $B_n(\alpha,a) \leqq B_{n+1}(\alpha,a)$. Since R_n is a regular subregion of R, we see similarly that $B_n(\alpha,a) \leqq B(\alpha,a)$. On the other hand, since $g_n(a,z)$ converge increasingly to $g(a,z)$, we have $R(\alpha,a) = \cup_{n=1}^{\infty} R_n(\alpha,a)$, which in turn implies the equation

$$B(\alpha,a) = \lim_{n\to\infty} B_n(\alpha,a). \tag{9}$$

In fact, let $C_1,\ldots, C_k$ with $k \leqq B(\alpha,a)$ be any finite set of independent 1-cycles in $H_1(R(\alpha,a))$. Since $R(\alpha,a) = \cup_{n=1}^{\infty} R_n(\alpha,a)$, these cycles are included in one of the regions $R_n(\alpha,a)$, say $R_{n'}(\alpha,a)$. Then they are independent in $R_{n'}(\alpha,a)$ and thus $B_{n'}(\alpha,a) \geqq k$. So (9) is valid. Since each $g_n(a,z)$ admits an at most finite number of critical values, (9) holds for all $\alpha > 0$ with a countable number of exceptions.

Combining (8), (9), Theorems 1C and 4C, and the Lebesgue monotone convergence theorem, we finally have

$$M(a) = \lim_{n\to\infty} M_n(a) = \lim_{n\to\infty} \exp[\textstyle\sum\{g_n(a,w)\colon w \in Z(a,R_n)\}]$$

$$= \lim_{n\to\infty} \exp\left(\int_0^\infty B_n(\alpha,a)d\alpha\right) = \exp\left(\int_0^\infty B(\alpha,a)d\alpha\right).$$

This establishes the theorem in the case $1 \leqq p < \infty$.

The case $p = \infty$ is entirely similar but much easier, for the argument concerning harmonic majorants does not occur here. □

§3. PROOF OF WIDOM'S THEOREM (II)

The objective of this section is to prove the implication (c) ⇒ (a) in Widom's theorem.

6. Review of Principal Operators

6A. Given any regular region G in R we denote by $P = P(G)$ any partition of $R \setminus Cl(G)$ into mutually disjoint nonvoid open sets. Since $R \setminus Cl(G)$ has finitely many components, $P(G)$ consists of a finite number of parts, say $R_1, \ldots, R_\ell$. $\beta_j(P)$ denotes the relative boundary of R_j in R, which is oriented negatively with respect to the part R_j. The union $\cup_j \beta_j(P)$ is the positively oriented boundary of G, which is denoted by $\beta(G)$ in what follows.

Two partitions $P(G)$ and $P(G')$ are called equivalent if there exists a regular region G'' containing $Cl(G)$ and $Cl(G')$ such that $\{R_j \setminus Cl(G''): R_j \in P(G)\}$ and $\{R'_k \setminus Cl(G''): R'_k \in P(G')\}$ define the same partition of $R \setminus Cl(G'')$. This relation is an equivalence relation and each equivalence class, $\mathbb{P}$, defines a <u>partition of the ideal boundary</u>, $\beta(R)$, of R.

Let $\mathbb{P} = \{P(G)\}$ be a partition of $\beta(R)$. Then each member $P(G)$ in $\mathbb{P}$ has the same number of parts, say ℓ, and, if we arrange parts in every $P(G)$ in some suitable order, we may always assume, using the above notation, that $R_j \setminus Cl(G'') = R'_j \setminus Cl(G'')$ for any $j = 1, \ldots, \ell$. For every fixed j, $\beta_j(\mathbb{P})$ denotes the collection of $\beta_j(P)$ with $P \in \mathbb{P}$ and $\beta(\mathbb{P}) = (\beta_1(\mathbb{P}), \ldots, \beta_\ell(\mathbb{P}))$.

If u is a harmonic function defined in some $R_j \in P(G)$ with $P(G) \in \mathbb{P}$, then the integrals $\int_{\beta_j(P)} {}^*du$ have the same value for any $P = P(G') \in \mathbb{P}$ with $Cl(G) \subseteqq G'$, where ${}^*du$ denotes the conjugate differential for du: ${}^*du = -(\partial u/\partial y)dx + (\partial u/\partial x)dy$. The common value is called the <u>flux</u> of u over the cycle $\beta_j(\mathbb{P})$ and is denoted by the symbol $\int_{\beta_j(\mathbb{P})} {}^*du$.

Finally, $\mathbb{P}$ is called the <u>identity partition</u> of $\beta(R)$ if each member $P(G)$ of $\mathbb{P}$ consists of a single part.

6B. Let W be an open set in R such that $R \setminus W$ is a compact set whose boundary consists of a finite number of nonintersecting analytic curves. We fix a partition $\mathbb{P}$ of the ideal boundary of R with $\beta(\mathbb{P}) = (\beta_1(\mathbb{P}), \ldots, \beta_\ell(\mathbb{P}))$. By a normal operator associated with W we mean a linear operator L which assigns to each real-valued continuous function f on ∂W a function Lf, harmonic in W and continuous on $\mathrm{Cl}(W)$, such that

(B1) $L1 = 1$,

(B2) $f \geqq 0$ implies $Lf \geqq 0$,

(B3) $Lf = f$ on ∂W,

(B4) the flux of Lf over every $\beta_j(\mathbb{P})$ vanishes.

Then we have the following

<u>Theorem</u>. Let L be a normal operator on W and let s be a function, harmonic in W and continuous on $\mathrm{Cl}(W)$, such that

$$\int_{\beta(\mathbb{P}_0)} {}^*ds = 0,$$

where $\mathbb{P}_0$ denotes the identity partition of $\beta(R)$. Then there exists a harmonic function p on R which satisfies

$$p - s = L((p - s)|\partial W)$$

on $\mathrm{Cl}(W)$. The function p is unique up to an additive constant. ([AS], Ch. III, 3A)

6C. We need normal operators of special kind, called principal operators. In order to construct such operators, let us first take a compact bordered surface $\overline{W}$ whose border, $\beta(W)$, consists of a finite number of nonintersecting closed analytic curves. Supposing that the border $\beta(W)$ is not connected, we divide it into at least two nonvoid parts $\alpha, \beta_1, \ldots, \beta_\ell$, each of which is a union of component curves. Every β_j is oriented positively with respect to W whereas α is oriented negatively with respect to W.

<u>Lemma</u>. For any continuous function f on α there exists a unique function u harmonic in W and continuous on $\overline{W}$ with the following properties:

(a) $u = f$ on α,

(b) $u = c_j$ on β_j, c_j being a constant, for $j = 1, \ldots, \ell$,

(c) $\int_{\beta_j} {}^*du = 0$ for $j = 1,\dots,\ell$.
The function u is positive if f is positive. ([AS], Ch. III, 5)

6D. We return to the situation considered in 6B. Namely, let W be an open set in R such that $R \setminus W$ is a compact set whose boundary consists of a finite number of nonintersecting closed analytic curves and let $\mathbb{P}$ be a fixed partition of $\beta(R)$, which has a representative of the form $P = P(G)$, where G is a fixed regular region containing $R \setminus W$. We use the notation described in 6A. Let $G_1 \subsetneqq G_2 \subsetneqq \cdots$ be a regular exhaustion of R with $\mathrm{Cl}(G) \subsetneqq G_1$. α denotes the negatively oriented boundary of W.

For each n let us denote by $L^{(n)}$ the normal operator on the set $W \cap G_n$ satisfying the following conditions: for every continuous function f on α, $L^{(n)}f$ denotes the function harmonic in $W \cap G_n$ and continuous on $\mathrm{Cl}(W \cap G_n)$ such that $(L^{(n)}f)|\alpha = f$, $(L^{(n)}f)|\beta_j(G_n) =$ a constant and $\int_{\beta_j(G_n)} {}^*d(L^{(n)}f) = 0$ for $j = 1,\dots,\ell$, where $\beta_j(G_n) = \beta(G_n) \cap R_j$. The preceding lemma shows that $L^{(n)}$ are well-defined normal operators on $W \cap G_n$ corresponding to the partition P.

In order to extend $L^{(n)}$ to the whole W, we consider continuous real-valued functions f on α such that the solution $H[f;W]$ for the Dirichlet problem on W is harmonic up to the boundary α. The set of such functions on α is denoted by $\Sigma(\alpha,W)$. Take any such f. We set

$$u_n = \begin{cases} L^{(n)}f & \text{on } \mathrm{Cl}(W \cap G_n) \\ c_j & \text{on } R_j \setminus \mathrm{Cl}(G_n), \end{cases}$$

where c_j, $j = 1,\dots,\ell$, are constants which make u_n continuous on W. The assumption on f implies that each u_n is harmonic up to the boundary α. Clearly, du_n is defined on $\mathrm{Cl}(W) \setminus \beta(G_n)$. Writing $\|\omega\|_W^2 = \iint_W \omega\bar{\omega}^*$ for a differential ω on W, we have

$$\|du_n\|_W^2 = \iint_{W\cap G_n} du_n{}^*du_n = \int_{\beta(G_n)} u_n{}^*du_n - \int_\alpha u_n{}^*du_n$$

$$= -\int_\alpha u_n{}^*du_n.$$

For $n < n'$ we have

$$\|du_{n'} - du_n\|_W^2 = \iint_{W\cap G_n} d(u_{n'} - u_n)^*d(u_{n'} - u_n) + \iint_{G_{n'}\setminus G_n} du_{n'}{}^*du_{n'}$$

$$= \int_\alpha u_{n'} {*du_{n'}} - \int_\alpha u_n {*du_n}$$

$$= \|du_n\|_W^2 - \|du_{n'}\|_W^2.$$

Thus, $\|du_n\|_W$ are monotonically decreasing and therefore

$$\|du_n\|_W^2 - \|du_{n'}\|_W^2 \to 0$$

as $n, n' \to \infty$. It follows that $\|du_{n'} - du_n\|_W \to 0$. This means that in each parametric disk V the derivatives $\partial u_n/\partial x$ (resp. $\partial u_n/\partial y$) form an L^2-Cauchy sequence, i.e.

$$\iint_V \{|\tfrac{\partial}{\partial x}(u_{n'} - u_n)|^2 + |\tfrac{\partial}{\partial y}(u_{n'} - u_n)|^2 dxdy \to 0$$

as $n, n' \to \infty$. Since $\partial u_n/\partial x$ and $\partial u_n/\partial y$ are harmonic, we see that $\{\partial u_n/\partial x\}$ and $\{\partial u_n/\partial y\}$ converge almost uniformly on V. As V is arbitrary, we conclude that the sequence of differentials $\{du_n\}$ converges almost uniformly on $Cl(W)$. Moreover, since $u_n|\alpha = f$ for all n, the sequence $\{u_n\}$ converges almost uniformly to a function, say u, which is easily seen to be harmonic on $Cl(W)$ with $u|\alpha = f$. A simple computation shows that $du \in \Gamma^1$ on $Cl(W)$ and thus $du \in \Gamma_h^1(Cl(W))$. Since u is uniquely determined by f independently of exhaustions $\{G_n\}$, we can define an operator L by setting $Lf = u$.

<u>Theorem</u>. The linear operator L maps the space $\Sigma(\alpha,W)$ into the space of bounded harmonic functions on $Cl(W)$ and satisfies the conditions (B1)-(B4) as well as

(D1) $d(Lf) \in \Gamma_{h0}(Cl(W))$, where the class $\Gamma_{h0}(Cl(W))$ consists of all harmonic differentials ω on $Cl(W)$ (harmonic up to the boundary α) with $\|\omega\|_W < \infty$ such that for any $dh \in \Gamma_e^1(Cl(W))$

$$(\omega,{*dh})_W \ (= - \iint_W \omega\cdot d\bar{h}\ ) = - \int_\alpha \bar{h}\omega.$$

The operator L is called the <u>(second) principal operator</u> on W associated with the partition $\mathbb{P}$ and is denoted sometimes by $(\mathbb{P})L_1$. For further details we refer the reader to Ahlfors and Sario [AS], Ch. III, §§1 and 2.

7. <u>Modified Green Functions</u>

7A. Let R' be a region in R such that the boundary $\partial R'$ in R consists of a finite number of nonintersecting closed analytic curves and $Cl(R')$ is noncompact. Such a region R' is called <u>hyperbolic</u> if the complement of $\partial R'$ in the ideal boundary of R' has positive harmonic measure or, equivalently, R' carries a nonconstant bounded harmonic function which vanish identically on $\partial R'$. We denote by α' the boundary $\partial R'$ oriented negatively with respect to R'.

<u>Lemma</u>. Let G be any regular region in R such that $\partial R' \subseteqq G$ and let $\mathcal{U}$ be the collection of positive functions, harmonic in $G \cap R'$ and continuous on $Cl(G \cap R')$, such that $u = 0$ on α' and $\int_{\alpha'} *du \leqq 1$. Then, for any compact subset K of $G \cap Cl(R')$, there exists a positive constant C such that $\sup_{z \in K} u(z) \leqq C$ for every $u \in \mathcal{U}$.

<u>Proof</u>. It is enough to show that, for any regular region G' in R with $\partial R' \subseteqq G' \subseteqq Cl(G') \subseteqq G$, there exists a constant $C' > 0$ such that

$$\max\{u(z): z \in Cl(G' \cap R')\} \leqq C'$$

for any $u \in \mathcal{U}$. Let G' be such a region and set $\beta' = \beta(G') \cap R'$. We denote by $h(z)$ the solution of the Dirichlet problem for the region $G' \cap R'$ with the boundary data equal to 0 on α' and to 1 on β'. Since the local harmonic conjugate of h increases along α', we have $\int_{\alpha'} *dh > 0$. Choose any fixed point a in $G' \cap R'$. Then, by use of Harnack's inequality we can find a constant $c > 0$ such that

$$c^{-1} \max_{z \in \beta'} v(z) \leqq v(a) \leqq c \min_{z \in \beta'} v(z) \tag{10}$$

for any positive harmonic function v on $G \cap R'$.

We now take any $u \in \mathcal{U}$. If $u = 0$ at some point in $G \cap R'$, then u vanishes identically and there is nothing to prove. So we assume $u > 0$ everywhere on $G \cap R'$. Since the second half of (10) implies

$$u(a)h(z) \leqq cu(z) \tag{11}$$

on β' and since $u(z) = h(z) = 0$ on α', we see that (11) holds everywhere on $Cl(G' \cap R')$. It follows that $0 \leqq u(a)\int_{\alpha'} *dh \leqq c \times \int_{\alpha'} *du \leqq c$. Hence $u(a) \leqq c/(\int_{\alpha'} *dh)$. Next we use the first half of (10) and get

$$\max_{z \in \beta'} u(z) \leqq c\,u(a) \leqq c^2 \Big(\int_{\alpha'} *dh\Big)^{-1}.$$

By the maximum principle $u(z)$ is majorized everywhere on $Cl(G' \cap R')$ by the constant $c^2/(\int_{\alpha'} *dh)$, as was to be proved. □

7B. We use the notations given above.

Lemma. (a) There is a positive harmonic function u on $Cl(R')$ such that $\int_{\alpha'} *du > 0$.

(b) If R' is hyperbolic, then there exist bounded positive harmonic functions v and v' on $Cl(R')$ such that $dv, dv' \in \Gamma_{h0}(Cl(R'))$ (cf. 6D) and

$$\int_{\alpha'} *dv > 0 \quad \text{and} \quad \int_{\alpha'} *dv' < 0.$$

Proof. Let $G_1 \subseteqq G_2 \subseteqq \cdots$ be a regular exhaustion of R with $\partial R' \subseteqq G_1$ and set $\beta_n' = \beta(G_n) \cap R'$ for each n. Let v_n be the solution of the Dirichlet problem for $R' \cap G_n$ with the boundary data equal to 0 on α' and to 1 on β_n'. Then v_n is harmonic on $Cl(R' \cap G_n)$ and the flux d_n of v_n over α', i.e. $\int_{\alpha'} *dv_n$, is positive. We set $u_n = v_n/d_n$. Then u_n is positive, vanishes on α' and its flux over α' is equal to 1. Thus the preceding lemma shows that the sequence $\{u_n : n = 1, 2, \ldots\}$ is uniformly bounded on any compact subset of $Cl(R')$. By the diagonal process we can find a subsequence $\{u_{n(k)} : k = 1, 2, \ldots\}$ convergent almost uniformly on $Cl(R')$. Its limit u clearly satisfies the assertion (a).

When R' is hyperbolic, the limit v of $\{v_n : n = 1, 2, \ldots\}$ is nonzero, $\leqq 1$ and $\int_{\alpha'} *dv > 0$. Setting $v' = 1 - v$, we see that $0 \leqq v' \leqq 1$ and $\int_{\alpha'} *dv' < 0$. The final assertion follows now from the fact that

$$\|dv\|_{R'}^2 = \lim_{n\to\infty} \|dv_n\|_{R' \cap G_n}^2 = \lim_{n\to\infty} \int_{\alpha'} *dv_n$$

$$= \int_{\alpha'} *dv < \infty. \quad \square$$

7C. Returning to the notations in 6A, we get the following

Theorem. Let G be a regular region in R and let $P = P(G)$ be any partition associated with G, whose parts are denoted by $R_1, \ldots, R_\ell$. The negatively oriented boundary of R_j is denoted by $\beta_j(P)$. Then, given any real numbers a_j, $j = 1, \ldots, \ell$, with $\sum_{j=1}^{\ell} a_j \equiv 0 \pmod{2\pi}$, there exists a nonnegative harmonic function u on R such that

$$\int_{\beta_j(P)} *du \equiv a_j \pmod{2\pi} \tag{12}$$

for every j. Moreover, the functions u for all possible choices of

$\{a_j\}$ can be found in such a way that the set of values $\{u(a)\}$ for every fixed $a \in R$ is bounded.

Proof. If $\ell = 1$, then $u \equiv 0$ satisfies the condition. So we suppose that $\ell > 1$. Since R is hyperbolic, one of the R_j's at least contains a hyperbolic component. We assume R_1 to be such a part. By Lemma 7B there is a nonnegative harmonic function u_1 on $Cl(R_1)$ such that $\int_{\beta_1(P)} {*du_1} = -1$. For each $j = 2, \ldots, \ell$ there exists a positive harmonic function u_j on $Cl(R_j)$ such that $\int_{\beta_j(P)} {*du_j} = 1$. We define a function s_j on $R \setminus G$ for each $j = 2, \ldots, \ell$ by setting $s_j = u_1$ on $Cl(R_1)$, $= u_j$ on $Cl(R_j)$ and $= 0$ on $Cl(R_i)$ with $i \neq 1, j$. Then, since $\int_{\beta(G)} {*ds_j} = 0$, we can apply Theorem 6B to the open set $R \setminus Cl(G)$, the function s_j and the operator L corresponding to $W = R \setminus Cl(G)$. So there exist harmonic functions p_j, $j = 2, 3, \ldots, \ell$, on R such that

$$p_j - s_j = L((p_j - s_j)|\partial G)$$

on $R \setminus G$. Since $L((p_j - s_j)|\partial G)$ are bounded on $R \setminus G$ in view of (B1) and (B2) in 6B and since s_j are nonnegative there, the functions p_j are seen to be bounded below on R and therefore, by adding a constant if necessary, we may assume that they are positive on R. Our assumption on s_j and the property (B4) in 6B imply the following:

$$\int_{\beta_j(P)} {*dp_k} = \int_{\beta_j(P)} {*ds_k} = \begin{cases} -1 & \text{if } j = 1 \\ 1 & \text{if } j = k \\ 0 & \text{if } j \neq 1, k \end{cases}$$

for $k = 2, \ldots, \ell$.

Finally let b_j be real numbers with $0 \leqq b_j < 2\pi$ and $b_j \equiv a_j \pmod{2\pi}$, and set $u = \sum_{k=2}^{\ell} b_k p_k$. Then u is nonnegative on R and clearly satisfies (12). Moreover, for any $a \in R$ we have the bound

$$u(a) = \sum_{k=2}^{\ell} b_k p_k(a) \leqq 2\pi \sum_{k=2}^{\ell} p_k(a),$$

as desired. □

7D. Here we will construct modified Green functions.

Theorem. Let $g(\cdot,\zeta)$ be the Green function for R with pole $\zeta \in R$. Then there exists a function $g_0(\cdot,\zeta)$ satisfying the following conditions:

(a) $g_0(\cdot,\zeta) - g(\cdot,\zeta)$ is a positive harmonic function on R with $d(g_0(\cdot,\zeta) - g(\cdot,\zeta)) \in \Gamma_{h0}(R)$;

(b) $\int_{\beta_j(\mathbb{P})} {*dg_0(\cdot,\zeta)} \equiv 0 \pmod{2\pi}$ for $j = 1,\ldots,\ell$, where $\mathbb{P}$ is a given partition of the ideal boundary $\beta(R)$;

(c) $g_0(z,\zeta)$ is bounded for z fixed and ζ lying in any compact set not containing z.

Proof. Let G be any regular region such that $\mathbb{P}$ contains a partition of the form $P = P(G) = \{R_1,\ldots, R_\ell\}$. We may suppose that $R_1, \ldots, R_{\ell'}$ are hyperbolic while $R_{\ell'+1},\ldots, R_\ell$ are not. Since R is hyperbolic, we have $\ell' \geqq 1$. By use of Lemma 7B we take bounded positive harmonic functions u_j on $Cl(R_j)$ with $du_j \in \Gamma_{h0}(Cl(R_j))$, $j = 1,\ldots, \ell'$, such that $\int_{\beta_1(P)} {*du_1} = -1$ and $\int_{\beta_j(P)} {*du_j} = 1$ for $j = 2,\ldots, \ell'$. We take any $\zeta \in G$ and set

$$a_j = a_j(\zeta) = - \int_{\beta_j(P)} {*dg(\cdot,\zeta)} \tag{13}$$

for $j = 1,\ldots, \ell$. Then the following holds: (i) $a_j > 0$ for $j = 1, \ldots, \ell'$; (ii) $a_j = 0$ for $j = \ell' + 1,\ldots, \ell$; and (iii) $\sum_{j=1}^{\ell} a_j = 2\pi$. Assuming these for a moment, we proceed further. If $\ell' = 1$, then we have only to set $g_0(\cdot,\zeta) = g(\cdot,\zeta)$ and the conditions (a)-(c) are obviously satisfied in view of the facts just stated. So assume $\ell' > 1$. Then we define s_j, $j = 2,\ldots, \ell'$, by setting $s_j = u_j$ on $Cl(R_j)$, $= u_1$ on $Cl(R_1)$ and $= 0$ on $Cl(R_i)$ with $i \neq 1, j$. Then

$$\int_{\beta(G)} {*ds_j} = \int_{\beta_j(P)} {*du_j} + \int_{\beta_1(P)} {*du_1} = 0,$$

so that an application of Theorem 6B gives us harmonic functions p_j on R for $j = 2,\ldots, \ell'$ such that $p_j - s_j = L((p_j - s_j)|\partial G)$ on $R \setminus G$. Since each s_j is harmonic up to the boundary of $R \setminus G$, Theorem 6D shows that $L((p_j - s_j)|\partial G)$ is bounded and $d(L((p_j - s_j)|\partial G))$ belongs to $\Gamma_{h0}(R \setminus G)$. Moreover, each s_j is bounded and $ds_j \in \Gamma_{h0}(R \setminus G)$. Thus, for every $j = 2,\ldots, \ell'$, p_j is bounded and $dp_j \in \Gamma_{h0}(R)$. We may assume, adding a suitable constant if necessary, that p_j are positive on R. In view of the property (B4) in 6B of the operator L, we see that

$$\int_{\beta_k(P)} {*dp_j} = \int_{\beta_k(P)} {*ds_j} = \begin{cases} -1 & \text{if } k = 1 \\ 1 & \text{if } k = j \\ 0 & \text{otherwise.} \end{cases} \tag{14}$$

For $j = \ell' + 1, \ldots, \ell$ we set $p_j \equiv 0$ on R.

We now define g_0 by the formula

$$(15) \qquad g_0(\cdot,\zeta) = \sum_{j=2}^{\ell} a_j(\zeta)p_j + g(\cdot,\zeta).$$

The property (a) is then clear. The property (b) also holds, for a simple computation using (13) and (14) shows that

$$\int_{\beta_k(P)} {}^*dg_0(\cdot,\zeta) = \begin{cases} -2\pi & \text{if } k = 1 \\ 0 & \text{if } k > 1. \end{cases}$$

When ζ lies in G, the same p_j, $j = 2,\ldots,\ell$, can be used and therefore, in view of the properties (i)-(iii) of a_j, $g_0(\cdot,\zeta) - g(\cdot,\zeta)$ is bounded on R by the same constant C not larger than $2\pi\sum_{j=2}^{\ell} \|p_j\|_\infty$. Take a regular exhaustion $G_1 \subseteqq G_2 \subseteqq \cdots$ of R with $\mathrm{Cl}(G) \subsetneqq G_1$ and denote by C_n the constant C corresponding to G_n. Then the property (c) is satisfied if $g_0(\cdot,\zeta)$ is constructed for $\zeta \in G_n \setminus G_{n-1}$ so as to have $g_0(\cdot,\zeta) - g(\cdot,\zeta) \leqq C_n$.

Finally, we have to prove the properties (i)-(iii). Take any j with $1 \leqq j \leqq \ell$ and let f_j be the restriction of $g(\cdot,\zeta)$ to the set $\beta_j(P)$. Then $f_j > 0$ and, by Theorem 6A, Ch. I, $g(\cdot,\zeta) = H[f_j;R_j]$ on $\mathrm{Cl}(R_j)$. Take again a regular exhaustion $G_1 \subseteqq G_2 \subseteqq \cdots$ of R with $\mathrm{Cl}(G) \subseteqq G_1$, and define $G_{nj} = G_n \cap R_j$ and $\beta_j(G_n) = \beta(G_n) \cap R_j$, so that $\partial G_{nj} = \beta_j(G_n) - \beta_j(P)$. Further, let g_n (resp. ψ_n) be the solution of the Dirichlet problem for the open set G_{nj} with the boundary data equal to f_j (resp. 1) on $\beta_j(P)$ and to 0 on $\beta_j(G_n)$. Then g_n and its derivatives converge almost uniformly on $\mathrm{Cl}(R_j)$ to $g(\cdot,\zeta)$ and its corresponding derivatives, respectively. Thus

$$\begin{aligned} \int_{\beta_j(P)} {}^*dg_n &= \int_{\beta_j(P)} \psi_n {}^*dg_n - \int_{\beta_j(G_n)} \psi_n {}^*dg_n \\ &= - \iint_{G_{nj}} d\psi_n {}^*dg_n = - \iint_{G_{nj}} dg_n {}^*d\psi_n \\ &= \int_{\beta_j(P)} g_n {}^*d\psi_n - \int_{\beta_j(G_n)} g_n {}^*d\psi_n \\ &= \int_{\beta_j(P)} f_j {}^*d\psi_n . \end{aligned}$$

Letting $n \to \infty$, we see that

(16) $$\int_{\beta_j(P)} {}^*dg(\cdot,\zeta) = \int_{\beta_j(P)} f_j {}^*d\psi,$$

where $\psi = H[1;R_j]$. Since $0 \leqq \psi \leqq 1$, we see that ${}^*d\psi \leqq 0$ along $\beta_j(P)$. Moreover, ${}^*d\psi < 0$ on some component of $\beta_j(P)$ if and only if R_j is hyperbolic. Since $f_j > 0$, it follows from (16) that

$$\int_{\beta_j(P)} {}^*dg(\cdot,\zeta) < 0$$

if and only if R_j is hyperbolic. This proves (i) and (ii). (iii) follows at once from the local expression of the Green function given in Theorem 6A, Ch. I. □

7E. We need some further property of g.

Lemma. Let $\{V, z\}$ be a parametric disk, in which we indentify a point and its parametric coordinate, and let ζ_0, ζ_0' be distinct points in V such that $|\zeta_0|, |\zeta_0'| < r_1 < r_2 < 1$. Let c be an arc joining ζ_0 to ζ_0' within the disk $\{|z| < r_1\}$. Then for any 1-cycle γ not intersecting with c

$$\int_\gamma {}^*d(g(\cdot,\zeta_0') - g(\cdot,\zeta_0)) = -2\pi \int_c \sigma(\gamma),$$

where $\sigma(\gamma)$ is a real differential in Γ_{h0} given in Theorem 9C, Ch. I.

Proof. We use the notation in Ch. I, 9. Let $v(z)$ (resp. $s(z)$) be a single-valued branch of the function $\log\{(z-\zeta_0)/(z-\zeta_0')\}$ (resp. $\arg\{(z-\zeta_0)/(z-\zeta_0')\}$) in the annulus $r_1 \leqq |z| < 1$ and let $e(z)$ be a real-valued C^2-function on R such that $e \equiv 1$ on $\{|z| < r_1\}$ and $\equiv 0$ on $R \setminus \{|z| < r_2\}$. We also set $w(z) = \log|(z-\zeta_0)/(z-\zeta_0')|$ on $\{|z| \leqq 1\}$ and $k(z) = g(z,\zeta_0') - g(z,\zeta_0)$. Then $d\{(1-e)k\} \in \Gamma_{e0}(R)$. From the definition of the Green function follows that $k = w + h_1$ for $|z| < 1$, where h_1 is a harmonic function. Thus,

$${}^*dk = d\{\arg(z-\zeta_0) - \arg(z-\zeta_0')\} + dh_2$$

for $|z| < 1$, where h_2 is a harmonic conjugate of h_1 in $\{|z| < 1\}$. We first see that

(17) $$\begin{aligned} d(k - ew) &= d\{(1-e)k\} + d\{e(k-w)\} \\ &= d\{(1-e)k\} + d(eh_1) \in \Gamma_{e0}(R) \end{aligned}$$

and so ${}^*d(k - ew) \in \Gamma_{e0}{}^*$. On the other hand, $dk + i{}^*dk - d(ev) \in \Gamma_c$.

In fact, on V we have

$$dk + i{*}dk - d(ev) = d\{(1 - e)v\} + d(h_1 + ih_2),$$

which is closed. Next, on $R \setminus \{|z| \leq r_2\}$ we have

$$dk + i{*}dk - d(ev) = dk + i{*}dk,$$

which is analytic and so is closed.

Let γ be a 1-cycle contained in $R \setminus \{|z| < r_1\}$. Then there is a real harmonic differential $\sigma = \sigma(\gamma) \in \Gamma_{h0}$ such that $\int_\gamma \omega = (\omega, \sigma^*)$ for any closed differential $\omega \in \Gamma_c$ (Theorem 9B, Ch. I). Since $dk - d(ev)$ is exact on $R \setminus \{|z| < r_1\}$, we have

$$\int_\gamma {*}dk = -i \int_\gamma dk + i{*}dk - d(ev) = -i(dk + i{*}dk - d(ev), \sigma^*)$$
$$= A, \text{ say.}$$

Now, from (17) follows that

$$d(k - ew) + i{*}d(k - ew) \in \Gamma_{e0} + \Gamma_{e0}^*.$$

This differential is equal to

$$dk - d(ew) + i({*}dk - e\,{*}dw - w\,{*}de) = dk - d(ew) + i({*}dk - eds - w\,{*}de)$$
$$= dk + i{*}dk - d(ev) + i(sde - w\,{*}de).$$

Since $\Gamma_{h0}^* \perp (\Gamma_{e0} + \Gamma_{e0}^*)$ by (4) in Ch. I, 8E, we have

$$(d(k - ew) + i{*}d(k - ew), \sigma^*) = 0$$

and thus

$$A = -i(i(w\,{*}de - sde, \sigma^*) = (w\,{*}de - sde, \sigma^*)$$
$$= - \iint_{r_1<|z|<r_2} (w\,{*}de - sde)\sigma$$
$$= \iint_{r_1<|z|<r_2} wde\sigma^* + \iint_{r_1<|z|<r_2} sde\sigma$$
$$= \iint_{r_1<|z|<r_2} d(ew)\sigma^* - \iint_{r_1<|z|<r_2} edw\sigma^*$$
$$+ \iint_{r_1<|z|<r_2} d(es)\sigma - \iint_{r_1<|z|<r_2} eds\sigma$$

$$= - \int_{|z|=r_1} w\sigma^* - \iint_{r_1<|z|<r_2} edw\sigma^* - \int_{|z|= r_1} s\sigma - \iint_{r_1<|z|<r_2} eds\sigma.$$

To compute this, we add purely imaginary terms as follows:

$$-i\int_{|z|=r_1} s\sigma^* - i\iint_{r_1<|z|<r_2} eds\sigma^* + i\int_{|z|=r_1} w\sigma + i\iint_{r_1<|z|<r_2} edw\sigma.$$

The result is this:

$$-\int_{|z|=r_1} v\sigma^* - \iint_{r_1<|z|<r_2} edv\sigma^* + i\int_{|z|=r_1} v\sigma + i\iint_{r_1<|z|<r_2} edv\sigma$$

$$= i \int_{|z|=r_1} v(\sigma + i\sigma^*) + i \iint_{r_1<|z|<r_2} edv(\sigma + i\sigma^*)$$

$$= i \int_{|z|=r_1} vda$$

$$= -i \int_{|z|=r_1} a(z)\left(\frac{1}{z-\zeta_0} - \frac{1}{z-\zeta_0'}\right)dz$$

$$= -2\pi \int_{\zeta_0}^{\zeta_0'} da = -2\pi \int_c (\sigma + i\sigma^*),$$

where a is an analytic function on V with $da = \sigma + i\sigma^*$. Hence,

$$\int_\gamma {}^*dk = -2\pi \int_c \sigma. \quad \square$$

7F. Concerning the function g_0 we have the following

<u>Theorem</u>. Under the same condition as in Lemma 7E, we have

$$\int_\gamma {}^*d(g_0(\cdot,\zeta_0') - g_0(\cdot,\zeta_0)) = -2\pi \int_c \{\sigma(\gamma) - \sum_{j=2}^{\ell} K_{j,\gamma}\sigma(\beta_j(P))\},$$

where $K_{j,\gamma}$ is a constant dependent of j and γ but not of points ζ_0 and ζ_0'.

<u>Proof</u>. We set $k_0 = g_0(\cdot,\zeta_0') - g_0(\cdot,\zeta_0)$. Then by (15) we have

$$k_0 = k + \sum_{j=2}^{\ell} (a_j(\zeta_0') - a_j(\zeta_0))p_j,$$

where p_j are harmonic with $dp_j \in \Gamma_{he}^1(R)$ and $a_j(\zeta)$'s are given by the formula (13). Thus by Lemma 7E

$$a_j(\zeta_0') - a_j(\zeta_0) = - \int_{\beta_j(P)} {*dk} = 2\pi \int_c \sigma(\beta_j(P)).$$

It follows that, for any cycle γ in $R \setminus \{|z| < r_1\}$,

$$\int_\gamma {*dk_0} = \int_\gamma {*dk} + \sum_{j=2}^{\ell} (a_j(\zeta_0') - a_j(\zeta_0)) \int_\gamma {*dp_j}$$

$$= -2\pi \int_c \{\sigma(\gamma) - \sum_{j=2}^{\ell} (\int_\gamma {*dp_j})\sigma(\beta_j(P))\}. \quad \square$$

8. Proof of Sufficiency

8A. We are now in a position to prove the implication (c) ⇒ (a) in Theorem 2B. We do this by contradiction. Namely we suppose that R is a hyperbolic Riemann surface but not a PWS and therefore that

$$\int_0^\infty B(\alpha,a)d\alpha = \infty$$

for some (and hence all) $a \in R$.

We set

$$m(\xi,a) = \sup\{|f(a)|: f \in \mathcal{H}^1(R,\xi), \|f\|_{1,a} \leqq 1\}$$

for any line bundle ξ over R and then

$$m(a) = \inf\{m(\xi,a): \xi \in H^1(R;\mathbb{T})\}.$$

In view of Theorem 5A we have $m(a) = \exp(-\int_0^\infty B(\alpha,a)d\alpha) = 0$. We should show that $m(\xi,a) = 0$ for some line bundle ξ. For this end we still have to prove two more facts.

8B. Let G_0 be a canonical region in R (Ch. I, 1C), $\{R_1,\ldots, R_\ell\}$ the set of all connected components of $R \setminus Cl(G_0)$, and α_j, $1 \leqq j \leqq \ell$, the boundary contour of R_j, which is oriented positively with respect to G_0. Then a homology basis for $Cl(G_0)$ consists of $\alpha_1,\ldots, \alpha_{\ell-1}$ and certain nondividing cycles $\gamma_1,\ldots, \gamma_N$.

Lemma. Let V be any parametric disk included in G_0 such that $Cl(V) \subsetneqq G_0$ and all γ_n's lie off V. Then, for any real numbers $b_1, \ldots, b_N$ there exist points $\zeta_1,\ldots, \zeta_{N'}$ in V such that

$$\sum_{k=1}^{N'} \int_{\gamma_n} {*dg_0(\cdot,\zeta_k)} \equiv b_n \pmod{2\pi}$$

for $n = 1, \ldots, N$. The number N' is independent of the numbers b_n.

Proof. Let ζ_0 be the center of V and let $\sigma_n = \sigma(\gamma_n)$, $n = 1, \ldots, N$, and $\tau_j = \sigma(\beta_j(P))$, $j = 2, \ldots, \ell$, be the harmonic differentials on R representing γ_n and $\beta_j(P)$, respectively, where $P = P(G_0)$. We write $\sigma_n = u_n dx + v_n dy$ and $\tau_j = u_j' dx + v_j' dy$ in V. Then, u_n, $n = 1, \ldots, N$, and u_j', $j = 2, \ldots, \ell$, are linearly independent on V. In fact, if $\sum_{n=1}^N c_n u_n + \sum_{j=2}^\ell c_j' u_j' = 0$ on V, then

$$0 = \sum_{n=1}^N c_n u_n + \sum_{j=2}^\ell c_j' u_j' = \frac{\partial}{\partial x} \int_{\zeta_0}^{\zeta} \left(\sum_{n=1}^N c_n \sigma_n + \sum_{j=2}^\ell c_j' \tau_j \right)$$

on V. This means that

$$\int_{\zeta_0}^{\zeta} \left(\sum_{n=1}^N c_n \sigma_n + \sum_{j=2}^\ell c_j' \tau_j \right)$$

is constant on V and therefore $\sum_{n=1}^N c_n \sigma_n + \sum_{j=2}^\ell c_j' \tau_j$ vanishes identically on R. Since σ_n, τ_j with $n = 1, \ldots, N$ and $j = 2, \ldots, \ell$ are linearly independent, we have $c_n = c_j' = 0$ for all n and j, as was to be shown. It follows that

$$u_n^{\#} = u_n - \sum_{j=2}^\ell K_{j,\gamma_n} u_j'$$

for $n = 1, \ldots, N$ are linearly independent on V. So we can find, by induction, points $\zeta_1', \ldots, \zeta_N' \in V$ such that

$$\Delta(\zeta_1', \ldots, \zeta_N') = \begin{vmatrix} u_1^{\#}(\zeta_1') & \ldots & u_1^{\#}(\zeta_N') \\ \vdots & & \vdots \\ u_N^{\#}(\zeta_1') & \ldots & u_N^{\#}(\zeta_N') \end{vmatrix} \neq 0.$$

We set $\sigma_n^{\#} = \sigma_n - \sum_{j=2}^\ell K_{j,\gamma_n} \tau_j$ and consider the mapping

$$\Phi(x_1, \ldots, x_N) = \left(\sum_{k=1}^N \int_{\zeta_k'}^{\zeta_k' + x_k} \sigma_1^{\#}, \ldots, \sum_{k=1}^N \int_{\zeta_k'}^{\zeta_k' + x_k} \sigma_N^{\#} \right)$$

defined in a neighborhood of 0 in $\mathbb{R}^N$ given by $U = \{(x_1, \ldots, x_N): |x_n| < r \text{ for } n = 1, \ldots, N\}$ with $r = \min\{1 - |\zeta_n'|: n = 1, \ldots, N\}$. Since the Jacobian of Φ at 0 is equal to $\Delta(\zeta_1', \ldots, \zeta_N') \neq 0$, $\Phi(U)$ contains an open set in $\mathbb{R}^N$. For a large integer $M > 0$, $(2\pi M)\Phi(U)$ includes an N-cube with sides of length 2π. On the other hand, we have, by Theorem 7F,

$$M\left(\sum_{k=1}^N \int_{\gamma_1} {*dg_0}(\cdot, \zeta_k' + x_k), \ldots, \sum_{k=1}^N \int_{\gamma_N} {*dg_0}(\cdot, \zeta_k' + x_k) \right)$$

$$= M\left(\sum_{k=1}^{N}\int_{\gamma_1} {}^*dg_0(\cdot,\zeta_k'),\ldots,\sum_{k=1}^{N}\int_{\gamma_N} {}^*dg_0(\cdot,\zeta_k')\right) - 2\pi M\Phi(x_1,\ldots,x_N).$$

In view of the definition of M, the right-hand side covers an N-cube with sides of length 2π when $(x_1,\ldots,x_N)$ ranges over U. Hence, for any $(b_1,\ldots,b_N) \in \mathbb{R}^N$ there exists a point $(x_1,\ldots,x_N) \in U$ such that

$$M\sum_{k=1}^{N}\int_{\gamma_n} {}^*dg_0(\cdot,\zeta_k' + x_k) \equiv b_n \pmod{2\pi}$$

for $n = 1,\ldots, N$. If $\zeta_1,\ldots, \zeta_{MN}$ are the points $\zeta_n' + x_n$, $n = 1,\ldots, N$, each repeated M times, then ζ_k's belongs to V and

$$\sum_{k=1}^{MN}\int_{\gamma_n} {}^*dg_0(\cdot,\zeta_k) \equiv b_n \pmod{2\pi}$$

for $n = 1,\ldots, N$. □

8C. Using the same notations as in the preceding, we have

<u>Theorem</u>. Suppose $a \in G_0$. Then there exists a constant $m > 0$ such that for any line bundle ξ_0' over G_0 there exist an extension ξ to R and an $f \in \mathcal{H}^\infty(R,\xi)$ satisfying $\|f\|_\infty \leq 1$ and $|f(a)| \geq m$.

<u>Proof</u>. As noted in Ch. II, 2B, the bundle ξ_0' is given by a character, say θ, of the fundamental group $F_O(\overline{G}_0)$ with initial point O, where $\overline{G}_0 = \mathrm{Cl}(G_0)$. Since θ is identically equal to 1 on the commutator subgroup $[F_O(\overline{G}_0)]$ of $F_O(\overline{G}_0)$, θ is determined by its values at α_1, $\ldots$, $\alpha_{\ell-1}$ and $\gamma_1,\ldots, \gamma_N$, for these elements generate the homology group $H_1(\overline{G}_0,\mathbb{Z}) = F_O(\overline{G}_0)/[F_O(\overline{G}_0)]$. Here, the paths are supposed to issue from O by performing obvious deformations. We set $\theta(\alpha_j) = \exp(-2\pi i a_j)$ for $j = 1,\ldots, \ell$ and $\theta(\gamma_n) = \exp(-2\pi i b_n)$ for $n = 1$, $\ldots$, N, where a_j and b_n are real numbers. Since $\sum_{j=1}^{\ell} \alpha_j$ bounds the region G_0, we should have $\sum_{j=1}^{\ell} a_j \equiv 0 \pmod{2\pi}$. By Theorem 7C there exists a nonnegative harmonic function u on R such that

$$\int_{\alpha_j} {}^*du \equiv a_j \pmod{2\pi}$$

for every j, where $u(a) \leq C < \infty$, C being a constant independent of a_j. Next we use Lemma 8B to choose a finite number of points $\zeta_1,\ldots, \zeta_{MN}$ from a fixed compact set K lying in $G_0 \setminus \{a\}$ such that

$$\sum_{k=1}^{MN}\int_{\gamma_n} {}^*dg_0(\cdot,\zeta_k) \equiv b_n - \int_{\gamma_n} {}^*du \pmod{2\pi}$$

for $n = 1,\ldots, N$. We set $v = u + \sum_{k=1}^{MN} g_0(\cdot,\zeta_k)$ and $f = \exp(-v - i\tilde{v})$, where $\tilde{v}$ denotes the harmonic conjugate of v. We see first that f is holomorphic on R. Since the definition of g_0 (Theorem 7D) shows that

$$\int_{\alpha_j} *dg_0(\cdot,\zeta) \equiv 0 \pmod{2\pi}$$

for any $\zeta \in G_0$ and any $j = 1,\ldots, \ell$, we see that

$$-\int_{\alpha_j} *dv \equiv -a_j \pmod{2\pi}.$$

We also have

$$-\int_{\gamma_n} *dv = -\int_{\gamma_n} *du - \sum_{k=1}^{MN} \int_{\gamma_n} *dg_0(\cdot,\zeta_k) \equiv -b_n \pmod{2\pi}.$$

Thus, f is a section of a line bundle ξ over R, so that the bundle ξ_0' is the restriction of ξ to G_0. Moreover,

$$|f(a)| = \exp(-v(a) - \sum_{k=1}^{MN} g_0(a,\zeta_k)) \geqq m > 0,$$

where

$$m = \exp(-C - MN \sup_{\zeta\in K} g_0(a,\zeta))$$

is independent of the choice of ξ. □

8D. For each line bundle ξ over R we denote by $\rho_0(\xi)$ the restriction of ξ to the region G_0. Then

Theorem. Suppose $m(a) = 0$. Then, for any line bundle ξ_0 over $\overline{G}_0$, we have

$$\inf\{m(\xi,a)\colon \rho_0(\xi) = \xi_0\} = 0.$$

Proof. Since $m(a) = 0$, we see that for any $\varepsilon > 0$ there exists a line bundle ξ' over R with $m(\xi',a) < \varepsilon$. We set $\xi_0' = \rho_0(\xi')^{-1}\xi_0$. By the preceding theorem we find an extension, say ξ'', of ξ_0' to R and a section $f_0 \in \mathcal{H}^\infty(R,\xi''^{-1})$ such that $\|f_0\|_\infty \leqq 1$ and $|f_0(a)| \geqq m > 0$, where m is a constant independent of ξ_0'. Set $\xi = \xi'\xi''$. Then, $\rho_0(\xi) = \rho_0(\xi')\rho_0(\xi'') = \xi_0$. If $h \in \mathcal{H}^1(R,\xi)$ with $\|h\|_{1,a} \leqq 1$, then f_0h belongs to $\mathcal{H}^1(R,\xi\xi''^{-1}) = \mathcal{H}^1(R,\xi')$ and $\|f_0h\|_{1,a} \leqq 1$, so that $|(f_0h)(a)| \leqq m(\xi',a) < \varepsilon$. Since $|f_0(a)| \geqq m$, we have $|h(a)| \leqq m^{-1}\varepsilon$, which means that $m(\xi,a) \leqq m^{-1}\varepsilon$. As ε is arbitrary, we get the desired result. □

8E. To finish the proof of (c) ⇒ (a) in Theorem 2B, we suppose that $m(a) = 0$ for some (and hence all) $a \in R$. We take a relatively compact open subset V of R and a countable dense subset S of V. We then choose a regular exhaustion $\{G_n: n = 1, 2, \ldots\}$ of R by canonical regions (Ch. I, 1C) such that $\mathrm{Cl}(V) \subseteqq G_1$. We denote by ρ_n (resp. ρ_{jk} with $j \leqq k$) the restriction to G_n (resp. G_j) of line bundles over R (resp. G_k). Finally we set

$$m_n(\xi_n,a) = \sup\{|f(a)|: f \in \mathcal{H}^1(G_n,\xi_n), \|f\|_{1,a} \leqq 1\}$$

for any line bundle ξ_n over G_n and any $a \in G_n$.

We fix a line bundle ξ_j over G_j and take any $b \in G_j$. We set

$$\tilde{m}_n = \inf\{m_n(\xi_n,b): \xi_n \in H^1(G_n;\mathbb{T}), \rho_{jn}(\xi_n) = \xi_j\}$$

for $n \geqq j$ and

$$\tilde{m} = \inf\{m(\xi,b): \xi \in H^1(R;\mathbb{T}), \rho_j(\xi) = \xi_j\}.$$

We claim

$$\lim_{n\to\infty} \tilde{m}_n = \tilde{m}. \tag{18}$$

This can be shown in the same way as in the proof of (8) (see 5A). In fact, if $\xi_{n+1} \in H^1(G_{n+1};\mathbb{T})$ with $\rho_{j,n+1}(\xi_{n+1}) = \xi_j$, then $\rho_{n,n+1}(\xi_{n+1})$ belongs to $H^1(G_n;\mathbb{T})$ and therefore

$$m_n(\rho_{n,n+1}(\xi_{n+1}),b) \geqq m_{n+1}(\xi_{n+1},b) \geqq \tilde{m}_{n+1}.$$

Since every line bundle over G_j is the restriction of one over R, we see that $\tilde{m}_n \geqq \tilde{m}_{n+1} \geqq \tilde{m}$ and hence $\lim_{n\to\infty} \tilde{m}_n \geqq \tilde{m}$.

In order to show the reverse inequality, we may assume that $\tilde{m} < \infty$. We take $\xi \in H^1(R;\mathbb{T})$ with $\rho_j(\xi) = \xi_j$ and $m(\xi,b) < \infty$. We assume as before that ξ is defined by means of a countable open covering $\{V'_\alpha\}$ in the form $(\{\xi_{\alpha\beta}\}, \{V'_\alpha\})$ such that all V'_α as well as all nonempty $V'_\alpha \cap G_n$ are simply connected. The definition of $m_n(\rho_n(\xi),b)$ shows that there exists an $f_n \in \mathcal{H}^1(G_n,\rho_n(\xi))$ with the properties $\|f_n\|_{1,b} \leqq 1$ and $|f_n(b)| \geqq m_n(\rho_n(\xi),b) - 1/n$. Let $\{f_{n\alpha}\}_\alpha$ be a representative of f_n with respect to the covering $\{V'_\alpha \cap G_n\}_\alpha$ of G_n and the representative $(\{\xi_{\alpha\beta}\}, \{V'_\alpha \cap G_n\}_\alpha)$ of $\rho_n(\xi)$. Let $\{V_\alpha\}$ be a refinement of the covering $\{V'_\alpha\}$ such that $\mathrm{Cl}(V_\alpha)$ is a compact subset of V'_α for every α. Since $\|f_n\|_{1,b} \leqq 1$, there exists a harmonic function, h_n, on G_n with $|f_n| \leqq h_n$ on G_n and $h_n(b) \leqq 1$. Take any α. If G_n includes $Cl(V_\alpha)$, then the Harnack inequality shows that h_n is bounded on the set $Cl(V_\alpha)$ by a constant C_α independent of n. Thus, $f_{n\alpha}$ with $n \geqq 1$ are uniformly bounded on V_α for each α. So there is a subsequence $\{n_k: k = 1, 2, \dots\}$ such that $\{f_{n_k\alpha}\}_k$ is convergent uniformly on V_α for each α and such that $\{h_{n_k}\}_k$ is convergent uniformly on each V_α. We set

$$f_\alpha = \lim_{k\to\infty} f_{n_k\alpha} \quad \text{and} \quad h = \lim_{k\to\infty} h_{n_k}.$$

Then $\{f_\alpha\}$ defines a section of the bundle ξ, which we denote by f. Since $|f| \leqq h$, $f \in \mathcal{H}^1(R,\xi)$ and $\|f\|_{1,b} \leqq h(b) = \lim_{k\to\infty} h_{n_k}(b) \leqq 1$. Finally, we have

$$|f(b)| = \lim_{k\to\infty} |f_{n_k}(b)| \geqq \limsup_{k\to\infty}\{m_{n_k}(\rho_{n_k}(\xi),a) - n_k^{-1}\}$$

$$\geqq \limsup_{k\to\infty} \tilde{m}_{n_k} = \lim_{n\to\infty} \tilde{m}_n.$$

As ξ is arbitrary, we have $\tilde{m} \geqq \lim_{n\to\infty} \tilde{m}_n$ and therefore (18) holds.

By Theorem 8D we have $\tilde{m} = 0$ for any $\xi_j \in H^1(G_j;\mathbb{T})$ and any $b \in G_j$, and thus

(19) $$\lim_{n\to\infty}[\inf\{m_n(\xi_n,b): \xi_n \in H^1(G_n;T),\ \rho_{jn}(\xi_n) = \xi_j\}] = 0.$$

Let us take a sequence $\{b_j: j = 1, 2, \dots\}$ in which each point in S appears infinitely often. By use of (19) we find a sequence $\{n_j\}$ with $n_1 < n_2 < \cdots$ and line bundles ξ_{n_i} over G_{n_i} for $i = 1, 2, \dots$ such that

$$\rho_{n_i,n_{i+1}}(\xi_{n_{i+1}}) = \xi_{n_i} \quad \text{and} \quad m_{n_i}(\xi_{n_i},b_i) < i^{-1}.$$

Then there exists a unique line bundle ξ over R satisfying $\rho_{n_i}(\xi) = \xi_{n_i}$ for all i, so that we have $m(\xi,b_i) \leqq m_{n_i}(\xi_{n_i},b_i) < i^{-1}$. Take any $b \in S$. Then our assumption on $\{b_j\}$ says that we have $b_i = b$ for infinitely many i. So $m(\xi,b) = 0$ for all $b \in S$. This means that every $f \in \mathcal{H}^1(R,\xi)$ vanishes at all points in S and therefore $f \equiv 0$ on R. Hence we have $\mathcal{H}^1(R,\xi) = \{0\}$, as was to be proved. □

9. A Few Direct Consequences

9A. First we restate Theorem 5A as follows:

Theorem. Let R be any PWS. Then, for every p with $1 \leq p \leq \infty$,

$$\inf_{\xi}[\sup\{|f(a)|: f \in \mathcal{H}^p(R,\xi),\ \|f\|_{p,a} = 1\}]$$

$$= \exp\left(-\int_0^{\infty} B(\alpha,a)d\alpha\right) > 0,$$

where ξ ranges over all line bundles over R. If R is regular, then the common value is equal to

$$\exp[-\sum\{g(a,w): w \in Z(a;R)\}].$$

The last statement comes from Theorem 1C. Since this theorem is valid for any point $a \in R$, we have in particular the following

Corollary. If R is a PWS, then the functions in $\mathcal{H}^{\infty}(R,\xi)$ have no common zeros for any line bundle $\xi \in H^1(R;\mathbb{T})$.

9B. Theorem. Let R be a PWS. Then the following hold:

(a) For each pair of distinct points $a, b \in R$ there exists an $f \in H^{\infty}(R)$ with $f(a) \neq f(b)$.

(b) For every point $a \in R$ there exists an $f \in H^{\infty}(R)$ which has a simple zero at a.

Proof. Consider the multiplicative bounded analytic function

$$z \to \exp[-g(a,z) - i\tilde{g}(a,z))$$

on R and let ξ be the line bundle over R determined by this function. Take an $h \in \mathcal{H}^{\infty}(R,\xi^{-1})$ such that $|h(b)| \neq 0$. This is possible because of the preceding Corollary. We then choose a function element of $h(z)\exp[-g(a,z) - i\tilde{g}(a,z)]$ and extend it analytically to the whole surface R. As is easily seen, the resulting function, say f, belongs $H^{\infty}(R)$ and has the property: $f(a) = 0$ and $f(b) \neq 0$. This proves the statement (a).

To prove (b), we choose $h \in \mathcal{H}^{\infty}(R,\xi^{-1})$ with $h(a) \neq 0$. Then, $h(z)\exp[-g(a,z) - i\tilde{g}(a,z)]$ gives rise to a single-valued function f in $H^{\infty}(R)$. Since $h(a) \neq 0$, f has a simple zero at the point a. □

NOTES

For the definition of PWS we follow Widom [70], which contains his fundamental theorem. Most results in this chapter are taken from this brilliant paper. Parreau's definition is in Parreau [52]. Essential equivalence of these two definitions was shown by Hasumi [18]. As a matter of fact, I did not know Parreau's results when [18] was written. Afterwards, in 1976, Z. Kuramochi pointed out to me the relevance of Parreau's paper [52] in connection with my work [19]. The name "surfaces of Parreau-Widom type" was then coined by M. Hayashi.

All results except those in Subsections 3 and 6 are due to Widom [70], although some small modifications have been made. The results in Subsection 3 are adapted from Hasumi [18].

CHAPTER VI. GREEN LINES

Every bounded harmonic function in the unit disk has limits along almost all radii and can be restored from its radial limit function by means of the Poisson integral. This famous result, known as the Fatou theorem, has a striking direct generalization to all Riemann surfaces of Parreau-Widom type. The first main results in this direction were discovered by M. Parreau [52] already in 1958. It is indeed in that paper of Parreau that surfaces of Parreau-Widom type appeared for the first time. Combined with Widom's theorem (Theorem 2B, Ch. V), results of Parreau then permit us to give a completely affirmative answer, in the case of surfaces of Parreau-Widom type, to the Brelot-Choquet problem asking relations between Green lines and the Martin boundary.

In §1 an introductory discussion is given about Green lines on general hyperbolic Riemann surfaces. The other sections are devoted to the case of surfaces of Parreau-Widom type. We show in §2 that every positive harmonic function has a limit along almost every Green line issuing from any fixed point and that the Dirichlet problem on the space of Green lines is resolutive for any bounded data measurable with respect to the Green measure. We solve in §3 the Brelot-Choquet problem and show that the radial limits (along Green lines) and the fine limits are essentially the same for a class of functions including bounded harmonic functions.

In this chapter we consider only regular hyperbolic Riemann surfaces. As our ultimate objective is to investigate Green lines on surfaces of Parreau-Widom type, this regularity assumption is evidently legitimate in view of Theorem 3B, Ch. V.

§1. THE DIRICHLET PROBLEM ON THE SPACE OF GREEN LINES

1. Definition of Green Lines

1A. Let R be a regular hyperbolic Riemann surface with Green function $g(a,z)$. Let $a \in R$ be held fixed and define two functions $r(z) = r(a;z)$ and $\theta(z) = \theta(a;z)$ by equations

$$dr(z)/r(z) = -dg(a,z) \quad \text{and} \quad d\theta(z) = -{*}dg(a,z),$$

respectively. In the following we denote by $r(z)$ the special solution $r(z) = \exp(-g(a,z))$ for the first equation. A <u>Green line</u> issuing from the point a is by definition any maximal open arc issuing from a on which $d\theta(z) \neq 0$ and $\theta(z)$ is constant. $\mathbb{G}(R,a)$ $(= \mathbb{G}(a)$, for short) denotes the set of Green lines issuing from a. For a sufficiently large $\alpha > 0$ the closed region $Cl(R(\alpha,a)) = \{z \in R: g(a,z) \geq \alpha\}$ is conformally isomorphic with the closed unit disk under the map

$$z \to e^{\alpha}r(z)\exp(i\theta(z)).$$

Thus each Green line ℓ in $\mathbb{G}(a)$ is parametrized with $\theta \in [0, 2\pi)$ as $\ell = \ell_\theta$ in such a way that $\theta(z) \equiv \theta \pmod{2\pi}$ with $\{z\} = \ell \cap \partial R(\alpha,a)$. Using the correspondence $e^{i\theta} \to \ell_\theta$, we can identify $\mathbb{G}(a)$ with the unit circumference and introduce a measure $dm_a(\ell_\theta) = dm_a(\theta) = d\theta/2\pi$, which is called the <u>Green measure</u> on $\mathbb{G}(a)$. Let $\mathbb{E}_0(R,a) = \mathbb{E}_0(a)$ be the set of Green lines ℓ in $\mathbb{G}(a)$ such that the closure $Cl(\ell)$ of ℓ in R is a compact set. It is easy to see that a line $\ell \in \mathbb{G}(a)$ belongs to $\mathbb{E}_0(a)$ if and only if it ends in a critical point of the function $g(a,z)$. So $\mathbb{E}_0(a)$ is a countable set. A Green line ℓ is called <u>regular</u> if the infimum, in z, of $g(a,z)$ over ℓ is equal to zero. We denote by $\mathbb{L}(R,a) = \mathbb{L}(a)$ the set of regular Green lines in $\mathbb{G}(a)$. Since R is regular, we have

$$\mathbb{G}(a) = \mathbb{L}(a) \cup \mathbb{E}_0(a).$$

The union of the point a and all Green lines in $\mathbb{G}(a)$ is called the <u>Green star region</u> centered at a and is denoted by $\mathbb{G}'(R,a)$ (or $\mathbb{G}'(a)$). The region $\mathbb{G}'(a)$ is a simply connected region and we can use the function $z \to r(z)e^{i\theta(z)}$ as a global coordinate on it. For every $\ell \in \mathbb{L}(a)$ let $z(\ell;\alpha)$ denote the point z on ℓ for which $g(a,z) = \alpha$.

1B. We define the Dirichlet problem on the set $\mathbb{G}(a)$ of Green lines. Given an extended real-valued function f on the set $\mathbb{L}(a)$, we denote by $\overline{S}(f;\mathbb{G}(a))$ the set of superharmonic functions s on R such that

(B1) s is bounded below;

(B2) $\liminf_{\alpha\to 0} s(z(\ell;\alpha)) \geq f(\ell)$ for dm_a-a.e. on $\mathbb{L}(a)$.

We set $\underline{S}(f;\mathbb{G}(a)) = -\overline{S}(-f;\mathbb{G}(a))$. Suppose that both $\overline{S}(f;\mathbb{G}(a))$ and $\underline{S}(f;\mathbb{G}(a))$ are nonvoid and let $\overline{G}[f;\mathbb{G}(a)]$ (resp. $\underline{G}[f;\mathbb{G}(a)]$) be the pointwise infimum (resp. supremum) on R of functions in $\overline{S}(f;\mathbb{G}(a))$

(resp. $\underline{S}(f;\mathbb{G}(a))$). Since we have $s' \geqq s''$ for any $s' \in \overline{S}(f;\mathbb{G}(a))$ and $s'' \in \underline{S}(f;\mathbb{G}(a))$, both $\overline{S}(f;\mathbb{G}(a))$ and $\underline{S}(f;\mathbb{G}(a))$ are Perron families, so that $\overline{G}[f;\mathbb{G}(a)]$ and $\underline{G}[f;\mathbb{G}(a)]$ are harmonic on R and $\underline{G}[f;\mathbb{G}(a)] \leqq \overline{G}[f;\mathbb{G}(a)]$. If $\overline{S}(f;\mathbb{G}(a))$ and $\underline{S}(f;\mathbb{G}(a))$ are nonvoid and if $\underline{G}[f;\mathbb{G}(a)] = \overline{G}[f;\mathbb{G}(a)]$, then f is called <u>resolutive</u>. The common function is denoted by $G[f;\mathbb{G}(a)]$ and is called the <u>solution of the Dirichlet problem for the space</u> $\mathbb{G}(a)$ <u>of Green lines</u>. It is clear that $G[f;\mathbb{G}(a)]$ belongs to $HP'(R)$.

2. The Dirichlet Problem

2A. We begin with the following, which is an analogue of Theorem 5A, Ch. III.

<u>Theorem</u>. An extended real-valued function f on $\mathbb{L}(a)$ is resolutive if and only if there exist a harmonic function u and a positive superharmonic function s on R such that

$$(1) \qquad u + \varepsilon s \in \overline{S}(f;\mathbb{G}(a)) \quad \text{and} \quad u - \varepsilon s \in \underline{S}(f;\mathbb{G}(a))$$

for any $\varepsilon > 0$. If this is the case, then $u = G[f;\mathbb{G}(a)]$.

<u>Proof</u>. The proof is similar to that of Theorem 5A, Ch. III. In fact, if f is resolutive with solution $u = G[f;\mathbb{G}(a)]$, then there exists a sequence $\{s_n' : n = 1, 2, \ldots\}$ in $\overline{S}(f;\mathbb{G}(a))$ such that

$$\sum_{n=1}^{\infty} (s_n'(a^*) - u(a^*)) < \infty,$$

where a^* is any prescribed point in R. So the sum $s' = \sum_{n=1}^{\infty} (s_n' - u)$ is a positive superharmonic function on R. Then for any fixed integer $m \geqq 1$ we have

$$u(z) + m^{-1}s'(z) \geqq u(z) + m^{-1} \sum_{n=1}^{m} (s_n' - u)(z)$$

$$= m^{-1} \sum_{n=1}^{m} s_n'(z)$$

for every $z \in R$. It follows readily that $u + m^{-1}s' \in \overline{S}(f;\mathbb{G}(a))$. Similarly, there exists a positive superharmonic function s'' on R such that $u - m^{-1}s''$ belongs to $\underline{S}(f;\mathbb{G}(a))$ for any integer $m \geqq 1$. So the superharmonic function $s = s' + s''$ satisfies the conditions in (1).

Suppose, conversely, that we have a harmonic function u and a positive superharmonic function s satisfying (1). Then

$$u = \lim_{\varepsilon\to 0} (u+\varepsilon s) \geqq \overline{G}[f;\mathbb{G}(a)] \geqq \underline{G}[f;\mathbb{G}(a)] \geqq \lim_{\varepsilon\to 0} (u-\varepsilon s) = u,$$

wherever s is finite. Since s is infinite only on a polar set, we conclude that f is resolutive and $u = G[f;\mathbb{G}(a)]$. □

Using this, we can easily show the following as in the case of the usual Dirichlet problem.

Corollary. (a) If f_1, f_2 are resolutive functions on $\mathbb{L}(a)$ and α_1, α_2 are real numbers, then $\alpha_1 f_1 + \alpha_2 f_2$, $\max\{f_1,f_2\}$ and $\min\{f_1,f_2\}$ are resolutive and satisfy the following, in which we write $G[\cdot]$ in place of $G[\cdot;\mathbb{G}(a)]$:

$$G[\alpha_1 f_1 + \alpha_2 f_2] = \alpha_1 G[f_1] + \alpha_2 G[f_2],$$

$$G[\max\{f_1,f_2\}] = G[f_1] \vee G[f_2], \quad G[\min\{f_1,f_2\}] = G[f_1] \wedge G[f_2].$$

(b) Let $\{f_n: n = 1, 2,...\}$ be a monotone sequence of resolutive functions on $\mathbb{L}(a)$ with limit $f = \lim_{n\to\infty} f_n$. Then f is resolutive if and only if the sequence $\{G[f_n]: n = 1, 2,...\}$ converges. If this is the case, $G[f] = \lim_{n\to\infty} G[f_n]$.

2B. Let $\{\alpha_n: n = 1, 2,...\}$ be a sequence of positive numbers strictly decreasing to zero, such that $\partial R(\alpha_n,a) \cap Z(a;R) = \emptyset$. We set $R_n = R(\alpha_n,a)$, $n = 1, 2,...$. Since the surface R is regular, the family $\{R_n: n = 1, 2,...\}$ forms a regular exhaustion of R in the sense of Ch. I, 1A. For every n the Green function $g_n(a,z)$ for R_n with pole a is equal to $g(a,z) - \alpha_n$. So if f is an extended real-valued function on ∂R_n which is summable with respect to the arc-length measure and if u is the solution of the Dirichlet problem for R_n with the boundary function f, then

$$(2) \qquad u(a) = -\frac{1}{2\pi}\int_{\partial R_n} f(z)\,{*dg_n(a,z)} = -\frac{1}{2\pi}\int_{\partial R_n} f(z)\,{*dg(a,z)}$$

$$= -\frac{1}{2\pi}\int_0^{2\pi} f(z(\ell_\theta;\alpha_n))d\theta = \int_{\mathbb{L}(a)} f(z(\ell;\alpha_n))dm_a(\ell).$$

Lemma. Let s be a superharmonic function which is bounded below and u the greatest harmonic minorant of s on R. We set

$$\underline{s}_{(\alpha_n)}(\ell) = \liminf_{n\to\infty} s(z(\ell;\alpha_n)), \quad \underline{s}(\ell) = \liminf_{\alpha\to 0} s(z(\ell;\alpha)).$$

Then we have

$$u(a) \geqq \int_{\mathbb{L}(a)} \underline{s}_{(\alpha_n)}(\ell)dm_a(\ell) \geqq \int_{\mathbb{L}(a)} \underline{s}(\ell)dm_a(\ell).$$

Proof. Let u_n be the greatest harmonic minorant of s on R_n, i.e. $u_n = H[s;R_n]$. Then, $u_n \geqq u_{n+1} \geqq \cdots \geqq u$ on R_n and $u_n \to u$. By use of (2) we have

$$u(a) = \lim_{n\to\infty} u_n(a) = \lim_{n\to\infty} \int_{\mathbb{L}(a)} s(z(\ell;\alpha_n))dm_a(\ell)$$

$$\geqq \int_{\mathbb{L}(a)} \underline{s}_{(\alpha_n)}(\ell)dm_a(\ell) \geqq \int_{\mathbb{L}(a)} \underline{s}(\ell)dm_a(\ell). \quad \square$$

2C. We apply the preceding result to the Dirichlet problem defined in 1B.

Theorem. Let $f(\ell)$ be an extended real-valued function on $\mathbb{L}(a)$ such that $\overline{S}(f;\mathbb{C}(a))$ and $\underline{S}(f;\mathbb{C}(a))$ are nonvoid. Then we have

$$\overline{G}[f;\mathbb{C}(a)](a) \geqq \overline{\int}_{\mathbb{L}(a)} f(\ell)dm_a(\ell) \geqq \underline{\int}_{\mathbb{L}(a)} f(\ell)dm_a(\ell) \geqq \underline{G}[f;\mathbb{C}(a)](a),$$

where $\overline{\int}$ and $\underline{\int}$ denote the lower and the upper integrals with respect to the Green measure dm_a, respectively.

Proof. If $s \in \overline{S}(f;\mathbb{C}(a))$ and if u is the greatest harmonic minorant of s, then Lemma 2B implies that

$$u(a) \geqq \int_{\mathbb{L}(a)} \underline{s}(\ell)dm_a(\ell) \geqq \overline{\int}_{\mathbb{L}(a)} f(\ell)dm_a(\ell).$$

Since s is arbitrary,

$$\overline{G}[f;\mathbb{C}(a)](a) \geqq \overline{\int}_{\mathbb{L}(a)} f(\ell)dm_a(\ell).$$

The remaining inequality can be obtained similarly. As the upper integral is always larger than the lower integral, we get the desired inequalities. $\square$

Corollary. If an extended real-valued function $f(\ell)$ on $\mathbb{L}(a)$ is resolutive, then f is dm_a-summable and

$$G[f;\mathbb{C}(a)](a) = \int_{\mathbb{L}(a)} f(\ell)dm_a(\ell).$$

§2. THE SPACE OF GREEN LINES ON A SURFACE OF PARREAU-WIDOM TYPE

In what follows in this chapter we assume that R is a regular Riemann surface of Parreau-Widom type.

3. The Green Star Regions

3A. We choose a point $a \in R$, which is held fixed, and use the notations given in 1A. We map the Green star region $\mathbb{G}'(a)$ onto a region $D = D(a)$ in the open unit disk $\mathbb{D}_w$ in the complex w-plane conformally and univalently by means of the global coordinate function

$$w = \Phi(z) = r(z)e^{i\theta(z)},$$

where $r(z) = r(a;z)$ and $\theta(z) = \theta(a;z)$. Let Ω_0 be the set of all $\theta \in [0, 2\pi)$ with $\ell_\theta \in \mathbb{E}_0(a)$. Since R is regular, we have $\mathbb{G}(a) = \mathbb{L}(a) \cup \mathbb{E}_0(a)$ and so

$$D = \mathbb{D}_w \setminus \cup\{S_\theta : \theta \in \Omega_0\}, \tag{3}$$

where S_θ, $\theta \in \Omega_0$, denotes the slit $\{re^{i\theta} : r(z(\theta)) \leqq r \leqq 1\}$, $z(\theta)$ being the end point of the Green line ℓ_θ other than a. Each $z(\theta)$, $\theta \in \Omega_0$, should be a critical point of $g(a,z)$, i.e. $z(\theta) \in Z(a;R)$. As Theorem 1C, Ch. V, shows, $\sum\{g(a,z_j): z_j \in Z(a;R)\} < \infty$ and therefore

$$\sum\{1 - \exp(-g(a,z_j)): z_j \in Z(a;R)\} < \infty.$$

In view of the fact $r(z) = \exp(-g(a,z))$, this implies that the total length of the slits S_θ, $\theta \in \Omega_0$, is finite. By the Riemann mapping theorem there exists a function Ψ which maps the open unit disk $\mathbb{D}_\zeta$ in the complex ζ-plane onto D conformally and univalently. Clearly, the topological boundary ∂D of D is the union of the circle $\partial\mathbb{D}_w$ and all the slits S_θ, $\theta \in \Omega_0$. The above observation shows that ∂D is rectifiable.

3B. We are going to mention some useful properties of the function Ψ. First we prove the following

<u>Theorem</u>. Ψ can be extended continuously to the closed unit disk $\mathrm{Cl}(\mathbb{D}_\zeta)$.

<u>Proof</u>. Ψ is a bounded holomorphic function on $\mathbb{D}_\zeta$. Let Ω be the set of $\theta \in [-\pi, \pi)$ such that the radial limit $\lim_{r\to 1} \Psi(re^{i\theta})$ exists. By Theorem 2A, Ch. IV, the set Ω has a negligible complement in the

interval $[-\pi, \pi)$. If we denote by c_θ the radius $\{re^{i\theta}: 0 \leq r < 1\}$ in $\mathbb{D}_\zeta$ and by c'_θ the image of c_θ under Ψ, then c'_θ converges to a point in ∂D for every $\theta \in \Omega$ as $r \to 1$. In order to prove the theorem, let c be any curve in $\mathbb{D}_\zeta$ converging to a point $\zeta_0 \in \partial\mathbb{D}_\zeta$. We have only to show that $c' = \Psi(c)$ converges to a point in ∂D. Set $\zeta_0 = 1$ for the sake of simplicity. Suppose on the contrary that c' oscillates in D and let $E \subseteqq \partial D$ be the set of accumulation points of c' in ∂D. Then E is a continuum included in ∂D. We then take two sequences $\{\theta'(n)\}$ and $\{\theta''(n)\}$ in Ω such that $\theta'(1) < \theta'(2) < \cdots \to 0$ and $\theta''(1) > \theta''(2) > \cdots \to 0$. We take any n and set

$$G_n = \{z \in \mathbb{D}_\zeta: \ \theta'(n) < \arg z < \theta''(n)\}$$

and

$$G'_n = \Psi(G_n).$$

Then G'_n is a region bounded by two convergent curves $c'_{\theta'(n)}$, $c'_{\theta''(n)}$ and the part of the boundary ∂D corresponding to the arc $\{e^{i\theta}: \theta'(n) < \theta < \theta''(n)\}$ on $\partial\mathbb{D}_\zeta$. Since the curve c eventually comes in the region G_n, the oscillating part of the image c' is contained in the region G'_n, so that the continuum E lies on the boundary of G'_n. Let $u(w)$ be the solution of the Dirichlet problem for D with the boundary values $f(w) = 1$ if $w \in E$ and $= 0$ if $w \in \partial D \setminus E$. Since E is a continuum, every point in E is a regular point for the Dirichlet problem and therefore $u > 0$ on D. We set $v = u \circ \Psi$, so that v is bounded and harmonic on $\mathbb{D}_\zeta$. If θ belongs to either $[-\pi, \theta'(n))$ or $(\theta''(n), \pi)$, then the curve c'_θ tends to $\partial D \setminus \partial G'_n$ and thus $u(w)$ tends to zero as w goes to the boundary along c'_θ. So $v(\zeta)$ tends to zero as ζ goes to $\partial\mathbb{D}_\zeta$ along c_θ. It follows that $v(0) \leq (\theta''(n) - \theta'(n))/2\pi \to 0$ as $n \to \infty$. This contradicts with the fact $u > 0$. Hence, c' should converge to a point in ∂D, as desired. □

3C. We denote the continuous extension of Ψ to $\mathrm{Cl}(\mathbb{D}_\zeta)$ by the same letter Ψ. So $\Psi|\partial\mathbb{D}_\zeta$ maps $\partial\mathbb{D}_\zeta$ continuously onto ∂D. Then we have the following

<u>Theorem</u>. The derivative $\Psi'(\zeta)$ belongs to the Hardy class $H^1(\mathbb{D}_\zeta)$ and its radial limit is almost everywhere equal to $\partial\Psi/\partial\zeta$, which is defined by

$$\frac{\partial\Psi}{\partial\zeta}(e^{i\theta}) = \lim_{t\to 0} (\Psi(e^{i(\theta+t)}) - \Psi(e^{i\theta}))/(e^{i(\theta+t)} - e^{i\theta}).$$

Proof. Since ∂D is rectifiable, the function $\Psi|\partial\mathbb{D}_\zeta$ is of bounded variation. We set $d\mu(e^{it}) = (ie^{it})^{-1}d\Psi(e^{it})$ on $\partial\mathbb{D}_\zeta$. Then

$$\int_{\partial\mathbb{D}_\zeta} e^{it}d\mu(e^{it}) = i^{-1}\int_{\partial\mathbb{D}_\zeta} d\Psi(e^{it}) = 0$$

and for $n = 2, 3, \ldots$

$$\int_{\partial\mathbb{D}_\zeta} e^{nit}d\mu(e^{it}) = i^{-1}\int_{\partial\mathbb{D}_\zeta} e^{(n-1)it}d\Psi(e^{it})$$

$$= (n-1)\int_{\partial\mathbb{D}_\zeta} \Psi(e^{it})e^{(n-1)it}dt = 0,$$

because $\Psi(\zeta)$ is holomorphic in $\mathbb{D}_\zeta$ and continuous up to the boundary. By the F. and M. Riesz theorem (see Appendix A.3.2 and A.3.3) $d\mu$ is absolutely continuous with respect to the arc-length measure and

$$u(re^{i\theta}) = \frac{1}{2\pi}\int_0^{2\pi} P(r,\theta - t)d\mu(e^{it})$$

belongs to the Hardy class $H^1(\mathbb{D}_\zeta)$. Setting $d\mu(e^{it}) = f(e^{it})dt$, we have by the Fatou theorem (Appendix A.1.2)

$$f(e^{it}) = \frac{\partial\Psi}{\partial\zeta}(e^{it}) \quad \text{a.e.}$$

So the radial limit of $u(\zeta)$ is equal to $\partial\Psi/\partial\zeta$ a.e. on $\partial\mathbb{D}_\zeta$. On the other hand, since $d\mu$ is orthogonal to e^{nit}, $n = 1, 2, \ldots$, we have

$$\int_{\partial\mathbb{D}_\zeta} \frac{e^{-it} + \bar{\zeta}}{e^{-it} - \bar{\zeta}}d\mu(e^{it}) = \int_{\partial\mathbb{D}_\zeta} d\mu(e^{it})$$

and therefore

$$u(\zeta) = \frac{1}{2\pi}\int_{\partial\mathbb{D}_\zeta} \frac{1}{2}\left(\frac{e^{it} + \zeta}{e^{it} - \zeta} + \frac{e^{-it} + \bar{\zeta}}{e^{-it} - \bar{\zeta}}\right)d\mu(e^{it})$$

$$= \frac{1}{4\pi}\int_{\partial\mathbb{D}_\zeta}\left(\frac{e^{it} + \zeta}{e^{it} - \zeta} + 1\right)d\mu(e^{it}) = \frac{1}{2\pi}\int_{\partial\mathbb{D}_\zeta} \frac{e^{it}}{e^{it} - \zeta}\,d\mu(e^{it})$$

$$= \frac{1}{2\pi i}\int_{\partial\mathbb{D}_\zeta} \frac{1}{e^{it} - \zeta}\,d\Psi(e^{it}) = \frac{1}{2\pi i}\int_{\partial\mathbb{D}_\zeta} \frac{\Psi(e^{it})}{(e^{it} - \zeta)^2}\,d(e^{it})$$

$$= \Psi'(\zeta),$$

which implies the desired result. □

Combined with the Fatou theorem, the preceding theorem shows:

Corollary. If $\partial\Psi/\partial\zeta$ exists at a point $\zeta_0 \in \partial\mathbb{D}_\zeta$, then $\Psi'(\zeta)$ converges to the value $(\partial\Psi/\partial\zeta)(\zeta_0)$ uniformly as ζ tends to ζ_0 through any fixed Stolz region in $\mathbb{D}_\zeta$ with vertex ζ_0.

3D. We now have to modify the definition of Stolz regions given in Appexdix A.1.2. For $\zeta_0 \in \partial\mathbb{D}$, $0 < \alpha < \pi/2$ and $0 < \rho < 1$ we denote by $S(\zeta_0;\alpha,\rho)$ the connected component of the set

$$\{z \in \mathbb{D}:\ -\alpha \leqq \arg(1 - z\bar{\zeta}_0) \leqq \alpha,\ \rho \leqq |z| < 1\},$$

to which the point ζ_0 is adherent. In order to make the new Stolz region $S(\zeta_0;\alpha,\rho)$ simply connected, we always assume that

$$\rho \geqq \sin\alpha.$$

Theorem. (a) $(\partial\Psi/\partial\zeta)(\zeta_0)$ exists and is different from zero at almost every angle $\theta_0 \in [0, 2\pi) \setminus \Omega_0$, where $\zeta_0 = \Psi^{-1}(\exp(i\theta_0))$.

(b) Let $\theta_0 \in [0, 2\pi) \setminus \Omega_0$ satisfy the condition in (a) and set $w_0 = \exp(i\theta_0) = \Psi(\zeta_0)$. Then ∂D (see (3) for the definition of D) and $\partial\mathbb{D}_w$ have the same tangent at the point w_0. Consequently, for any α with $0 < \alpha < \pi/2$ the region D contains a Stolz region of the form $S(w_0;\alpha,\rho)$, where ρ depends on w_0 and α. Under the map Ψ^{-1} the Stolz region $S(w_0;\alpha,\rho)$ is sent to a curvilinear Stolz region in $\mathbb{D}_\zeta$ with vertex ζ_0 and angular measure 2α.

In the statement (b) above we mean by a curvilinear Stolz region with vertex ζ_0 and angular measure 2α $(< \pi)$ in the disk $\mathbb{D}_\zeta$ any closed curvilinear triangle with smooth sides which, except the vertex ζ_0, lies in $\mathbb{D}_\zeta$ and such that the tangents at ζ_0 to the two adjacent sides have directions $\zeta_0 e^{i(\pi-\alpha)}$ and $\zeta_0 e^{i(\pi+\alpha)}$, respectively.

Proof. We showed in Theorem 3C that $\partial\Psi/\partial\zeta$ exists a.e. and Ψ' is the Poisson integral of $\partial\Psi/\partial\zeta$. Since $\Psi'(\zeta)$ is not identically zero, its boundary function $\partial\Psi/\partial\zeta$ can vanish only on a set of measure zero by Theorem 2A, Ch. IV. Hence the statement (a) holds for almost every point in $\partial\mathbb{D}_\zeta$.

Suppose that $\theta_0 \in [0, 2\pi) \setminus \Omega_0$ satisfies the condition in (a). Draw a smooth curve c which, except the initial point ζ_0, lies in the disk $\mathbb{D}_\zeta$ and has a tangent at ζ_0. We suppose that c is not tangential to the circumference $\partial\mathbb{D}_\zeta$. Let ζ_1, ζ_2 be two points on c distinct from ζ_0. We have

$$\Psi(\zeta_1) - \Psi(\zeta_2) = \int_{\zeta_2}^{\zeta_1} \Psi'(\zeta)\,d\zeta,$$

where the right-hand side is the integral along the curve c. Fix ζ_1 and let ζ_2 tend to the point ζ_0 along c. Then $\Psi(\zeta_2) \to \Psi(\zeta_0)$, so that

$$\Psi(\zeta_1) - \Psi(\zeta_0) = \int_{\zeta_0}^{\zeta_1} \Psi'(\zeta)d\zeta.$$

Since $\Psi'(\zeta)$ tends to $(\partial\Psi/\partial\zeta)(\zeta_0)$ as $\zeta \in c$ tends to ζ_0, we have shown that the derivative of Ψ at ζ_0 along the curve c exists and is equal to $(\partial\Psi/\partial\zeta)(\zeta_0)$. So the tangent to c at ζ_0 is rotated by $\arg\{\frac{\partial\Psi}{\partial\zeta}(\zeta_0)\}$ under the map $\zeta \to \Psi(\zeta)$ and therefore $\Psi(\zeta)$ preserves the angle between two curves c_1 and c_2 issuing from ζ_0 nontangentially with respect to the circumference $\partial\mathbb{D}_\zeta$. As $(\partial\Psi/\partial\zeta)(\zeta_0) \neq 0$, the boundary ∂D has a tangent at $w_0 = \Psi(\zeta_0)$, which is obtained from the tangent to $\partial\mathbb{D}_\zeta$ at ζ_0 under a rotation by the angle $\arg\{\frac{\partial\Psi}{\partial\zeta}(\zeta_0)\}$. Hence the angle between the circle $\partial\mathbb{D}_\zeta$ and the curve c is equal to that between ∂D and $\Psi(c)$. Since $\Psi(\zeta_0) = w_0 \in \partial\mathbb{D}_w$ and ∂D is rectifiable, we can find a sequence $\{\zeta_n^*: n = 1, 2, \dots\} \subseteq \partial\mathbb{D}_\zeta$ such that $\zeta_n^* \neq \zeta_0$, $\zeta_n^* \to \zeta_0$ and $\Psi(\zeta_n^*)$ ($= w_n$, say) $\in \partial\mathbb{D}_w$. Then, $w_n \to w_0$ and

$$\frac{\partial\Psi}{\partial\zeta}(\zeta_0) = \lim_{n\to\infty} \frac{\Psi(\zeta_n^*) - \Psi(\zeta_0)}{\zeta_n^* - \zeta_0} = \lim_{n\to\infty} \frac{w_n - w_0}{\zeta_n^* - \zeta_0}.$$

We thus have

$$\arg\{\frac{\partial\Psi}{\partial\zeta}(\zeta_0)\} = \arg w_0 - \arg\zeta_0 \pmod{2\pi}.$$

This shows in particular that ∂D and $\partial\mathbb{D}_w$ have the same tangent at w_0. The first half of (b) is now clear.

Let us now consider a Stolz region $S(w_0;\alpha,\rho)$ included in D and let c' be any one of its sides issuing from w_0: for instance

$$\arg(1 - w\overline{w}_0) = \alpha$$

for $w \in c'$. Set $c = \Psi^{-1}(c' \setminus \{w_0\})$, so that c is a smooth curve in $\mathbb{D}_\zeta$. We should show that c converges to ζ_0 and has a tangent at ζ_0, for which we have $\arg(1 - \zeta\overline{\zeta}_0) = \alpha$. To see this, we take any α_1, α_2 with $0 < \alpha_1 < \alpha < \alpha_2 < \pi/2$ and draw two segments c_1 and c_2 in $\mathbb{D}_\zeta$ with the common initial point ζ_0 such that $\arg(1 - \zeta_i\overline{\zeta}_0) = \alpha_i$ for $\zeta_i \in c_i$, $i = 1, 2$. And further we set $c_i' = \Psi(c_i)$. As shown above, c_i' converge to w_0 and have tangents at w_0, which satisfy the equations $\arg(1 - w_i w_0) = \alpha_i$ for $w_i \in c_i'$, respectively. So the curve c' lies in a Stolz region with vertex w_0 determined by c_1' and c_2', and consequently c lies in a Stolz region with vertex ζ_0 determined by c_1 and c_2. Since α_1 and α_2 can be taken arbitrarily close to α, we have shown that c converges to ζ_0 and has a tangent at ζ_0 satisfying the equation $\arg(1-\zeta\bar{\zeta}_0) = \alpha$. This proves the latter half of the statement (b). □

4. Limit along Green Lines

4A. Let u be a function on R with values in a compact space X. For a given Green line $\ell \in \mathbb{L}(a)$ we set

$$u^{\vee}(\ell) = \bigcap_{\alpha>0} \mathrm{Cl}[u(\{z(\ell;\beta) : 0 < \beta < \alpha\})].$$

If $u^{\vee}(\ell)$ consists of only one point, then the point is called the radial limit of u along the Green line ℓ and is denoted by $\check{u}(\ell)$.

Theorem. Let $a \in R$ and let u be a positive harmonic function on the Green star region $\mathbb{G}'(a)$. Then for almost every $\ell \in \mathbb{L}(a)$ the limit $\check{u}(\ell)$ exists and is finite. The function $\ell \to \check{u}(\ell)$ is dm_a-measurable on $\mathbb{L}(a)$. If u is bounded holomorphic, then u vanishes identically whenever $\check{u}(\ell)$ vanishes on a set of positive dm_a-measure.

Proof. (i) Let u be a bounded holomorphic function on $\mathbb{G}'(a)$ and set $v = u \circ \Phi^{-1} \circ \Psi$. Then v is a bounded holomorphic function on $\mathbb{D}_\zeta$, so that by the Fatou theorem $v(\zeta)$ tends for almost every point ζ_0 in $\partial\mathbb{D}_\zeta$ to a finite limit as ζ tends to ζ_0 through any fixed Stolz region in $\mathbb{D}_\zeta$ with vertex ζ_0. Let E be the totality of such ζ_0 in $\partial\mathbb{D}_\zeta$. Since $\Psi: \partial\mathbb{D}_\zeta \to \partial D$ preserves sets of measure zero, we see that, for almost every $\theta_0 \in [0, 2\pi) \setminus \Omega_0$, $\Psi^{-1}(\exp(i\theta_0))$ $(= \zeta_0$, say) belongs to E (see 3A for the definition of Ω_0). Let θ_0 possess this property. Then the image of the Green line ℓ_{θ_0} under the map $\Psi^{-1} \circ \Phi$ is contained in any Stolz region in $\mathbb{D}_\zeta$ with vertex in E. Hence $\check{u}(\ell_{\theta_0})$ exists. So $\check{u}(\ell)$ exists for almost every $\ell \in \mathbb{L}(a)$. If $u \not\equiv 0$, then $v \not\equiv 0$ on $\mathbb{D}_\zeta$, so that its radial limit vanishes only on a set of measure zero. Hence $\check{u}(\ell) \neq 0$ for almost every $\ell \in \mathbb{L}(a)$.

(ii) Let u be any positive harmonic function on $\mathbb{G}'(a)$. Since $\mathbb{G}'(a)$ is simply connected, the harmonic conjugate $\tilde{u}$ of u exists uniquely under the condition $\tilde{u}(a) = 0$. Since u is positive,

$$h(z) = \exp(-u(z) - i\tilde{u}(z))$$

is a bounded holomorphic function on $\mathbb{G}'(a)$ and therefore by (i) $\check{h}(\ell)$ exists for almost every $\ell \in \mathbb{L}(a)$. Since $h \circ \Phi^{-1} \circ \Psi$ does not vanish, its radial limit can vanish only on a set of measure zero. It follows that $\check{h}(\ell) \neq 0$ for almost every $\ell \in \mathbb{L}(a)$. As $u(z) = -\log|h(z)|$, $\check{u}(\ell) = -\log|\check{h}(\ell)|$ exists and is finite a.e. □

4B. In the case of a PWS, Theorem 2C is much strengthened. In fact, we have:

Theorem. Let $f(\ell)$ be a bounded measurable function on $\mathbb{L}(a)$. Then f is resolutive and the radial limit $\check{G}[f;\mathbb{G}(a)](\ell)$ exists and is equal to $f(\ell)$ for almost every $\ell \in \mathbb{L}(a)$.

Proof. Consider an exhaustion $\{R_n: n = 1, 2, \ldots\}$ of R with $R_n = R(\alpha_n, a)$, as given in 2B. For each n we consider the Dirichlet problem for R_n with the boundary data f_n, where

$$f_n(z(\ell;\alpha_n)) = f(\ell)$$

for $\ell \in \mathbb{L}(a)$. Then f_n is a bounded measurable function on ∂R_n, which is defined a.e. We set $u_n = H[f_n;R_n]$, which is the solution of the ordinary Dirichlet problem. Then $\{u_n: n = 1, 2, \ldots\}$ is uniformly bounded. It is easy to see that $\{u_{n+1}, u_{n+2}, \ldots\}$ forms a uniformly bounded equicontinuous family on $\mathrm{Cl}(R_n)$. Consequently, passing to a subsequence if necessary, we can assume that $\{u_n\}$ converges to a harmonic function, say u, almost uniformly on R.

On the other hand, we consider the conformal map Φ of the Green star region $\mathbb{G}'(a)$ onto the region $D = D(a)$, which we defined in 3A. We set $R_n' = \mathbb{G}'(a) \cap R_n$ and $\Phi_n(z) = \rho_n \Phi(z)$ for $z \in R_n'$, where $\rho_n = \exp \alpha_n$. Let D_n be the image of Φ_n, i.e. $D_n = (\rho_n D) \cap \mathbb{D}_w$. It is then clear that $D_1 \supseteq D_2 \supseteq \cdots \supseteq D_n \supseteq \cdots \supseteq D$.

We are going to solve Dirichlet problems for the region D. For this purpose, we regard each slit S_θ, $\theta \in \Omega_0$, as the union of its two edges S_θ^- and S_θ^+: a point $w \in S_\theta$ is defined to belong to S_θ^- (resp. S_θ^+) if w is regarded as the limit of $w' \in D$ satisfying $\mathrm{Im}(w'w^{-1}) < 0$ (resp. $\mathrm{Im}(w'w^{-1}) > 0$). The only common point of S_θ^- and S_θ^+ is the vertex of the slit S_θ, i.e. $\exp(-g(a,z(\theta)) + i\theta)$. Then the region D can be viewed in an obvious sense as a Jordan region with rectifiable boundary L, where

$$L = \cup\{S_\theta^+ \cup S_\theta^-: \theta \in \Omega_0\} \cup \{e^{i\theta}: \theta \in [0, 2\pi) \setminus \Omega_0\}.$$

Now let v_n be the restriction of $u_n \circ \Phi_n^{-1}$ to the region D. Then v_n can clearly be extended continuously to the boundary L, the boundary values being denoted by f_n^*. If $\ell = \ell_\theta \in \mathbb{L}(a)$, then we have

$f_n^*(e^{i\theta}) = f(\ell)$. If $\theta \in \Omega_0$ and $w \in S_\theta^-$ (or S_θ^+), then there exists a point $z_n(w) \in R$ such that (i) $f_n^*(w) = u_n(z_n(w))$ and (ii) there is a sequence $\{z_j: j = 1, 2, \ldots\} \subseteqq R_n'$ with the properties $z_j \to z_n(w)$ and $\Phi_n(z_j) \to w$ in D. Such a point $z_n(w)$ is uniquely determined by continuity. For each such w the sequence $\{z_n(w): n = 1, 2, \ldots\}$ stays in a compact subset of R and converges to a point $z_0(w) \in R$. In fact, if $w' \in D$ and $w' \to w$, then $\Phi^{-1}(\rho_n^{-1}w') \to z_0(w)$. Thus, if $w \in S_\theta^-$ (or S_θ^+), then $f_n^*(w) = u_n(z_n(w)) \to u(z_0(w))$. Namely, $\{f_n^*\}$ converges pointwise to a well-defined function f^*, i.e.

$$f^*(e^{i\theta}) = f(\ell)$$

if $\ell = \ell_\theta \in \mathbb{L}(a)$, and

$$f^*(w) = u(z_0(w))$$

if $z \in S_\theta^+ \cup S_\theta^-$. For any given $w \in D$ let $d\omega_w$ be the harmonic measure of D supported on L. Since L is rectifiable, it is not hard to see that $d\omega_w$ is absolutely continuous with respect to the arc-length of L. As we have

$$(u_n \circ \Phi_n^{-1})(w) = v_n(w) = \int_L f_n^* d\omega_w,$$

so we deduce, by letting $n \to \infty$ and using the Lebesgue dominated convergence theorem:

$$(u \circ \Phi^{-1})(w) = \int_L f^*\, d\omega_w.$$

Namely, $u \circ \Phi^{-1}$ is exactly the solution of the Dirichlet problem for D corresponding to the function f^*.

As we have seen in Theorem 3D, almost every radius $\{re^{i\theta}: 0 \leqq r < 1\}$ with $\theta \in [0, 2\pi) \setminus \Omega_0$ of $\mathbb{D}_w$ is orthogonal to ∂D. So the Fatou theorem shows that

$$(u \circ \Phi^{-1})(re^{i\theta}) \to f^*(e^{i\theta}) = f(\ell_\theta)$$

as $r \to 1$ for almost every $\theta \in [0, 2\pi) \setminus \Omega_0$. This means that

$$\lim_{\alpha \to 0} u(z(\ell;\alpha)) = f(\ell)$$

for almost every $\ell \in \mathbb{L}(a)$, as was to be shown. □

As a consequence we can show:

<u>Corollary</u>. Every dm_a-summable function on $\mathbb{L}(a)$ is resolutive.

§3. THE GREEN LINES AND THE MARTIN BOUNDARY

5. Convergence of Green Lines

5A. Let us consider the Martin compactification R^* of a hyperbolic Riemann surface R and let Δ (resp. Δ_1) be the Martin boundary (resp. the set of minimal points of Δ). Let k_b, $b \in \Delta$, be the Martin function with pole at b. We denote by 0 the origin of R, so that $k_b(0) = 1$ for any $b \in \Delta$. The basic properties of k_b have been given in Ch. III. Let $a \in R$ be fixed and consider Green lines $\ell \in \mathbb{L}(a)$. For $\ell \in \mathbb{L}(a)$ let e_ℓ denote the end of ℓ in R^*, i.e.

$$e_\ell = \mathrm{Cl}(\ell) \setminus (\ell \cup \{a\}),$$

where $\mathrm{Cl}(\ell)$ means the closure of ℓ in R^*. Since R^* is compact and $\ell \cup \{a\}$ is non-compact in R, we see that e_ℓ is a nonvoid subset of Δ. We say that a Green line $\ell \in \mathbb{L}(a)$ is convergent if e_ℓ consists of a single point. Here is a long standing problem due to Brelot and Choquet asking whether almost every Green line in $\mathbb{L}(a)$ (with respect to the Green measure dm_a) is convergent. See Brelot [5] for a detailed exposition. We shall show that the problem has an affirmative solution at least in the case of PWS's.

5B. We assume in what follows that R is a regular PWS. We take another point $a' \in R$ and set

$$P(a,a';z) = \delta g(a',z)/\delta g(a,z)$$

for $z \in R$, where $\delta g(a,z) = 2(\partial g(a,z)/\partial z)dz$. It is easy to see that $P(a,a';z)$ is a meromorphic function on R. If $a \neq a'$, then it has poles in the set $Z(a;R) \cup \{a'\}$. Since R is a PWS, we have by Theorem 1C, Ch. V, $\sum\{g(a,z_j): z_j \in Z(a;R)\} < \infty$. So, if we set

$$s^{(a)}(z) = \sum\{g(z,z_j): z_j \in Z(a;R)\},$$

then the Harnack inequality shows that $s^{(a)}(z) < \infty$ for $z \notin Z(a;R)$ and therefore that $s^{(a)}$ is a positive superharmonic function. Set $g^{(a)}(z) = \exp(-s^{(a)}(z))$ and

$$u(a,a';z) = g^{(a)}(z)\exp(-g(a',z)). \tag{5}$$

We define $s(z) = s^{(a)}(z) + g(a',z)$, which is also a positive superharmonic function on R. In fact, it is a potential. We take an open covering $\{V_\alpha\}$ of R consisting of simply connected open sets V_α

such that, whenever $V_\alpha \cap V_\beta \neq \emptyset$, $V_\alpha \cup V_\beta$ is included in a parametric disk which contains at most one singularity of $s(z)$. With each V_α we associate a function of the form $s(z) + i\tilde{s}_\alpha(z)$, $z \in V_\alpha$, where $\tilde{s}_\alpha$ denotes a harmonic conjugate of $s|V_\alpha$. To be more precise, if V_α contains no singularities of s, then $s(z)$ is harmonic on V_α and so $\tilde{s}_\alpha$ is a usual harmonic conjugate of s on V_α. If V_α contains a critical point, say z_j, of multiplicity m_j, then

$$s(z) = -m_j \log|z - z_j| + u(z)$$

for $z \in V_\alpha$, where u is harmonic, and so

$$\tilde{s}_\alpha(z) = -m_j \arg(z - z_j) + \tilde{u}(z),$$

where $\tilde{u}$ is a harmonic conjugate of u on V_α. Then we set

$$f_\alpha(z) = \exp(-(s(z) + i\tilde{s}_\alpha(z)))$$

for $z \in V_\alpha$, which is a holomorphic function on V_α. It is easy to see that, for each pair α, β with $V_\alpha \cap V_\beta \neq \emptyset$, $f_\alpha(z) f_\beta(z)^{-1}$ is equal to a constant $\xi_{\alpha\beta}$ of modulus one on the intersection $V_\alpha \cap V_\beta$; for the union $V_\alpha \cup V_\beta$ is included in a simply connected set. So the pair $(\{V_\alpha\}, \{\xi_{\alpha\beta}\})$ defines a line bundle, say $\xi_{a,a'}$, over R, for which $(\{V_\alpha\}, \{f_\alpha\})$ forms a holomorphic section, say $f^{(a,a')}$. Since $|f_\alpha| \leq 1$ for each α, we see that $f^{(a,a')}$ belongs to $\mathcal{H}^\infty(R,\xi_{a,a'})$. By the Widom theorem (Theorem 2B, Ch. V) we know $\mathcal{H}^\infty(R,\xi_{a,a'}^{-1}) \neq \{0\}$. Let us choose any nonzero element, say $h^{(a,a')}$, of $\mathcal{H}^\infty(R,\xi_{a,a'})$ with norm less than 1. Then one can find a representative of this element associated with the covering $\{V_\alpha\}$, i.e. $(\{V_\alpha\}, \{h_\alpha\})$. It is immediate that the pair $(\{V_\alpha\}, \{f_\alpha h_\alpha\})$ defines a nontrivial single-valued holomorphic function on R, which we denote by $F(a,a';z)$. The property of F that we need is this:

$$|F(a,a';z)| \leq u(a,a';z) \quad \text{on } R. \tag{6}$$

Let $\{\alpha_n : n = 1, 2, \ldots\}$ be a sequence of positive numbers strictly decreasing to zero with the property $\partial R(\alpha_n,a) \cap Z(a;R) = \emptyset$ and set $R_n = R(\alpha_n,a)$. We may suppose here that $a' \in R_1$. Let $g_n(a',z)$ be the Green function for R_n with pole a'. Since $g_a - \alpha_n$ is the Green function for R_n with pole a, the Harnack inequality shows that there exists a constant c depending only on a, a' and R, such that $0 < \delta g_n(a',z)/\delta g(a,z) < c$ on ∂R_n, for $-(2\pi i)^{-1}\delta g_n(a',z)$ (resp. $-(2\pi i)^{-1}\delta g(a,z)$) along ∂R_n represents the harmonic measure of R_n at the point a' (resp. a). Since $u(a,a';z) \leq 1$ on R, we have

(7) $$0 \leq u(a,a';z)(\delta g_n(a',z)/\delta g(a,z)) \leq c$$

on ∂R_n. Since $u(a,a';z)$, defined by (5), vanishes at every pole of the meromorphic function $\delta g_n(a',z)/\delta g(a,z)$ with the same multiplicity and since (6) holds, the function

$$F(a,z';z)(\delta g_n(a',z)/\delta g(a,z))$$

is a holomorphic function on R_n and is continuous up to the boundary. It follows from (7) that

$$|F(a,a';z)(\delta g_n(a',z)/\delta g(a,z))| \leq c$$

on $Cl(R_n)$. Since $\delta g_n(a',z)$ converge to $\delta g(a',z)$ almost uniformly on R, we have

$$|F(a,a';z)P(a,a';z)| \leq c$$

on R. Since both $F(a,a';z)$ and $F(a,a';z)P(a,a';z)$ are bounded and holomorphic on R, the real and the imaginary parts of these functions are bounded harmonic functions on R. So by Theorem 4A both $F(a,a';z)$ and $F(a,a';z)P(a,a';z)$ have nonzero radial limits along almost every Green line $\ell \in \mathbb{L}(a)$. $P(a,a';z)$ thus has a nonzero finite radial limit $\check{P}(a,a';\ell)$ for almost every $\ell \in \mathbb{L}(a)$.

A regular Green line $\ell \in \mathbb{L}(a)$ is said to be <u>convergent</u> if $z(\ell;\alpha)$ converge to a point in Δ with respect to the Martin topology as $\alpha \to 0$. We denote by $\Lambda(R;a) = \Lambda(a)$ the set of convergent Green lines in $\mathbb{L}(a)$. For every $\ell \in \Lambda(a)$ let b_ℓ denote the limit of ℓ in Δ. We sometimes write b_θ for b_ℓ when $\ell = \ell_\theta$ according to the parametrization of Green lines defined in 1A.

<u>Theorem</u>. The set $\Lambda(a)$ of convergent Green lines issuing from a forms a measurable subset of $\mathbb{L}(a)$ of Green dm_a-measure one.

<u>Proof</u>. Take any $a' \in R$ with $a' \neq a$. As we have seen above, the function $P(a,a';z)$ has a nonzero finite radial limit $\check{P}(a,a';\ell)$ for almost every $\ell \in \mathbb{L}(a)$. We look at any $\ell \in \mathbb{L}(a)$ for which the radial limit $\check{P}(a,a';\ell)$ exists. At each $z \in \ell$ we take a local coordinate $z = x + iy$ in such a way that

$$dx = dg(a,z) \quad \text{and} \quad dy = {}^*dg(a,z).$$

Along ℓ we then have

$$\delta g(a,z) = (\partial_x g(a,z))dx = dx$$

and

$$\delta g(a',z) = (\partial_x g(a',z) + i\partial_x \tilde{g}(a',z))dx,$$

where $\tilde{g}(a',z)$ denotes the harmonic conjugate of $g(a',z)$, determined uniquely up to an additive constant factor. We may assume that $x = g(a,z)$ and $y = y_0$ = constant along ℓ. Then we have on ℓ

(8) $$\operatorname{Re}(P(a,a';z)) = (dg(a',x+iy_0)/dx)/(dg(a,x+iy_0)/dx).$$

Since $\check{P}(a,a';\ell)$ exists, the left-hand side of (8) has a limit as x tends to zero. Since both $g(a,x+iy_0)$ and $g(a',x+iy_0)$ tend to zero as $x \to 0$ in view of the regularity of the surface R, l'Hospital's rule shows that

(9) $$\begin{aligned}\operatorname{Re}(\check{P}(a,a';\ell)) &= \lim_{x\to 0} \{dg(a',x+iy_0)/dx)/(dg(a,x+iy_0)/dx\} \\ &= \lim_{x\to 0} \{g(a',x+iy_0)/g(a,x+iy_0)\}.\end{aligned}$$

Let $b \in e_\ell$. Then there exists a sequence of points z_n in ℓ with the coordinates $x_n + iy_0$ such that $x_n \to 0$ and $z_n \to b$. So the final member of (9) is equal to $k_b(a')/k_b(a)$. Since b is arbitrary in e_ℓ, the quotient $k_b(a')/k_b(a)$ is constant on e_ℓ as a function in b.

Now let a' run through a countable dense subset A of R. Then we see that there exists a measurable subset A of $\mathbb{L}(a)$ such that (i) $m_a(A) = 1$; (ii) for each $a' \in A$ and $\ell \in A$, $\check{P}(a,a';\ell)$ exists and is finite. So, if $b, b' \in e_\ell$ with $\ell \in A$, then

$$k_b(a')/k_b(a) = k_{b'}(a')/k_{b'}(a)$$

for every $a' \in A$. Since the function k_b with $b \in \Delta$ is continuous on R, the density of A in R implies that k_b and $k_b{}'$ are proportional. As shown in Ch. III, 2B, this means that $b = b'$. Namely, e_ℓ consists of a single point for each $\ell \in A$. □

6. Green Lines and the Martin Boundary

6A. We shall establish further connection between the space of Green lines and the Martin boundary.

Theorem. Let $a \in R$ be fixed. Then we have the following:

(a) The map $\pi: \ell \to b_\ell$ of $\Lambda(a)$ into Δ is measurable. For any $f \in L^1(d\chi)$ the function $f \circ \pi$, defined dm_a-a.e. on $\Lambda(a)$, is resolutive and $H[f] = G[f \circ \pi; \mathfrak{G}(a)]$. In particular, $f \circ \pi$ is dm_a-summable and

$$\int_\Delta f(b)k_b(a)d\chi(b) = \int_{\mathbb{L}(a)} f(b_\ell)dm_a(\ell).$$

(b) If u is a positive harmonic function on R, then $\hat{u}(b_\ell) = \check{u}(\ell)$ for almost every $\ell \in \Lambda(a)$. The same is true of every meromorphic function u of bounded characteristic, i.e. $\log|u| \in SP'(R)$.

(c) $\hat{P}(O,a;b)$ exists a.e. on Δ_1 and

$$\hat{P}(O,a;b) = k_b(a) \quad \text{a.e. on } \Delta_1.$$

(d) The set $\Lambda(a)$ can be chosen in such a way that the map π in (a) is injective.

(e) The map π is an isomorphism between the measure spaces $(\Lambda(a), dm_a)$ and $(\Delta, d\chi_a)$ with $d\chi_a(b) = k_b(a)d\chi(b)$.

Proof. (a) We take a sequence $\{\alpha_n\}$ satisfying the properties given in 5B. Then π is the limit of continuous functions $\ell \to z(\ell;\alpha_n)$ as n goes to infinity, so that it is measurable. Secondly, let f be any $d\chi$-summable function on Δ. Put $g = f \circ \pi$. By comparing the definitions of the Dirichlet problems given respectively in 1B and Ch. III, 3A, we see that $\overline{S}(f) \subseteq \overline{S}(g;\mathbb{G}(a))$ and therefore $\overline{H}[f] \geqq \overline{G}[g;\mathbb{G}(a)]$. We also have $\underline{H}[f] \leqq \underline{G}[g;\mathbb{G}(a)]$. Since f is assumed to be $d\chi$-summable, it is resolutive by Corollary 3D, Ch. III, i.e. $\overline{H}[f] = \underline{H}[f]$ and

$$H[f](a) = \int_{\Delta_1} f(b)k_b(a)d\chi(b).$$

It follows that $\overline{G}[f;\mathbb{G}(a)] = \underline{G}[g;\mathbb{G}(a)] = H[f]$. By Corollary 2C we have

$$G[g;\mathbb{G}(a)](a) = \int_{\mathbb{L}(a)} g(\ell)dm_a(\ell).$$

From these the statement (a) follows at once.

(b) Suppose first that u is a singular positive harmonic function on R, i.e. $u \in HP(R) \cap I(R)$. Then, by Theorem 5E, Ch. III, we have $\hat{u}(b) = 0$ a.e. on Δ_1. On the other hand, by Theorem 4A, the radial limit $\check{u}(\ell)$ exists as a nonnegative dm_a-measurable function which is defined a.e. on $\mathbb{L}(a)$. We claim that $\check{u}(\ell) = 0$ dm_a-a.e. on $\mathbb{L}(a)$. Suppose on the contrary that there exist a positive number $\delta > 0$ and a measurable set $\mathcal{B} \subseteq \mathbb{L}(a)$ of positive dm_a-measure such that $\check{u}(\ell) \geqq \delta$ for every $\ell \in \mathcal{B}$. Then the characteristic function $c_{\mathcal{B}}$ of $\mathcal{B}$ is dm_a-summable so that, by Corollary 4B, $c_{\mathcal{B}}$ is resolutive on $\mathbb{L}(a)$. Since $\check{u}(\ell) \geqq \delta c_{\mathcal{B}}(\ell)$ for any $\ell \in \mathbb{L}(a)$, we have $u(z) \geqq G[\delta c_{\mathcal{B}};\mathbb{G}(a)](z)$ on R. This is clearly a contradiction, for u is singular while the function $G[\delta c_B;\mathbb{C}(a)]$ is quasibounded and strictly positive. The property (a) implies in particular that, if $A \subseteq \Delta$ is $d\chi$-negligible, $\pi^{-1}(A)$ is dm_a-negligible. Hence we have shown that $\check{u}(\ell) = \hat{u}(b_\ell) = 0$ for almost every $\ell \in \Lambda(a)$.

Suppose next that u is a quasibounded positive harmonic function. We define $f(b) = \hat{u}(b)$ for $b \in \Delta_1$, whenever $\hat{u}(b)$ exists; otherwise, $f(b) = 0$. Then f is $d\chi$-summable on Δ. Set $g(\ell) = f(b_\ell)$ for every $\ell \in \Lambda(a)$. By (a) above and Theorem 5E, Ch. III, we have $u = H[f] = G[g;\mathbb{C}(a)]$. As an analogue of Theorem 2A is true for the Dirichlet problem for the Martin compactification, there exists a positive superharmonic function s on R satisfying $u + \varepsilon s \in \overline{S}(f)$ and $u - \varepsilon s \in \underline{S}(f)$ for any $\varepsilon > 0$. We already know that $\overline{S}(f) \subseteq \overline{S}(g;\mathbb{C}(a))$, so that

$$\liminf_{\alpha\to 0} (u + \varepsilon s)(z(\ell;\alpha)) \geqq g(\ell) = f(b_\ell)$$

for any $\ell \in \Lambda(a)$. By Theorem 4A the radial limit $\check{u}(\ell)$ exists dm_a-a.e. on $\mathbb{L}(a)$. For any $\ell \in \Lambda(a)$ for which $\check{u}(\ell)$ exists, we have

$$\check{u}(\ell) + \varepsilon \liminf_{\alpha\to 0} s(z(\ell;\alpha)) \geqq f(b_\ell).$$

By Lemma 2B, $\liminf_{\alpha\to 0} s(z(\ell;\alpha))$ is finite a.e.; so we have $\check{u}(\ell) \geqq f(b_\ell)$ dm_a-a.e. on $\Lambda(a)$. The reverse inequality is obtained from the fact $u - \varepsilon s \in \underline{S}(f)$ and hence

$$\check{u}(\ell) = f(b_\ell) = \hat{u}(b_\ell)$$

for almost every $\ell \in \Lambda(a)$. Finally if u is a meromorphic function of bounded characteristic, then it can be expressed as a quotient of bounded holomorphic functions, in view of Widom's theorem (Theorem 2B, Ch. V); so the desired result follows from the above consideration.

(c) In the proof of Theorem 5B we showed that

$$\text{(10)} \qquad \mathrm{Re}(\check{P}(0,a;\ell)) = k(b_\ell,a)/k(b_\ell,0) = k(b_\ell,a)$$

for every $\ell \in \Lambda(a)$. Since $P(0,a;z)$ is expressed as a quotient of bounded holomorphic functions, the property (b) implies that

$$\text{(11)} \qquad \check{P}(0,a;\ell) = \hat{P}(0,a;b_\ell)$$

for almost every $\ell \in \Lambda(a)$. The property (a) implies in particular that $\pi(\Lambda(a))$ has a $d\chi$-negligible complement in the space Δ. So it follows from (10) and (11) that

$$\text{(12)} \qquad \mathrm{Re}(\hat{P}(0,a;b)) = k(b,a) \quad \text{a.e. on } \Delta_1.$$

Changing the role of $\mathcal{O}$ and a, we get

$$\text{(13)} \qquad \mathrm{Re}(1/\hat{P}(\mathcal{O},a;b)) = \mathrm{Re}(\hat{P}(a,\mathcal{O};b)) = 1/k(b,a) \quad \text{a.e. on } \Delta_1.$$

It is easy to see that $\hat{P}(\mathcal{O},a;b)$ is real and is equal to $k(b,a)$, when $b \in \Delta_1$ satisfies both (12) and (13). Hence $\hat{P}(\mathcal{O},a;b) = k(b,a)$ a.e. on Δ_1.

(d) Let f be a bounded measurable function on $\mathbb{L}(a)$. Then it is resolutive by Theorem 4B and we have

$$\text{(14)} \qquad \check{G}[f;\mathbb{G}(a)](\ell) = f(\ell)$$

for almost every $\ell \in \mathbb{L}(a)$. Of course (14) is true for almost every $\ell \in \Lambda(a)$. In view of (c), we have for almost every $\ell \in \Lambda(a)$

$$\hat{G}[f;\mathbb{G}(a)](b_\ell) = f(\ell).$$

In particular, we look at the function $f_0(\ell) = \theta$ if $\ell = \ell_\theta$ according to the parametrization mentioned in 1A. Then we have

$$\text{(15)} \qquad \hat{G}[f_0;\mathbb{G}(a)](b_\ell) = \theta$$

for almost every $\ell \in \Lambda(a)$ with $\ell = \ell_\theta$. Let $\Lambda'(a)$ be the set of $\ell \in \Lambda(a)$ for which (15) holds. Then the set $\Lambda'(a)$ has a dm_a-negligible complement in $\Lambda(a)$ and the correspondence $\pi\colon \ell \to b_\ell$ is one-to-one on $\Lambda'(a)$.

(e) This is an immediate consequence of (a) and (d). □

6B. Applying Theorem 6A to multiplicative analytic functions, we get the following

<u>Theorem</u>. Let u be an l.m.m. of bounded characteristic on R (see Ch. II, 5D). Then the following hold:

(a) $\hat{u}(b)$ exists a.e. on Δ_1.

(b) $\check{u}(\ell)$ exists a.e. on $\Lambda(a)$.

(c) $\hat{u}(b_\ell) = \check{u}(\ell)$ a.e. for $\ell \in \Lambda(a)$.

<u>Proof</u>. We set as in Ch. IV, 4B $v = \log u$, $v_i = \mathrm{pr}_I(v)$, $v_q = \mathrm{pr}_Q(v)$, $u_I = \exp(v_i)$ and $u_Q = \exp(v_q)$. Then $\hat{u}_I = 1$ and $\hat{u} = \hat{u}_Q$ a.e. on Δ_1 by Corollary 4B, Ch. IV.

Suppose now that $v \geqq 0$. So $v_i \geqq 0$ and $v_q \geqq 0$. Since v_q is harmonic on R by Theorem 5B, Ch. II, the statement (b) of the preceding theorem says that

$$\text{(16)} \qquad \check{v}_q(\ell) = \hat{v}_q(b_\ell)$$

for almost every $\ell \in \Lambda(a)$. On the other hand, we set

$$v_i^{\#}(b_\ell) = \limsup_{\alpha \downarrow 0} v_i(z(\ell;\alpha))$$

for $\ell \in \Lambda(a)$. Since $v_i \geqq 0$, we have $v_i^{\#}(b_\ell) \geqq 0 = \hat{v}_i(b_\ell)$ a.e., the last equality sign being true by Theorem 4B, Ch. IV. We look at the bounded l.a.m. $h = \exp(-v_i - v_q)$ and denote by η the line bundle defined by h (Ch. II, 2C). Then we take any nonzero l.a.m. $k \leqq 1$, whose line bundle is η^{-1}. This is possible in view of Widom's theorem (Theorem 2B, Ch. V). We set $-\log k = v_i' + v_q'$ with $v_i' = \mathrm{pr}_I(-\log k) \geqq 0$ and $v_q' = \mathrm{pr}_Q(-\log k) \geqq 0$, and see as above that

$$\check{v}_q'(\ell) = \hat{v}_q'(b_\ell) \tag{17}$$

for almost every $\ell \in \Lambda(a)$.

Since h and k correspond to mutually inverse line bundles, there exists a nonzero $f \in H^\infty(R)$ such that $hk = |f|$ on R. We thus have $\check{f}(\ell) = \hat{f}(b_\ell)$ for almost every $\ell \in \Lambda(a)$. This means that for almost every $\ell \in \Lambda(a)$

$$\lim_{\alpha \downarrow 0} h(z(\ell;\alpha))k(z(\ell;\alpha))$$

exists and is equal to

$$\begin{aligned} |\check{f}(\ell)| &= |\hat{f}(b_\ell)| = \hat{h}(b_\ell)\hat{k}(b_\ell) \\ &= \exp(-\hat{v}_q(b_\ell))\exp(-\hat{v}_q'(b_\ell)) \end{aligned} \tag{18}$$

by means of Corollary 4B, Ch. IV. From (16) and (17) follows that, for almost every $\ell \in \Lambda(a)$,

$$\lim_{\alpha \downarrow 0} \exp[-v_q(z(\ell;\alpha)) - v_q'(z(\ell;\alpha))]$$

exists and is equal to $\exp[-\hat{v}_q(b_\ell) - \hat{v}_q'(b_\ell)] \neq 0$. Therefore

$$\lim_{\alpha \downarrow 0} \exp[-v_i(z(\ell;\alpha)) - v_i'(z(\ell;\alpha))]$$

exists and is equal to $|\hat{f}(b_\ell)|\exp[\hat{v}_q(b_\ell) + \hat{v}_q'(b_\ell)]$ for almost every $\ell \in \Lambda(a)$. On the other hand, since v_i and v_i' are nonnegative, the value (18) is majorized by

$$\liminf_{\alpha \downarrow 0} \exp[-v_i(z(\ell;\alpha))] = \exp[-v_i^{\#}(b_\ell)] \leqq 1.$$

Combined with (18), this gives

$$1 \leqq |\hat{f}(b_\ell)|\exp[\hat{v}_q(b_\ell) + \hat{v}_q'(b_\ell)] \leqq \exp[-v_i^{\#}(b_\ell)] \leqq 1.$$

Hence, $v_i^{\#}(b_\ell) = 0$ a.e., which means that $\check{v}_i(\ell)$ exists and is equal to 0 for almost every $\ell \in \Lambda(a)$. Thus

$$\check{v}_i(\ell) = \hat{v}_i(b_\ell) = 0$$

a.e. on $\Lambda(a)$. This establishes our assertions when $\log u \geqq 0$.

The general case can be shown by applying the above consideration to the positive and the negative parts of $\log u$. □

7. Boundary Behavior of Analytic Maps

7A. Let R be a regular PWS. We first define Stolz regions in R with vertices in the Martin boundary Δ_1. Let $a, a' \in R$ be fixed. We denote by $\Omega(a,a')$ the set of $\theta \in [0, 2\pi)$ satisfying the following conditions:

(i) $\ell_\theta \in \Lambda(a)$;

(ii) $(\partial\Psi/\partial\zeta)(\zeta)$ exists and is different from 0 for $\zeta = \Psi^{-1}(\exp(i\theta))$;

(iii) both $\check{P}(a,a';\ell_\theta)$ and $\hat{P}(a,a';b_\theta)$ exist, are finite and satisfy

$$\check{P}(a,a';\ell_\theta) = \hat{P}(a,a';b_\theta) = k(b_\theta,a')/k(b_\theta,a);$$

(iv) both $F_0 \circ \Phi^{-1}$ and $F_1 \circ \Phi^{-1}$ have nonvanishing finite radial limits at $e^{i\theta}$, which we denote by $(F_0 \circ \Phi^{-1})(e^{i\theta})$ and $(F_1 \circ \Phi^{-1})(e^{i\theta})$, respectively, where $F_0(z) = F(a,a';z)$ and $F_1(z) = F(a,a';z)P(a,a';z)$. Moreover, the functions $F_0 \circ \Phi^{-1}(w)$ and $F_1 \circ \Phi^{-1}(w)$ tend uniformly to $(F_0 \circ \Phi^{-1})(e^{i\theta})$ and $(F_1 \circ \Phi^{-1})(e^{i\theta})$, respectively, as w tends to $e^{i\theta}$ through any Stolz region $S(e^{i\theta};\alpha,\rho)$ with $0 < \alpha < \pi/2$.

In these statements, Φ and Ψ were defined in 3A, while $P(a,a';z)$, $F(a,a';z)$ and $\Lambda(a)$ were given in 5B. Our previous observations show the following

<u>Lemma</u>. The set $\Omega(a,a')$ has negligible complement in $[0, 2\pi)$.

7B. Let $\theta \in \Omega(a,a')$. We consider any Stolz region $S(e^{i\theta};\alpha,\rho)$ included in the set $D(a)$ (Theorem 3D), where $0 < \alpha < \pi/2$ and $0 < \rho < 1$, and set

$$S(b_\theta;\alpha,\rho|a) = S(b_\theta;\alpha,\rho) = \Phi^{-1}(S(e^{i\theta};\alpha,\rho)).$$

This may be called a Stolz region in R if it clusters only at the point b_θ. To see this, we denote by $e(b_\theta,\alpha)$ the set of cluster points in Δ of the set $S(b_\theta;\alpha,\rho)$. It is easy to see that $e(b_\theta,\alpha)$ is independent of the choice of ρ. By an argument similar to that used in the proof of Theorem 5B we have

Lemma. For any $b' \in e(b_\theta,\alpha)$ we have

(19) $$k(b',a')/k(b',a) = k(b_\theta,a')/k(b_\theta,a).$$

7C. We take a fixed countable dense subset A of R and set

$$\Omega_1(a) = \cap\{\Omega(a,a'): a' \in A\}.$$

Then Lemma 7A shows that $\Omega_1(a)$ is a measurable subset of $[0, 2\pi)$ with negligible complement. Now let $\theta \in \Omega_1(a)$. For any point $b' \in e(b_\theta,\alpha)$ we have (19) for any $a' \in A$. Since A is dense in R, we see that the equation (19) holds for all $a' \in R$. Since Martin functions with distinct poles are never proportional (Ch. III, 2B), we should have $b' = b_\theta$. Namely, the set $e(b_\theta,\alpha)$ consists of a single point b_θ for any $\theta \in \Omega_1(a)$ and any α with $0 < \alpha < \pi/2$. We thus have:

Theorem. Let $a \in R$ be fixed and let $\Delta_1(a)$ be the set of points b_θ in Δ_1 with $\theta \in \Omega_1(a)$. Then the set $\Delta_1(a)$ is a measurable subset of Δ_1 of $d\chi$-measure one and satisfies the following: For any $b_\theta \in \Delta_1(a)$ with $\theta \in \Omega_1(a)$ and any $0 < \alpha < \pi/2$ there exists a positive number $\rho = \rho(b_\theta,\alpha)$ with $0 < \rho < 1$ such that the system

(20) $$r(a;z) = c \quad \text{and} \quad \theta(a;z) \equiv d \mod 2\pi$$

has a unique solution z in $\mathbb{C}'(R,a)$ for any pair of numbers (c, d) satisfying

(21) $$\rho \leqq c < 1 \quad \text{and} \quad -\alpha \leqq \arg(1 - ce^{i(d-\theta)}) \leqq \alpha,$$

where $r(a;z) = \exp(-g(a,z))$. The set of solutions z for the system (20) and (21) is equal to the set $S(b_\theta;\alpha,\rho)$, which we defined in 7B, and has only one cluster point b_θ in Δ.

We call the set $S(b_\theta;\alpha,\rho)$ (or, more exactly, $S(b_\theta;\alpha,\rho|a)$) a Stolz region in R with vertex b_θ and angular measure 2α (with origin a). We can generalize this definition of Stolz regions as follows. Let $\theta \in \Omega_1(a)$. Then, for any α_1, α_2 with $-\pi/2 < \alpha_1 < \alpha_2 < \pi/2$, there exists a positive number ρ, $0 < \rho < 1$, such that the set

$$S(e^{i\theta};\alpha_1,\alpha_2,\rho) = \{z \in \mathbb{D}: \alpha_1 \leqq \arg(1 - ze^{-i\theta}) \leqq \alpha_2\}$$

is included in the region $D(a)$. We set

$$S(b_\theta;\alpha_1,\alpha_2,\rho|a) = S(b_\theta;\alpha_1,\alpha_2,\rho) = \Phi^{-1}(S(e^{i\theta};\alpha_1,\alpha_2,\rho)).$$

Then, $S(b_\theta;\alpha_1,\alpha_2,\rho)$ has a unique cluster point b_θ in Δ and may be viewed as a kind of Stolz regions. It is not hard to see that distinct choices of a give rise to essentially equivalent Stolz regions.

7D. We can extend some results on boundary behavior of analytic maps appearing in Chapter 19 of Constantinescu and Cornea [CC]. We will describe a few of them here. Let f be a continuous map of R into a compact metric space X. Let $a \in R$ be fixed, so that $S(b;\alpha_1,\alpha_2,\rho)$ always means $S(b;\alpha_1,\alpha_2,\rho|a)$. Take any $b \in \Delta_1(a)$ and set, for any α_1, α_2 with $-\pi/2 < \alpha_1 < \alpha_2 < \pi/2$,

$$f(b;\alpha_1,\alpha_2) = \cap \operatorname{Cl}(f(S(b;\alpha_1,\alpha_2,\rho))),$$

where the intersection on the right-hand side is taken for all possible ρ with $0 < \rho < 1$. We further put

$$f^{\Delta}(b) = \cap\{f(b;\alpha_1,\alpha_2): -\pi/2 < \alpha_1 < \alpha_2 < \pi/2\},$$

$$f^{\nabla}(b) = \cup\{f(b;\alpha_1,\alpha_2): -\pi/2 < \alpha_1 < \alpha_2 < \pi/2\}.$$

<u>Theorem</u>. $f^{\wedge}(b) \subseteqq f^{\Delta}(b)$ for almost every $b \in \Delta_1(a)$.

<u>Corollary</u>. (a) $f^{\Delta}(b)$ is nonempty for almost every $b \in \Delta_1(a)$.

(b) Let f be a nonconstant analytic map of R into a Riemann surface R', R'^* a metrizable compactification of R' and A' a polar subset of R'^*. If A denotes the set of $b \in \Delta_1(a)$ for which $f^{\Delta}(b) \subseteqq A'$, then A is $d\chi$-negligible.

7E. In order to speak of f^{∇}, we define $\Delta_2(a)$ to be the set of points b_θ with $\theta \in \Omega_1(a)$ such that $R \setminus \mathbb{G}'(R;a)$ is thin at b_θ and $\mathbb{D}_w \setminus D(a)$ is thin at $e^{i\theta}$. Then $\Delta_2(a)$ is seen to be a subset of the set $\Delta_1(a)$ of $d\chi$-measure one. We then have for instance the following

<u>Theorem</u>. Let f be an analytic map of R into the extended complex plane $\overline{\mathbb{C}}$. If there exists a point $w_0 \in \overline{\mathbb{C}}$ such that $\sum\{g(a,z): f(z) = w_0\} < \infty$, where the summation is repeated according to multiplicity, then $f^{\nabla}(b) = \{\hat{f}(b)\}$ for almost every b in $\mathcal{D}(f) \cap \Delta_2(a)$.

<u>Corollary</u>. If f is a meromorphic function of bounded characteristic on R, then $f^{\nabla}(b) = \{\hat{f}(b)\}$ for almost every point $b \in \Delta_2(a)$.

NOTES

The general discussion in §1 on Green lines and the corresponding Dirichlet problem is adapted from Brelot and Choquet [6]. See also Sario and Nakai [62; pp. 199-209].

The results in §2 are due to Parreau [52], although some of them were rediscovered by Hasumi [17, 18].

The Brelot-Choquet problem--the main topic of §3--can be found in Brelot [5; §14]. The result in this section were obtained by Hasumi [17, 18], except Theorem 6A, (c). The latter is an important complement due to Hayashi [27]. Stolz regions on PWS's and related things sketched in Subsection 7 are discussed at length in Hasumi [19]. Niimura [48] contains a cluster value theorem on PWS's.

CHAPTER VII. CAUCHY THEOREMS

The Cauchy integral theorem and its inverse will be discussed in this chapter. If these were taken in the classical sense, their validity would be a matter of triviality. But, in fact, we are thinking of them in a form more fitted to our present situation. So the problem is far from being evident and deserves rather detailed analysis. The inverse Cauchy theorem in its most general form holds for any surfaces of Parreau-Widom type. As we shall see in Ch. IX, this is indeed characteristic of surfaces of Parreau-Widom type. On the other hand, the direct Cauchy theorem--an utmost generalization of the Cauchy integral theorem--is not always valid and will be discussed again in Ch. IX.

In §1 we prove the inverse Cauchy theorem. We state in §2 the direct Cauchy theorem and prove its weaker variation, which is true for any hyperbolic Riemann surface. A few applications are given in §3. Among other things, we shall show that the algebra $H^\infty(d\chi)$ is a maximal weak-star closed subalgebra of $L^\infty(d\chi)$.

Unless otherwise stated, R denotes a surface of Parreau-Widom type (= PWS), which is regular in the sense of potential theory.

§1. THE INVERSE CAUCHY THEOREM

1. Statement of Results

1A. Let R be a regular PWS with Green function $g(a,z)$. Let $a \in R$ be fixed and let $\{z_1, z_2, \ldots\}$ be a fixed enumeration of members of $Z(a;R)$, the critical points of $g(a,\cdot)$, which we repeat according to multiplicity. We use the following notations:

$$s^{(a)}(z) = \sum_{j=1}^{\infty} g(z,z_j), \quad s_n^{(a)}(z) = \sum_{j=n+1}^{\infty} g(z,z_j),$$

$$g^{(a)}(z) = \exp(-s^{(a)}(z)), \quad g_n^{(a)}(z) = \exp(-s_n^{(a)}(z)),$$

$$S^{(a)} = \exp(-s^{(a)}(z) - i\tilde{s}^{(a)}(z))$$

and

$$S_n^{(a)}(z) = \exp(-s_n^{(a)}(z) - i\tilde{s}_n^{(a)}(z)),$$

where $\tilde{u}$ denotes the harmonic conjugate of u normalized by the condition $\tilde{u}(a) = 0$. $S^{(a)}$ (resp. $S_n^{(a)}$) is a bounded multiplicative analytic function on R. We denote by $\xi^{(a)}$ (resp. $\xi_n^{(a)}$) the line bundle which $S^{(a)}$ (resp. $S_n^{(a)}$) defines over R.

1B. Our objective in this section is to prove the following, which is called the <u>inverse Cauchy theorem</u> and will be proved in 2D below.

<u>Theorem</u>. Let $a \in R$ be fixed. If $u \in L^1(d\chi)$ satisfies

$$\int_{\Delta_1} \hat{h}(b)u(b)k(b,a)d\chi(b) = 0$$

for any h, meromorphic on R, such that $|h|g^{(a)}$ is bounded on R and $h(a) = 0$, then there exists an $f \in H^1(R)$ such that $\hat{f} = u$ a.e. on Δ_1.

1C. By the way we wish to state without proof the inverse Cauchy theorem for an arbitrary PWS.

<u>Theorem</u>. Let R be a (not necessarily regular) PWS with origin O and $R^\dagger$ the regularization of R in the sense of Theorem 3B, Ch. V. Let $g^\dagger(a,z)$ be the Green function for $R^\dagger$ and $Z(O;R^\dagger)$ the set of critical points in $R^\dagger$ of the function $z \to g^\dagger(O,z)$, which we repeat according to multiplicity. Set

$$g^{\dagger(O)}(z) = \exp(-\sum \{g^\dagger(z,w) : w \in Z(O;R^\dagger)\})$$

for $z \in R$. If $u \in L^1(d\chi)$ satisfies $\int_{\Delta_1} \hat{h}u d\chi = 0$ for any h, meromorphic on R, such that $h(O) = 0$ and $|h|g^{\dagger(O)}$ is bounded on R, then there exists an $f \in H^1(R)$ such that $u = \hat{f}$ a.e. on Δ_1.

This follows easily from Theorem 3B, Ch. V and the preceding theorem.

2. <u>Proof of Theorem 1B</u>

2A. To prove Theorem 1B we need some preliminary observation, which we state in three lemmas. The first is the following

Lemma. $\hat{g}^{(a)}(b) = 1$ a.e. on Δ_1.

Proof. Since $s(z) = s^{(a)}(z)$ is a positive superharmonic function on R, F. Riesz's theorem (Theorem 6F, Ch. I) shows that $s = u + U$, where u is a nonnegative harmonic function and U is a potential. By Theorem 5A, Ch. III, s is a Wiener function on R and by Theorem 5B, Ch. III, $h[s] = u$. Since s is continuous as an extended real-valued function, it follows from Theorem 5D, Ch. III, that $\hat{s}$ exists a.e. on Δ_1 and the quasibounded part of the harmonic function $h[s]$ is given by $\int_{\Delta_1} \hat{s}(b) d_b d\chi(b)$. For $n = 1, 2, \ldots$ we set $s_n = s_n^{(a)}$ and $s_n' = s - s_n$. Since s_n' is a potential, $h[s_n'] = 0$ and so $h[s] = h[s_n]$ for $n = 1, 2, \ldots$. Thus, $u = h[s] \leqq s_n$, $n = 1, 2, \ldots$. Since $s(z)$ is finite except at z_j, s_n converge to zero on R except at z_j. Hence $u = 0$ on R and consequently $\hat{s} = 0$ a.e. on Δ_1. □

2B. Next we investigate the function

$$P(a,a';z) = \delta g(a',z)/\delta g(a,z),$$

appearing already in Ch. VI, 5B.

Lemma. Let $V = \{|\zeta| < 1\}$ be a parametric disk in R and let a be fixed. Then there exists a constant C such that

$$|P(a,\zeta';z) - P(a,\zeta'';z)| g^{(a)}(z) \leqq C|\zeta' - \zeta''|$$

for any $\zeta', \zeta'' \in V' = \{\zeta \in V: |\zeta| < 1/4\}$ and any $z \in R \setminus Cl(V)$.

Proof. We consider a regular exhaustion $\{R_n: n = 1, 2, \ldots\}$, as defined in Ch. VI, 5B, i.e. $R_n = R(\alpha_n, a)$ for a strictly decreasing sequence $\{\alpha_n: n = 1, 2, \ldots\}$ of positive numbers tending to 0 with $\partial R(\alpha_n, a) \cap Z(a;R) = \emptyset$. We assume further that $Cl(V) \subsetneqq R_1$. Let $\zeta', \zeta'' \in V'$ and let $g_n(\zeta',z)$ and $g_n(\zeta'',z)$ be the Green functions for R_n with poles ζ' and ζ'', respectively. Then, for any real quasibounded harmonic function h on R_n, we have

$$(1)\quad h(\zeta') - h(\zeta'') = -\frac{1}{2\pi i}\int_{\partial R_n} h(z)\left(\frac{\delta g_n(\zeta',z)}{\delta g(a,z)} - \frac{\delta g_n(\zeta'',z)}{\delta g(a,z)}\right)\delta g(a,z).$$

Put $h^+ = h \vee 0$ and $h^- = (-h) \vee 0$. Then, h^+ and h^- are positive quasibounded harmonic functions on R_n. By the Harnack inequality, we have

$$c(r)^{-1} \leqq h^+(\zeta')/h^+(\zeta'') \leqq c(r),$$

where $r = |\zeta' - \zeta''|$ and $c(r) = (3+4r)/(3-4r)$. The same is true of the function h^-. So

$$(2) \qquad \begin{aligned} |h(\zeta') - h(\zeta'')| &\leqq |h^+(\zeta') - h^+(\zeta'')| + |h^-(\zeta') - h^-(\zeta'')| \\ &\leqq 8r(h^+(\zeta'') + h^-(\zeta'')) \\ &= 8r\left(-\frac{1}{2\pi i}\right)\int_{\partial R_n} |h(z)|\frac{\delta g_n(\zeta'',z)}{\delta g(a,z)}\,\delta g(a,z) \\ &\leqq 8r\lambda\left(-\frac{1}{2\pi i}\right)\int_{\partial R_n} |h(z)|\,\delta g(a,z), \end{aligned}$$

where λ is a constant depending only on a, V and R, and not on n. Combining (1) and (2), we have

$$\left|\frac{\delta g_n(\zeta',z)}{\delta g(a,z)} - \frac{\delta g_n(\zeta'',z)}{\delta g(a,z)}\right| \leqq 8\lambda r$$

on ∂R_n. We set

$$v(z) = g^{(a)}(z)\exp(-g(\zeta',z) - g(\zeta'',z)).$$

Then, $v(z)$ is the modulus of a multiplicative analytic function on R with $v \leqq 1$. It follows that

$$(3) \qquad \left|\frac{\delta g_n(\zeta',z)}{\delta g(a,z)} - \frac{\delta g_n(\zeta'',z)}{\delta g(a,z)}\right|\cdot v(z) \leqq 8\lambda r$$

on ∂R_n. Since the left-hand member of (3) is the modulus of a multiplicative analytic function on $\mathrm{Cl}(R_n)$ so that the inequality sign remains to be valid when z ranges over R_n. Letting $n \to \infty$, we have

$$|P(a,\zeta';z) - P(a,\zeta'';z)|v(z) \leqq 8\lambda r$$

on R. Since $\mathrm{Cl}(V')$ is a compact subset of V, the set of functions

$$\exp(g(\zeta',z) + g(\zeta'',z))$$

with $\zeta', \zeta'' \in V'$ forms a uniformly bounded family on the region $R \setminus \mathrm{Cl}(V)$. Hence, there exists a desired constant C. □

2C. <u>Lemma</u>. Let V be a parametric disk whose closure lies in $R \setminus Z(a;R)$ and J any closed rectifiable curve contained in $\{\zeta \in V: |\zeta| < 1/4\}$, where $a \in R$ is fixed. Set

$$P_J(z) = \int_J P(a,\zeta;z)\,d\zeta$$

for $z \in R \setminus (Z(a;R) \cup Cl(V))$. Then we have the following:

(a) P_J is holomorphic on $R \setminus (Z(a;R) \cup Cl(V))$ and can be extended analytically to $R \setminus Z(a;R)$.

(b) $P_J(a) = 0$.

(c) P_J is meromorphic with poles in $Z(a;R)$ counting multiplicity such that $|P_J|g^{(a)}$ is bounded.

(d) $\hat{P}_J(b)$ exists a.e. on Δ_1 and

(4) $$\hat{P}_J(b) = \int_J (k_b(\zeta)/k_b(a))d\zeta \quad \text{a.e. on } \Delta_1.$$

Proof. Since the poles of $P(a,\zeta;z)$ are contained in the union $\{\zeta\} \cup Z(a;R)$, the function P_J is analytic on $R \setminus (Z(a;R) \cup \{\zeta \in V: |\zeta| \leqq 1/4\})$. If $\zeta, \zeta' \in V$, then

$$g(\zeta,\zeta') = -\log|\zeta - \zeta'| + h(\zeta,\zeta')$$

for $\zeta \neq \zeta'$, where $h(\zeta,\zeta')$ is symmetric in ζ and ζ', is harmonic in ζ' and has a removable singularity at $\zeta' = \zeta$. So we have

$$\delta_{\zeta'}g(\zeta,\zeta') = -(\zeta - \zeta')^{-1}d\zeta' + \delta_{\zeta'}h(\zeta,\zeta'),$$

where $\delta_{\zeta'}h(\zeta,\zeta')$ is an analytic differential in $\zeta' \in V$. For $\zeta' \in V$ with $1/4 < |\zeta'| < 1$, we have

$$\int_J P(a,\zeta;\zeta')d\zeta = \int_J \left(\frac{\delta_{\zeta'}h(\zeta,\zeta')}{d\zeta'} \Big/ \frac{\delta_{\zeta'}g(a,\zeta')}{d\zeta'}\right)d\zeta,$$

the right-hand member being analytic throughout V. Hence, the function P_J can be continued analytically to the whole V, so that it can be regarded analytic on $R \setminus Z(a;R)$. This proves the statement (a).

Since $\delta g(a,z)$ has a pole at a, $P_J(a) = 0$. The poles of P_J are contained in $Z(a;R)$ counting multiplicity. Since J is compact, the discussion in Ch. VI, 5B shows that there exists a constant c depending only on a, J and R with

(5) $$|P_J(z)|g^{(a)}(z) \leqq c$$

on R. In fact, let $\zeta \in J$. Then, as we saw in Ch. VI, 5B,

$$g^{(a)}(z)|P(a,\zeta;z)| \leqq c(a,\zeta)e^{g(\zeta,z)}$$

on R, where $c(a,\zeta)$ depends only on a and ζ and is bounded when a is fixed and ζ ranges over any compact set, e.g. J. In view of the fact that $J \subseteqq \{\zeta \in V: |\zeta| < 1/4\}$, $g(\zeta,z)$ is bounded above, by c' say, when ζ and z range over J and $R \setminus \{z' \in V: |z'| < 1/2\}$, respectively. Therefore

$$g^{(a)}(z)|P_J(z)| \leqq \{\max_{\zeta\in J} c(a,\zeta)\}\{\text{length}(J)\}\cdot e^{C'},$$

where z ranges over the outside of $\{z' \in V: |z'| < 1/2\}$. On the other hand, P_J is analytic on $Cl(V)$ and so is bounded there. Thus $g^{(a)}|P_J|$ is bounded everywhere on R, which shows the inequality (5). Hence (c) is proved.

The inequality (5) also shows that P_J is a meromorphic function of bounded characteristic, i.e. $\log|P_J| \in SP'(R)$, and thus $\hat{P}_J$ exists a.e. on Δ_1. In order to show the equation (4), let $\gamma: [0, 1) \to J$ be a fixed parametrization of the curve J. Since $a' \to P(a,a';z)$ is continuous on J for any fixed $z \in R \setminus (Z(a;R) \cup Cl(V))$, we have for such z,

$$P_J(z) = \lim_{n\to\infty} \sum_{j=1}^{n} P(a,\zeta_{n,j};z)(\zeta_{n,j} - \zeta_{n,j-1}),$$

where $\zeta_{n,j} = \gamma(j/n)$, $j = 0, 1,\ldots, n-1$, and $\zeta_{n,n} = \zeta_{n,0}$. Let Δ' be a measurable subset of Δ_1 with $\chi(\Delta_1 \setminus \Delta') = 0$ such that, for each $b \in \Delta'$, we have $\hat{g}^{(a)}(b) = 1$ and $\hat{P}(a,\zeta_{n,j};b) = k_b(\zeta_{n,j})/k_b(a)$ for every n and j. Such a set Δ' exists in view of Theorem 6A, Ch. VI, and Lemma 2A.

Take any $b \in \Delta'$. Then, for any $0 < \varepsilon < 1$ and any n, there exists an open set $D_n \in G(b)$ (see Ch. III, 4A) such that D_n is included in $R \setminus Cl(V)$ and

$$|g^{(a)}(z)P(a,\zeta_{n,j};z) - k_b(\zeta_{n,j})/k_b(a)| < \varepsilon$$

for any $z \in D_n$ and any $j = 1,\ldots, n$. Thus, for $z \in D_n$,

$$\left|\sum_{j=1}^{n} g^{(a)}(z)P(a,\zeta_{n,j};z)(\zeta_{n,j} - \zeta_{n,j-1}) - \sum_{j=1}^{n} \frac{k_b(\zeta_{n,j})}{k_b(a)}(\zeta_{n,j} - \zeta_{n,j-1})\right|$$

$$\leqq \varepsilon\, \text{length}(J).$$

We take an integer $n_0 > 0$ in such a way that $\gamma([(j-1)/n, j/n])$ is contained in a disk of diameter ε for each $n \geqq n_0$ and $j = 1,\ldots, n$. Put $J_{n,j} = \gamma([(j-1)/n, j/n])$, $j = 1,\ldots, n$. Let $z \in D_n$ with $n \geqq n_0$. Then, since $|\zeta - \zeta_{n,j}| < \varepsilon$ for each $\zeta \in J_{n,j}$, we have, in view of Lemma 2B,

$$\left|\int_J g^{(a)}(z)P(a,\zeta;z)d\zeta - \sum_{j=1}^{n} g^{(a)}(z)P(a,\zeta_{n,j};z)(\zeta_{n,j} - \zeta_{n,j-1})\right|$$

$$\leq \sum_{j=1}^{n} \left|\int_{J_{n,j}} \{g^{(a)}(z)P(a,\zeta;z) - g^{(a)}(z)P(a,\zeta_{n,j};z)\}d\zeta\right|$$

$$\leq C\varepsilon\cdot\text{length}(J).$$

Since $a' \to k_b(a')$ is continuous on R, there exists an n_1 such that, for $n \geq n_1$,

$$\left|\sum_{j=1}^{n} \frac{k_b(\zeta_{n,j})}{k_b(a)}(\zeta_{n,j} - \zeta_{n,j-1}) - \int_J \frac{k_b(\zeta)}{k_b(a)}d\zeta\right| < \varepsilon.$$

Hence, for $n \geq \max\{n_0, n_1\}$ and for $z \in D_n$, we have

$$\left|g^{(a)}(z)P_J(z) - \int_J \frac{k_b(\zeta)}{k_b(a)}d\zeta\right| \leq \varepsilon\cdot\text{length}(J) + C\varepsilon\cdot\text{length}(J) + \varepsilon.$$

Thus we have shown that the fine boundary function for $g^{(a)}P_J$ exists a.e. on Δ_1 and is equal to $\int_J (k_b(\zeta)/k_b(a))d\zeta$. By Lemma 2A we have $\hat{g}^{(a)} = 1$ a.e. on Δ_1, so that we obtain the desired equation (4). □

2D. Proof (Theorem 1B). To prove Theorem 1B, we set

$$f(z) = \int_{\Delta_1} u(b)k(b,z)d\chi(b)$$

for $z \in R$. Then, f is a quasibounded harmonic function on R. Let V be any parametric disk included in $R \setminus Z(a;R)$, in which points and their coordinate variables are identified. Let J be any closed rectifiable curve contained in $\{z \in V: |z| < 1/4\}$. Then the Fubini theorem and Lemma 2C show that

$$\int_J f(z)dz = \int_{\Delta_1} \left(\int_J \frac{k(b,z)}{k(b,a)}dz\right)u(b)k(b,a)d\chi(b)$$

$$= \int_{\Delta_1} \hat{P}_J(b)u(b)k(b,a)d\chi(b) = 0.$$

By the Morera theorem f is analytic on $R \setminus Z(a;R)$. Since f is continuous on R, every point in $Z(a;R)$ is a removable singularity and hence f is everywhere holomorphic on R. Clearly, $|f|$ has a harmonic majorant, so that $f \in H^1(R)$. In view of Theorem 5E, Ch. III, $\hat{f} = u$ a.e. on Δ_1. □

§2. THE DIRECT CAUCHY THEOREM

3. Formulation of the Condition

3A. First we explain what we mean by the direct Cauchy theorem. To do this, we begin with two simple lemmas.

Lemma. Let f be a meromorphic function on R of bounded characteristic. Then f is a quotient of bounded holomorphic functions on R.

Proof. We may suppose that f does not vanish identically. We have, by definition, $\log|f| \in SP'(R)$. As we saw in Ch. II, 5A, there exist positive elements u_1 and u_2 in $SP'(R)$ such that $\log|f| = u_2 - u_1$. Thus we have

$$f = c\cdot\exp(-u_1 - i\tilde{u}_1)/\exp(-u_2 - i\tilde{u}_2),$$

where c is a constant of modulus one. Since f is single-valued, the functions $\exp(-u_j - i\tilde{u}_j)$, $j = 1, 2$, generate the same line bundle, ξ_0 say. Let h be a nonzero element in $\mathcal{H}^\infty(R,\xi_0^{-1})$, which exists because of Widom's theorem (Theorem 2B, Ch. V). Setting $f_1 = ch\cdot\exp(-u_1 - i\tilde{u}_1)$ and $f_2 = h\cdot\exp(-u_2 - i\tilde{u}_2)$, we get a desired expression $f = f_1/f_2$. □

3B. Lemma. If f is a meromorphic function on R such that $|f|g^{(a)}$ has a harmonic majorant on R, then $\hat{f}$ exists a.e. on Δ_1 and belongs to $L^1(d\chi)$.

Proof. The hypothesis shows that f is of bounded characteristic. So, by Lemma 3A, it is a quotient of bounded analytic functions, e.g. $f = f_1/f_2$ with f_j bounded. By Theorem 5E, Ch. III, both f_1 and f_2 have fine boundary values a.e. on Δ_1 and therefore $\hat{f}$ exists a.e. on Δ_1. Since $\hat{g}^{(a)} = 1$ a.e. on Δ_1 by Lemma 2A, we have $|\hat{f}| \leqq \hat{u}$, where u is a harmonic majorant of $|f|g^{(a)}$. Since $\hat{u}$ is summable again by Theorem 5E, Ch. III, so is $\hat{f}$, as was to be proved. □

3C. We mean by the direct Cauchy theorem for a point $a \in R$ the following statement:

(DCT_a) If f is a meromorphic function on R such that $|f|g^{(a)}$ has a harmonic majorant on R, then

$$f(a) = \int_{\Delta_1} \hat{f}(b)k_b(a)d\chi(b).$$

As indicated in the introduction, the statement (DCT_a) is not a theorem but a condition on PWS's. In the next chapter it will be shown that, for a PWS, (DCT_a) are simultaneously true or false for all $a \in R$ and so we may write (DCT) without referring to any particular a. Various conditions equivalent to (DCT) will be given in Ch. X.

4. The Direct Cauchy Theorem of Weak Type

4A. Although (DCT) does not always hold for PWS's, there is a weaker statement which is valid for any hyperbolic Riemann surface. In the remaining part of this section, we denote by R any hyperbolic Riemann surface, which is regular in the sense of potential theory. Let $a \in R$ be fixed arbitrarily and choose a strictly decreasing sequence $\{\alpha_n : n = 1, 2, \ldots\}$ of positive numbers tending to zero such that the level curves $\{z \in R: g(a,z) = \alpha_n\}$ do not contain any critical points of the function $g(a,\cdot)$. We set $R_n = R(\alpha_n, a) = \{z \in R: g(a,z) > \alpha_n\}$, $n = 1, 2, \ldots$. Then, R_n are Jordan regions, $Cl(R_n) \subsetneqq R_{n+1}$ and $\bigcup_{n=1}^{\infty} R_n = R$. We also set $\delta g(a,z) = 2(\partial g(a,z)/\partial z)dz$.

<u>Lemma</u>. Let F be a continuous Wiener function on R such that $|F| \leqq u$ on R for some quasibounded harmonic function u on R. Then we have

$$-\lim_{n\to\infty} \frac{1}{2\pi i} \int_{\partial R_n} F(z)\delta g(a,z) = \int_{\Delta_1} \hat{F}(b)k_b(a)d\chi(b). \tag{6}$$

<u>Proof</u>. We may suppose that F is nonnegative. By Lemma 5C and Theorem 4B in Ch. III, $\hat{F}$ exists a.e. on Δ_1 and is measurable. Since we have $0 \leqq F \leqq u$, we have $0 \leqq \hat{F} \leqq \hat{u}$ a.e. on Δ_1. Since $\hat{u}$ is seen to be summable, $\hat{F}$ is also summable.

In order to show the convergence of (6), we first assume that F is bounded. Then $h[F]$ is bounded and a fortiori is quasibounded. So by Theorem 5D, Ch. III, we have

$$h[F] = \int_{\Delta_1} \hat{F}(b)k_b d\chi(b).$$

By Theorem 5A, Ch. III, there exists a potential U on R such that, for any $\varepsilon > 0$,

$$h[F] - \varepsilon U \leqq F \leqq h[F] + \varepsilon U \tag{7}$$

outside a compact subset K_ε of R. Since F is finite everywhere, the proof of Theorem 5D, Ch. III, shows that we may assume $U(a) < \infty$.

Given any $\varepsilon > 0$, we take n so large that $K_\varepsilon \subseteqq R_n$ and integrate the inequality (7) with respect to $d\mu_n$, which is the restriction of the differential $-(2\pi i)^{-1}\delta g(a,z)$ to ∂R_n. Since we have

$$h[F](a) = \int_{\partial R_n} h[F](z)d\mu_n(z) = \int_{\Delta_1} \hat{F}(b)k_b(a)d\chi(b)$$

and

$$\int_{\partial R_n} U(z)d\mu_n(z) \leqq U(a),$$

we conclude that

$$\left|\int_{\partial R_n} F(z)d\mu_n(z) - \int_{\Delta_1} \hat{F}(b)k_b(a)d\chi(b)\right| \leqq \varepsilon U(a).$$

Since $U(a)$ is assumed to be finite, we have the identity (6) by letting $\varepsilon \to 0$.

Next we consider the general case. By Theorem 5B-(b), Ch. III, $F_m = \min\{F, m\}$ is a Wiener function for $m = 1, 2, \ldots$ with $h[F_m] = h[F] \wedge m$ and therefore $\hat{F}_m = \min\{\hat{F}, m\}$ a.e. on Δ_1. By the fact we have shown above, there exists, for any $m = 1, 2, \ldots$ and any $\varepsilon > 0$, a positive integer $n_0 = n_0(m,\varepsilon)$ such that

$$\left|\int_{\partial R_n} F_m(z)d\mu_n(z) - \int_{\Delta_1} \hat{F}_m(b)k_b(a)d\chi(b)\right| < \varepsilon$$

for $n \geqq n_0$. Since $\hat{F}$ is summable and $\hat{F}_m \to \hat{F}$ a.e., there exists, for any $\varepsilon > 0$, a number $m_0 = m_0(\varepsilon)$ such that

$$\int_{\Delta_1} \hat{F}(b)k_b(a)d\chi(b) < \int_{\Delta_1} \hat{F}_m(b)k_b(a)d\chi(b) + \varepsilon$$

for $m \geqq m_0$. On the other hand, we have $0 \leqq F \leqq u$ and therefore $F - F_m \leqq u - u_m$ on R, where $u_m = \min\{u, m\}$. It follows that

$$\begin{aligned} 0 &\leqq \int_{\partial R_n} F(z)d\mu_n(z) - \int_{\partial R_n} F_m(z)d\mu_n(z) \\ &\leqq \int_{\partial R_n} u(z)d\mu_n(z) - \int_{\partial R_n} u_m(z)d\mu_n(z) \\ &\leqq u(a) - (u \wedge m)(a). \end{aligned}$$

If we take $m \geqq m_0(\varepsilon)$ and $n \geqq n_0(m,\varepsilon)$, then we obtain

$$\left|\int_{\Delta_1} \hat{F}(b)k_b(a)d\chi(b) - \int_{\partial R_n} F(z)d\mu_n(z)\right| \leqq 2\varepsilon + u(a) - (u \wedge m)(a).$$

Since u is quasibounded, we have $(u \wedge m)(a) \to u(a)$ as $m \to \infty$, which implies the desired result. □

4B. By the direct Cauchy theorem of weak type we mean the following result:

Theorem. Let R be a regular hyperbolic Riemann surface and let z_1, ..., z_ℓ be any finite subset of $Z(a;R)$. We set

$$g_0(z) = \exp(-\sum_{j=1}^{\ell} g(z_j,z)).$$

If f is a meromorphic function on R such that $|f|g_0$ has a harmonic majorant on R, then $\hat{f}$ exists a.e. on Δ_1, is summable and

$$f(a) = \int_{\Delta_1} \hat{f}(b)k_b(a)d\chi(b). \tag{8}$$

Proof. Since $|f|g_0$ is an l.a.m. having a harmonic majorant on R, Theorem 4A, Ch. IV, shows that $u = \mathrm{LHM}(|f|g_0)$ is quasibounded. Our surface R being regular, there exists a positive number $c > 0$ and a compact set K in R such that $\mathrm{Int}(K)$ contains the points $z_1,\ldots,$ z_ℓ and $g_0 \geqq c$ on $R \setminus K$. So we have $|f| \leqq c^{-1}u$ on $R \setminus K$. We see that $c^{-1}u$ is quasibounded on $R \setminus K$. Since $\mathrm{Re}(f)$ (resp. $\mathrm{Im}(f)$) is harmonic on $R \setminus K$ and is majorized in modulus by $c^{-1}u$, it is quasibounded on $R \setminus K$. $\mathrm{Re}(f)$ (resp. $\mathrm{Im}(f)$) is thus a Wiener function on $R \setminus K$. We then modify $\mathrm{Re}(f)$ (resp. $\mathrm{Im}(f)$) on the set K so as to have a real continuous function f_1 (resp. f_2) on the surface R with the property $|f_1| \leqq c^{-1}u$ (resp. $|f_2| \leqq c^{-1}u$). By applying Lemma 5C, Ch. III, with $G = R \setminus K$, we see that $\Delta_1 \setminus \mathcal{D}(\mathrm{Re}(f))$ and $\Delta_1 \setminus \mathcal{D}(\mathrm{Im}(f))$ are negligible. It follows that f_1/u and f_2/u are bounded continuous functions on R with $\chi(\Delta_1 \setminus \mathcal{D}(f_j/u)) = 0$ for $j = 1, 2$. In view of Theorem 5D, Ch. III, $f_j = (f_j/u)u$, $j = 1, 2$, are Wiener functions on R. Since $|f_j| \leqq c^{-1}u$, the preceding lemma shows that

$$-\lim_{n\to\infty} \frac{1}{2\pi i}\int_{\partial R_n} f(z)\delta g(a,z)$$

$$= -\lim_{n\to\infty} \frac{1}{2\pi i}\int_{\partial R_n} (f_1 + if_2)(z)\delta g(a,z)$$

$$= \int_{\Delta_1} (\hat{f}_1 + i\hat{f}_2)(b)k_b(a)d\chi(b) = \int_{\Delta_1} \hat{f}(b)k_b(a)d\chi(b),$$

where $\{R_n: n = 1, 2, \ldots\}$ is an exhaustion of R mentioned in 4A. If n is so large that R_n includes K, then $f(z)\delta g(a,z)$ is a meromorphic differential in z on the closed region $Cl(R_n)$ with only one pole at the point a, whose residue is equal to $-2\pi i f(a)$. Hence, we get the formula (8). □

§3. APPLICATIONS

5. Weak-star Maximality of H^∞

5A. As an application of the inverse Cauchy theorem, we prove the following, which extends a fact well-known in the case of the unit disk (see Helson [33], p. 27).

Theorem. $H^\infty(d\chi)$ is maximal among weak* closed subalgebras of $L^\infty(d\chi)$.

Proof. Let $f^* \notin L^\infty(d\chi) \setminus H^\infty(d\chi)$ and let C be the weakly* closed subalgebra of $L^\infty(d\chi)$ generated by $H^\infty(d\chi)$ and the function f^*. Suppose that $s^* \in L^1(d\chi)$ is orthogonal to C. It is sufficient to show that $s^* = 0$. To see this, we take any meromorphic function u on R such that $|u|g^{(a)}$ is bounded on R, where a is any fixed point in R. Since R is a PWS, there exists a nonzero function $B \in H^\infty(R)$ such that $|B(z)| \leqq g^{(a)}(z)$ on R. Such a function B is obtained by multiplying $S^{(a)}$ and a nonzero element in $\mathcal{H}^\infty(R,(\xi^{(a)})^{-1})$. Thus Bu is bounded and analytic on R. Since C is closed under the multiplication by functions in $H^\infty(d\chi)$, we have

$$\int_{\Delta_1} \hat{B}(b)\hat{u}(b)(f^*(b))^n s^*(b)d\chi(b) = 0$$

for $n = 0, 1, \ldots$. By the inverse Cauchy theorem there exists, for each $n = 0, 1, \ldots$, a function $h_n \in H^1(R)$ such that $h_n(a) = 0$ and $\hat{h}_n = \hat{B}s^*(f^*)^n$ a.e. on Δ_1. Let $\phi_R\colon \mathbb{D} \to R$ be a universal covering map of R with $\phi_R(0) = a$. Then $h_n \circ \phi_R \in H^1(\mathbb{D})$, $h_n \circ \phi_R(0) = 0$ and, by Theorem 7B, Ch. III,

$$(h_n \circ \phi_R)^\wedge = ((\hat{B}s^*) \circ \hat{\phi}_R)(f^* \circ \hat{\phi}_R)^n \quad \text{a.e.}$$

on $\partial\mathbb{D}$. Thus $(\hat{B}s^*) \circ \hat{\phi}_R$ is orthogonal to the weakly* closed subalgebra C' of $L^\infty(d\sigma)$ generated by $H^\infty(d\sigma)$ and $f^* \circ \hat{\phi}_R$. Since $f^* \notin H^\infty(d\chi)$, it is easy to see, in view of Ch. III, 7D, that $f^* \circ \hat{\phi}_R \notin H^\infty(d\sigma)$. Since $H^\infty(d\sigma)$ is a maximal weakly* closed subalgebra of $L^\infty(d\sigma)$, we see that $C' = L^\infty(d\sigma)$ and therefore $(\hat{B}s^*) \circ \hat{\phi}_R = 0$ a.e. This shows that $\hat{B}s^* = 0$. As we know that $\hat{B} \neq 0$ a.e., we conclude $s^* = 0$, as was to be proved. □

6. Common Inner Factors

6A. We begin with the following

Lemma. Let η be a line bundle over R such that $\mathcal{H}^\infty(R,\eta)$ has no nonconstant common inner factors (see Ch. II, 5D). Then $\mathcal{H}^\infty(R,\eta^{-1})$ has no nonconstant common inner factor, too.

Proof. Let Q_0 be the greatest common inner factor of $\mathcal{H}^\infty(R,\eta^{-1})$ and let η_0 be the line bundle associated to Q_0. Then $\mathcal{H}^\infty(R,\eta^{-1}) = Q_0\mathcal{H}^\infty(R,\eta^{-1}\eta_0^{-1})$. Since $\mathcal{H}^\infty(R,\eta^{-1}\eta_0^{-1})$ has no nonconstant common inner factors and since

$$\mathcal{H}^\infty(R,\eta_0)\mathcal{H}^\infty(R,\eta^{-1}\eta_0^{-1}) \subseteqq \mathcal{H}^\infty(R,\eta^{-1}),$$

the inner factor of any element in $\mathcal{H}^\infty(R,\eta_0)$ should be divided by Q_0. Obviously, $\mathcal{H}^\infty(R,\eta_0)$ contains Q_0 and therefore $\mathcal{H}^\infty(R,\eta_0) = Q_o H^\infty(R)$. By means of induction we see that $\mathcal{H}^\infty(R,\eta_0^k) = Q_0^k H^\infty(R)$ for each $k = 1, 2, \ldots$. This means, in view of Theorem 9A, Ch. V, that

$$|Q_0(a)|^k \geqq \exp\left(-\int_0^\infty B(\alpha,a)d\alpha\right) > 0$$

for $k = 1, 2, \ldots$. It follows that $|Q_0(a)| = 1$ and so Q_0 is a constant function, as was to be shown. □

6B. Let us choose a point $a \in R$, which is held fixed, and set

$$J_n(a) = J_n = S_n^{(a)}\mathcal{H}^\infty(R,(\xi_n^{(a)})^{-1}),$$

$S_n = S_n^{(a)}$ and $\xi_n = \xi_n^{(a)}$ for $n = 1, 2, \ldots$ (see 1A). Then we have

Lemma. $\cup_{n=1}^\infty J_n$ has no nonconstant common inner factors.

Proof. First we look at each $\mathcal{H}^\infty(R,\xi_n)$, $n = 1, 2, \ldots$. If Q is a common inner factor of $\mathcal{H}^\infty(R,\xi_n)$, then Q should divide S_n. If Q were nonconstant, then it should vanish at some $z_j \in Z(a;R)$ with $j > n$. So $\mathcal{H}^\infty(R,\xi_n)$ would have a common zero, contradicting to Corollary

9A, Ch. V. This means that $\mathcal{H}^\infty(R,\xi_n)$ has no nonconstant common inner factors. By the preceding lemma $\mathcal{H}^\infty(R,\xi_n^{-1})$ has no nonconstant common inner factors. Thus the function $S_n^{(a)}$ is the greatest common inner factor of J_n. Since the common inner factors of the sequence $\{S_n^{(a)}: n = 1, 2, \dots\}$ are only constant functions, we are done. □

7. The Orthocomplement of $H^\infty(d\chi)$

7A. We want to determine the orthocomplement of $H^\infty(d\chi_a)$ in the space $L^1(d\chi_a)$. For this purpose we fix a point $a \in R$ and use the following notations besides those given in 1A. We set

$$S_n'(z) = S_n^{(a)}(z)S^{(a)}(z)^{-1}\exp(-g(a,z) - i\tilde{g}(a,z))$$

for $n = 1, 2, \dots$ and

$$S'(z) = S^{(a)}(z)^{-1}\exp(-g(a,z) - i\tilde{g}(a,z)),$$

whose line bundles are denoted by ξ_n' and ξ', respectively. We also set

$$K_n(a) = K_n = S_n'\mathcal{H}^1(R,\xi_n'^{-1})$$

for $n = 1, 2, \dots$ and

$$K'(a) = K' = S'\mathcal{H}^1(R,\xi'^{-1}).$$

Let $\hat{J}_n$ (resp. $\hat{K}_n$) be the set of fine boundary functions for elements in J_n (resp. K_n), whose existence has been shown in Theorem 5E, Ch. III (resp. Lemma 3B). We further denote by $\hat{J}(a) = \hat{J}$ (resp. $\hat{K}(a) = \hat{K}$) the weak* closure of $\cup_{n=1}^\infty \hat{J}_n$ in $L^\infty(d\chi_a)$ (resp. the closure of $\cup_{n=1}^\infty \hat{K}_n$ in $L^1(d\chi_a)$), where $d\chi_a = k_b(a)d\chi(b)$. Then we have the following result concerning orthocomplements, which will be used later and has its own interest.

Theorem. For any PWS R we have the following:

(a) $\hat{K} = H^\infty(d\chi_a)^\perp$,

(b) $\hat{J} = (\hat{K}')^\perp$,

where the orthocomplementation is taken with respect to the dual pair $(L^1(d\chi_a), L^\infty(d\chi_a))$.

Proof. (a) If $f \in K_n$, then f is meromorphic on R, $f(a) = 0$ and $|f(z)|\exp(-\sum_{j=1}^n g(z_j,z))$ has a harmonic majorant. By Theorem 4B we have

$$\int_{\Delta_1} \hat{f}(b)\hat{h}(b)d\chi_a(b) = f(a)h(a) = 0$$

for any $h \in H^\infty(R)$. So, $\hat{K}_n \subseteqq H^\infty(d\chi_a)^\perp$ for every n and therefore

(9) $$\hat{K} \subseteqq H^\infty(d\chi_a)^\perp.$$

Next we take any $f^* \in \hat{K}^\perp$ $(\subseteqq L^\infty(d\chi_a))$. Since $K'J_n \subseteqq K_n$ for $n = 1, 2, \ldots$, we have

$$\int_{\Delta_1} f^*\hat{h}\hat{k}\, d\chi_a = 0$$

for any $h \in K'$ and $k \in J_n$. Thus, $f^*\hat{k} \perp \hat{K}'$. By the inverse Cauchy theorem (Theorem 1B) there exists a function $u \in H^1(R)$ such that $\hat{u} = f^*\hat{k}$. The function $f^*\hat{k}$ being bounded on Δ_1, we see that $\hat{u}$ is also bounded and thus $u \in H^\infty(R)$. If $k \neq 0$, then $\hat{k} \neq 0$ a.e. and so $f^* = \hat{u}/\hat{k}$ a.e. Namely, f^* can be viewed as the boundary function for an analytic function, say f, of bounded characteristic. As seen for instance by using Theorem 7B, Ch. III, with $R' = \mathbb{D}$ and properties of analytic functions on the disk, analytic functions of bounded characteristic on R are completely determined by their boundary functions, so that f is determined uniquely by f^*. We write $f = f_1/f_2$ with f_1, $f_2 \in H^\infty(R)$ by use of Lemma 3A. So we have seen that $kf_1/f_2 \in H^\infty(R)$ for any $k \in \cup_{n=1}^\infty J_n$. Since $\cup_{n=1}^\infty J_n$ has, by Lemma 6B, no nonconstant common inner factors, the inner factor of f_2 must be an inner factor of f_1. It follows that $f = f_1/f_2$ belongs to $H^\infty(R)$. Hence, $f^* = \hat{f}$ belongs to $H^\infty(d\chi_a)$ and consequently $\hat{K}^\perp \subseteqq H^\infty(d\chi_a)$. Since $\hat{K}$ is closed in the space $L^1(d\chi_a)$, we conclude that

$$\hat{K} = \hat{K}^{\perp\perp} \supseteqq H^\infty(d\chi_a).$$

Combining this with (9), we get the desired identity.

(b) Take any $f \in K'$, so that f is meromorphic on R, $f(a) = 0$ and $|f|g^{(a)}$ has a harmonic majorant on R. If $u \in J_n$, then fu is meromorphic on R, $(fu)(a) = 0$ and $|(fu)(z)|\exp(-\sum_{j=1}^n g(z_j,z))$ has a harmonic majorant. By Theorem 4B we have

$$\int_{\Delta_1} \hat{f}\hat{u}\, d\chi_a = (fu)(a) = 0.$$

So $\hat{f} \in \hat{J}_n^\perp$ and, as n is arbitrary, $\hat{f} \in \hat{J}$. Thus $\hat{K}' \subseteqq \hat{J}^\perp$.

Next we take any $f^* \in \hat{J}^\perp$ $(\subseteqq L^1(d\chi_a))$. Since J_n is an ideal of

$H^\infty(R)$, $h \in J_n$ and $u \in H^\infty(R)$ imply $hu \in J_n$ and so

$$\int_{\Delta_1} f^* \hat{h}\hat{u}\, d\chi_a = 0.$$

Namely, $f^*\hat{h} \in H^\infty(d\chi_a)^\perp = \hat{K}$ in view of (a). Since $K_n \subseteqq K'$ for every $n = 1, 2, \ldots$ and since $\hat{K}'$ is closed in $L^1(d\chi_a)$, we have $\hat{K} \subseteqq \hat{K}'$. So there exists a function $u \in \mathcal{H}^1(R, \xi'^{-1})$ such that $f^*h = \hat{S}'\hat{u}$. By the argument used in (a), we deduce from this observation that u/h belongs to $\mathcal{H}^1(R,\xi'^{-1})$ and thus f^* belongs to $\hat{K}'$; namely, $\hat{J}^\perp \subseteqq \hat{K}'$. Hence, $\hat{J}^\perp = \hat{K}'$ or, equivalently, $\hat{J} = \hat{K}'^\perp$. □

7B. <u>Theorem</u>. Let R be a PWS for which (DCT_a) holds. Then

$$\hat{K}'(a) = \hat{K}(a) = H^\infty(d\chi_a)^\perp.$$

<u>Proof</u>. We have shown in the proof of the preceding theorem that $\hat{K}(a)$ is included in $\hat{K}'(a)$. Now we take any $f \in K'$. Then, $fh \in K'$ for any $h \in H^\infty(R)$ and thus, by (DCT_a), $\int_{\Delta_1} \hat{f}\hat{h}\, d\chi_a = 0$. So $\hat{f} \in H^\infty(d\chi_a)^\perp$, which is equal to $\hat{K}$. Hence, $\hat{K}' \subseteqq \hat{K}$, as desired. □

NOTES

The classical Cauchy-Read theorem is discussed in detail in Chapter IV of Heins [31]. The inverse Cauchy theorem was shown in Hasumi [18]. Neville [47] has a version of this depending on another compactification. It is possible to formulate our results in terms of Wiener compactification, as was indicated in [18]. The direct Cauchy theorem is false in its full generality. It was proved in Hasumi [17] under a strong restriction. A weaker version (Theorem 4B) is in Hasumi [17].

The weak-star maximality of $H^\infty(d\chi)$ in $L^\infty(d\chi)$ is in an unpublished note [20]. The results in Subsections 6 and 7 are due to Hayashi [27].

CHAPTER VIII. SHIFT-INVARIANT SUBSPACES

The study of shift-invariant subspaces has been one of the central themes in the modern theory of Hardy classes under the decisive influence of A. Beurling's paper [2], as seen in books of Helson, Gamelin and others. Although the study of invariant subspaces over Riemann surfaces began only recently, we have already found it very interesting. Our aim is to classify closed $H^\infty(d\chi)$-modules of $L^p(d\chi)$ defined on the Martin boundary of a given surface R of Parreau-Widom type. As usual, such modules are divided into two classes: doubly invariant subspaces and simply invariant ones. It turns out that the inverse Cauchy theorem is strong enough to determine all doubly invariant subspaces. As for simply invariant subspaces, on the other hand, the problem is not so simple. We have to see when and how Beurling's theorem extends to surfaces of Parreau-Widom type. Here the direct Cauchy theorem (DCT) plays a crucial role. As a matter of fact, (DCT) is equivalent to the condition that every β-closed ideal of $H^\infty(R)$ is generated by some (multiplicative) inner function. Further discussion will be given in Ch. X.

In §1 we define invariant subspaces and their liftings to the universal covering surface. The case of the unit disk is recalled here as a preliminary. Invariant subspaces over a PWS are discussed in §2. First, doubly invariant subspaces are determined. Assuming the direct Cauchy theorem, we next show that all simply invariant subspaces are of Beurling type. Finally, it is proved that the statements (DCT_a) with $a \in R$ are all equivalent.

In this chapter, let R denote a hyperbolic Riemann surface and $d\chi$ the harmonic measure on Δ_1 for the point O--the origin of R--as defined in Ch. III. We assume as before that R is regular in the sense of potential theory.

§1. PRELIMINARY OBSERVATIONS

1. Generalities

1A. We are going to consider L^p-spaces $L^p(d\chi)$ in what follows. By Theorem 3A, Ch. IV, the map $f \to \hat{f}$ gives an isometric linear injection of $H^p(R)$ into $L^p(d\chi)$ for every p with $0 < p \leqq \infty$, where the space $H^p(R)$ is equipped with the norm (the quasinorm, if $0 < p < 1$)

$$\|f\|_p = \begin{cases} ((\mathrm{LHM}(|f|^p))(0))^{1/p} & \text{if } 0 < p < \infty \\ \sup\{|f(z)|: z \in R\} & \text{if } p = \infty. \end{cases}$$

The space $H^p(R)$ is thus identified with a subspace of $L^p(d\chi)$, which is denoted by $H^p(d\chi)$. Moreover, we set $H_0^p(R) = \{f \in H^p(R): f(0) = 0\}$ and $H_0^p(d\chi) = \{\hat{f}: f \in H_0^p(R)\}$ for $0 < p \leqq \infty$.

A subspace M of $L^p(d\chi)$ is called an invariant subspace if it is closed (weakly* closed, if $p = \infty$) and if $H^\infty(d\chi)\cdot M \subseteqq M$. Such an invariant subspace M is said to be doubly invariant (resp. simply invariant) if $H_0^\infty(d\chi)\cdot M$ is dense (resp. not dense) in M. When $p = \infty$, the density should also be taken in the sense of the weak* topology $w(L^\infty(d\chi), L^1(d\chi))$ of $L^\infty(d\chi)$.

Let f be any measurable function on Δ_1. We say that a measurable subset E of Δ_1 supports f if $f = 0$ a.e. on $\Delta_1 \setminus E$. Then there exists a measurable subset, say E_f, of the smallest measure that supports f. Such a set is determined uniquely by f up to equivalence and is called the support of f. For a set M of measurable functions, the support of M is defined to be a measurable set $E_M \subseteqq \Delta_1$ of the smallest measure such that E_M supports every $f \in M$. The set E_M is again determined uniquely by M up to equivalence.

1B. Let $\phi = \phi_R$ be a universal covering map: $\mathbb{D} \to R$ with $\phi(0) = 0$ and let $T = T_R$ be the group of cover transformations for ϕ. By Theorem 7B, Ch. III, $\hat{\phi}$ is a measurable map defined a.e. on $\mathbb{T}$ with values in Δ_1 such that $\sigma(\hat{\phi}^{-1}(A)) = \chi(A)$ for any measurable set A in Δ_1. For $0 < p < \infty$, $[\cdot]_p$ denotes the closure operation in $L^p(d\chi)$ or in $L^p(d\sigma)$. $[\cdot]_\infty$ denotes the operation of taking weak* closure in $L^\infty(d\chi)$ or in $L^\infty(d\sigma)$.

Let M be a subspace of $L^p(d\chi)$. Then $M \circ \hat{\phi} = \{f^* \circ \hat{\phi}: f^* \in M\}$ is a subspace of $L^p(d\sigma)_T$, where $L^p(d\sigma)_T$ is the set of T-invariant elements of $L^p(d\sigma)$. Let $\{M\}_p$ be the invariant subspace of $L^p(d\sigma)$ which is generated by $M \circ \hat{\phi}$. $\{M\}_p$ is the closure (weak* closure, if $p = \infty$) in $L^p(d\sigma)$ of the linear envelope of $H^\infty(d\sigma)(M \circ \hat{\phi})$.

We note in passing that $H^\infty(d\chi) \circ \hat{\phi}$ is equal to the set $H^\infty(d\sigma)_T$ of T-invariant elements in $H^\infty(d\sigma)$.

Lemma. Let M be a doubly invariant subspace of $L^p(d\chi)$; then $\{M\}_p$ is a doubly invariant subspace of $L^p(d\sigma)$.

Proof. Since the map $f^* \to f^* \circ \hat{\phi}$ is an isometry of $L^p(d\chi)$ into $L^p(d\sigma)$ and since the linear envelope of $H_0^\infty(d\chi)\cdot M$ is dense in M, the linear envelope of $H_0^\infty(d\sigma)_T(M \circ \hat{\phi})$ is dense in $M \circ \hat{\phi}$. This implies that $M \circ \hat{\phi}$ is included in $e^{i\theta}\{M\}_p$. Since $e^{i\theta}\{M\}_p$ is invariant, it thus includes $\{M\}_p$ and therefore $e^{i\theta}\{M\}_p = \{M\}_p$. This proves the lemma for $p < \infty$.

Next we suppose $p = \infty$, so that $H_0^\infty(d\chi)\cdot M$ is $w(L^\infty(d\chi), L^1(d\chi))$-dense in M. The linear envelope $\mathrm{lin}(H_0^\infty(d\chi)\cdot M)$ of $H_0^\infty(d\chi)\cdot M$ is dense in $[M]_2$ with respect to the L^2-norm topology. To see this, we first note that $H_0^\infty(d\chi)\cdot M$ is $w(L^2(d\chi), L^2(d\chi))$-dense in M. If $[\cdot]_{2w}$ denotes the weak closure in $L^2(d\chi)$, then

$$M \subseteqq [H_0^\infty(d\chi)\cdot M]_{2w} \subseteqq [\mathrm{lin}(H_0^\infty(d\chi)\cdot M)]_{2w}.$$

Since the operations $[\cdot]_2$ and $[\cdot]_{2w}$ coincide for any convex subset of $L^2(d\chi)$ (Dunford and Schwartz [8], p. 422), we infer that $M \subseteqq [\mathrm{lin}(H_0^\infty(d\chi)\cdot M)]_2$. Passing to the universal covering surface and using Theorem 7B, Ch. III, we see that $M \circ \hat{\phi} \subseteqq [\mathrm{lin}(H_0^\infty(d\sigma)_T\cdot(M \circ \hat{\phi}))]_2$ and a fortiori

$$M \circ \hat{\phi} \subseteqq [\mathrm{lin}(H_0^\infty(d\sigma)\cdot(M \circ \hat{\phi}))]_2.$$

Clearly, the right-hand side is an invariant subspace of $L^2(d\sigma)$. So we have

$$\{M\}_\infty \subseteqq \{M\}_2 \subseteqq [\mathrm{lin}(H_0^\infty(d\sigma)\cdot\{M\}_2)]_2 = e^{i\theta}\{M\}_2.$$

This shows that $\{M\}_2$ is doubly invariant and also that $\{M\}_2$ is equal to the L^2-closure of $\{M\}_\infty$. In view of Lemma 2B below, the latter fact implies then that $\{M\}_\infty = \{M\}_2 \cap L^\infty(d\sigma)$. Hence, $\{M\}_\infty$ must be doubly invariant. □

1C. Here is another simple fact:

Lemma. Let M be any set of measurable functions on Δ_1 and let Σ (resp. S) be the support of M (resp. $M \circ \hat{\phi}$). Then $C_S = C_\Sigma \circ \hat{\phi}$ a.e., where C_Σ (resp. C_S) denotes the characteristic function of Σ (resp. S).

2. Shift-Invariant Subspaces on the Unit Disk

2A. A subspace M of $L^p(d\sigma)$ is invariant if and only if it is closed (weakly* closed, if $p = \infty$) and $e^{i\theta}M \subseteqq M$.

Lemma. If $M \subseteq L^p(d\sigma)$ is invariant, then $hM \subseteq M$ for any $h \in H^\infty(d\sigma)$.

Proof. Let $f \in M$ and $h \in H^\infty(d\sigma)$. By Appendix A.1.4 there is a sequence $\{h_n: n = 1, 2, \ldots\}$ in $P(\mathbb{T})$ such that $\|h_n\|_\infty \leqq \|h\|_\infty$ and h_n tend to h a.e. on $\mathbb{T}$. Since M is invariant, $h_n f \in M$ for $n = 1, 2, \ldots$. If $0 < p < \infty$, then

$$|hf - h_n f|^p = |h - h_n|^p |f|^p \leqq 2^p \|h\|_\infty^p |f|^p \in L^1(d\sigma)$$

for $n = 1, 2, \ldots$ and $hf - h_n f \to 0$ a.e. By Lebesgue's dominated convergence theorem we have $\int |hf - h_n f|^p d\sigma \to 0$. Since M is closed, hf belongs to M, as desired. If $p = \infty$, then, for any $k \in L^1(d\sigma)$,

$$|(hf - h_n f)k| \leqq 2\|h\|_\infty \|f\|_\infty |k| \in L^1(d\sigma)$$

for $n = 1, 2, \ldots$ and $(hf - h_n f)k \to 0$ a.e. So $\int (hf - h_n f)k d\sigma \to 0$ as $n \to \infty$. This means that $h_n f \to hf$ in the weak* topology of $L^\infty(d\sigma)$. Hence $hf \in M$. □

2B. Lemma. (a) If M is an invariant subspace of $L^p(d\sigma)$ with $0 < p < \infty$, then, for $p < q \leqq \infty$, $M \cap L^q(d\sigma)$ is an invariant subspace of $L^q(d\sigma)$ and M is the closure of $M \cap L^q(d\sigma)$ in $L^p(d\sigma)$.

(b) If M is an invariant subspace of $L^p(d\sigma)$ with $0 < p \leqq \infty$, then, for any $0 < q < p$, the closure N of M in $L^q(d\sigma)$ is an invariant subspace of $L^q(d\sigma)$ and $M = N \cap L^p(d\sigma)$.

Proof. (a) First we show that $M \cap L^q(d\sigma)$ is closed. If $q < \infty$, then the inequality $\|f\|_p \leqq \|f\|_q$ implies that $M \cap L^q(d\sigma)$ is closed in the space $L^q(d\sigma)$. To see that $M \cap L^\infty(d\sigma)$ is weakly* closed, take any net $\{f_\lambda\} \subset M \cap L^\infty(d\sigma)$ which is weakly* convergent to an $f \in L^\infty(d\sigma)$. When $p > 1$, the net converges to f in the weak topology $w(L^p(d\sigma), L^{p'}(d\sigma))$ with $p' = p/(p-1)$. The duality theory of Banach spaces shows that the weak closure of $M \cap L^\infty(d\sigma)$ in $L^p(d\sigma)$ is equal to its L^p-norm closure. Since M is closed in $L^p(d\sigma)$, we conclude that $f \in M$. Suppose now that $p \leqq 1$. The above argument shows that f belongs to the L^2-closure of $M \cap L^\infty(d\sigma)$. Since the L^p-closure always contains the L^2-closure and M is L^p-closed, we see that $f \in M$. In either case, f belongs to $M \cap L^\infty(d\sigma)$ and thus $M \cap L^\infty(d\sigma)$ is weakly* closed.

To see the latter half of the statement (a), take any $f \in M$. For each $n = 1, 2, \ldots$, let $u_n = \max\{\log^+|f| - \log n, 0\}$. Then, on the set $\{|f| \geqq n\}$, $0 \leqq u_n = \log(|f|/n) \leqq p^{-1}|f|^p/n^p$. Since $u_n = 0$ on the set $\{|f| < n\}$, we have $0 \leqq u_n \leqq p^{-1}|f|^p/n^p$ on $\mathbb{T}$ and thus

$$\int_{\mathbb{T}} u_n d\sigma \leq p^{-1} n^{-p} \|f\|_p^p \to 0$$

as $n \to \infty$. Let $\tilde{u}_n$ be the harmonic conjugate of u_n with $\tilde{u}_n(0) = 0$. Then by Appendix A.2.2 $\tilde{u}_n$ belongs to $L^s(d\sigma)$, $0 < s < 1$, and

$$(1) \qquad \|\tilde{u}_n\|_s \leq c_s \|u_n\|_1,$$

where $c_s = 2^{(s+1)/s}(\cos \pi s/2)^{-1/s}$. Fixing an s with $0 < s < 1$, we see that $\|\tilde{u}_n\|_s \to 0$ as $n \to \infty$. This means that $u_n + i\tilde{u}_n \to 0$ in measure. So, there exists a subsequence $\{u_{n(j)} + i\tilde{u}_{n(j)}: j = 1, 2, \dots\}$ converging a.e. to 0. We set $h_j = \exp(-u_{n(j)} - i\tilde{u}_{n(j)})$. Then we have $\|h_j\|_\infty \leq 1$ and $h_j \to 1$ a.e. Since $|h_j f| = |f| \exp(-u_{n(j)}) \leq n(j)$ and since $h_j f \in M$ by Lemma 2A, $h_j f \in M \cap L^\infty(d\sigma) \subseteq M \cap L^q(d\sigma)$. Moreover,

$$|f - h_j f|^p = |(1 - h_j) f|^p \leq 2^p |f|^p \in L^1(d\sigma)$$

for every j and also $f - h_j f \to 0$ a.e. By Lebesgue's dominated convergence theorem, $\int |f - h_j f|^p d\sigma \to 0$ as $j \to \infty$. Hence M is equal to the closure of $M \cap L^q(d\sigma)$ in $L^p(d\sigma)$.

(b) Clearly, N is an invariant subspace of $L^p(d\sigma)$. To see the latter half, we take any $f \in N \cap L^p(d\sigma)$, so that there exists a sequence $\{f_n: n = 1, 2, \dots\}$ in M with $\|f - f_n\|_q \to 0$. Since $f_n \to f$ in measure, we may assume, by passing to a subsequence if necessary, that $f_n \to f$ a.e. We set $u_n = \max\{\log^+|f_n| - \log|f|, 0\}$ if $1 \leq |f| \leq n$ and $= \log^+|f_n|$ if $|f| < 1$ or $|f| > n$. Then, $|f_n| \exp(-u_n) \leq \min\{|f|, n\} + 1$ for $n = 1, 2, \dots$ and

$$\int u_n d\sigma \leq \int \left| \log^+|f_n| - \log^+|f| \right| d\sigma + \int_{|f|>n} \log^+|f| d\sigma$$

$$= I_1 + I_2, \text{ say.}$$

Then $I_1 \leq \|f_n - f\|_q$ for $q \geq 1$ and $\leq q^{-1}\|f_n - f\|_q^q$ for $0 < q < 1$.

We now consider the case $p < \infty$. Then we see that

$$I_2 \leq p^{-1} \int_{|f|>n} |f|^p d\sigma.$$

Thus, $I_1 + I_2 \to 0$ as $n \to \infty$ and consequently $\int u_n d\sigma \to 0$. Let $\tilde{u}_n$ be the harmonic conjugate of u_n with $\tilde{u}_n(0) = 0$. Then by (1) we have $\|u_n\|_s \leq c_s \|u_n\|_1$ for $0 < s < 1$. As in (a), we find a subsequence $\{u_{n(j)} + i\tilde{u}_{n(j)}\}$, which converges to 0 a.e. on $\mathbb{T}$. We set $h_j =$

$\exp(-u_{n(j)} - i\tilde{u}_{n(j)})$ and see that $h_j f_{n(j)} \in M \cap L^\infty(d\sigma)$, $h_j f_{n(j)} \to f$ a.e. and $|h_j f_{n(j)} - f|^p \leqq 2|f|^p + 1 \in L^1(d\sigma)$. By Lebesgue's dominated convergence theorem, $h_j f_{n(j)} \to f$ in $L^p(d\sigma)$. Since M is closed in $L^p(d\sigma)$, we have $f \in M$ and therefore $N \cap L^p(d\sigma)$ is included in M. The converse inclusion being trivial, we have got the desired result.

We finally suppose $p = \infty$. So f is bounded. If n is large enough, then $I_2 = 0$ and thus $\int u_n d\sigma \to 0$ as $n \to \infty$. Proceeding as above, we see that $h_j f_{n(j)} \in M \cap L^\infty(d\sigma) = M$, $h_j f_{n(j)} \to f$ a.e. and $|h_j f_{n(j)} - f| \leqq 2|f| + 1 \leqq 2\|f\|_\infty + 1$. Again by the dominated convergence theorem, we conclude that $h_j f_{n(j)} \to f$ in $L^\infty(d\sigma)$ in the weak* topology $w(L^\infty(d\sigma), L^1(d\sigma))$. Since M is weakly* closed in $L^\infty(d\sigma)$, we have $f \in M$ and so $N \cap L^\infty(d\sigma) \subseteqq M$. The converse inclusion is true as before and we are done. □

2C. As is easily seen, an invariant subspace M of $L^p(d\sigma)$ is doubly invariant (resp. simply invariant) if $e^{i\theta}M = M$ (resp. $e^{i\theta}M \neq M$).

Theorem. Let M be a closed (weakly* closed, if $p = \infty$) subspace of $L^p(d\sigma)$, $0 < p \leqq \infty$.

(a) M is doubly invariant if and only if there exists a measurable subset S of $\mathbb{T}$ such that $M = C_S L^p(d\sigma)$, where C_S is the characteristic function of S. The set S is determined uniquely by M up to a set of Lebesgue measure zero.

(b) M is simply invariant if and only if there exists a function $q \in L^\infty(d\sigma)$ with $|q| = 1$ a.e. such that $M = qH^p(d\sigma)$. The function q is determined uniquely up to a constant factor of modulus one.

Proof. (i) We begin with the case $p = 2$. Let M be an invariant subspace of $L^2(d\sigma)$. Suppose first that $e^{i\theta}M = M$. Then let q be the orthogonal projection of the constant function $1 \in L^2(d\sigma)$ to the subspace M. So $1 - q \perp e^{in\theta}q$ for all integers n, i.e.

$$\int_{\mathbb{T}} (1 - q)\bar{q}e^{in\theta} d\sigma = 0$$

for all integers n. Since $\{e^{in\theta}\}$ forms an orthonormal basis of the Hilbert space $L^2(d\sigma)$, $(1 - q)\bar{q} = 0$ a.e., so that q is the characteristic function of a measurable subset, say S, of $\mathbb{T}$. Our assumption implies that $e^{-i\theta}M = M$ and therefore that $hq \in M$ for any h in $P(\mathbb{T}) + \overline{P(\mathbb{T})}$. Since M is closed in $L^2(d\sigma)$, Appendix A.1.4 shows that $hq \in M$ for any $h \in L^\infty(d\sigma)$ and consequently for any $h \in L^2(d\sigma)$. So we have $qL^2(d\sigma) \subseteqq M$. To see the reverse inclusion, take any $f \in M$

which is orthogonal to $qL^2(d\sigma)$. Then $fq = 0$ a.e. Moreover, we have $1 - q \perp e^{in\theta}f$ for all integers n and consequently $(1-q)f = 0$ a.e. Hence $f = 0$ a.e., which we had to show. The uniqueness of S, up to a null set, is almost obvious.

Suppose next that $e^{i\theta}M \neq M$. Since $e^{i\theta}M$ is a closed subspace of M, there exists a function $q \in M$, with $\|q\|_2 = 1$, which is orthogonal to $e^{i\theta}M$. Then $q \perp qe^{in\theta}$ for $n = 1, 2, \ldots$. This means that $|q|^2$ is equal to a constant a.e. Our assumption $\|q\|_2 = 1$ then implies that $|q| = 1$ a.e. Since $e^{i\theta}M \subseteqq M$, we have $qP(\mathbb{T}) \subseteqq M$. It follows that $qH^2(d\sigma) \subseteqq M$. Conversely, take any $f \in M$ which is orthogonal to the space $qH^2(d\sigma)$. Then $f \perp qe^{in\theta}$ for $n = 0, 1, \ldots$ and also $q \perp fe^{in\theta}$ for $n = 1, 2, \ldots$. So $f\bar{q}$ is orthogonal to $e^{in\theta}$ with all integral n and hence vanishes a.e. Since $|q| = 1$ a.e., $f = 0$ a.e. Hence $M \subseteqq qH^2(d\sigma)$ and consequently $M = qH^2(d\sigma)$. To see the uniqueness of q, we suppose that $q_1H^2(d\sigma) = q_2H^2(d\sigma)$ with $|q_1| = |q_2| = 1$ a.e. Then both $q_1\bar{q}_2$ and $\bar{q}_1q_2$ belong to $H^2(d\sigma)$ and hence $q_1\bar{q}_2$ is equal a.e. to a constant of modulus one, as desired.

Finally, it is easy to see that $C_SL^2(d\sigma)$ (resp. $qH^2(d\sigma)$ with $|q| = 1$ a.e.) gives a doubly (resp. simply) invariant subspace of $L^2(d\sigma)$. This proves the case $p = 2$.

(ii) Let $0 < p \leqq \infty$, $p \neq 2$, and let M be an invariant subspace of $L^p(d\sigma)$. Suppose first that $0 < p < 2$. Then, by Lemma 2B, the intersection $M \cap L^2(d\sigma)$ is an invariant subspace and M is the closure of $M \cap L^2(d\sigma)$ in $L^p(d\sigma)$. By (i) $M \cap L^2(d\sigma)$ is equal either to $C_SL^2(d\sigma)$ with a measurable $S \subseteqq \mathbb{T}$ or to $qH^2(d\sigma)$ with $|q| = 1$ a.e. By taking L^p-closure, we find that $M = C_SL^p(d\sigma)$ or $= qH^p(d\sigma)$. Next suppose that $2 < p \leqq \infty$ and let N be the L^2-closure of M. Then, by (i), $N = C_SL^2(d\sigma)$ with a measurable $S \subseteqq \mathbb{T}$ or $= qH^2(d\sigma)$ with $|q| = 1$ a.e. It follows from Lemma 2B that $M = N \cap L^p(d\sigma) = C_SL^p(d\sigma)$ or $= qH^p(d\sigma)$.

Finally we have to show that $C_SL^p(d\sigma)$ (resp. $qH^p(d\sigma)$) is a doubly (resp. simply) invariant subspace of $L^p(d\sigma)$ and also that C_S (resp. q) is unique in the sense described in the theorem. The result is obvious for $C_SL^p(d\sigma)$. As for $qH^p(d\sigma)$ the result follows from Lemma 2B and the case $p = 2$, which has already been established. □

§2. INVARIANT SUBSPACES

3. Doubly Invariant Subspaces

3A. We now consider the case of Riemann surfaces and use the notations mentioned in 1A. First we show the following

Theorem. Let Σ be a measurable subset of Δ_1 and let C_Σ denote the characteristic function of Σ. Then $M = C_\Sigma L^p(d\chi)$ is a doubly invariant subspace of $L^p(d\chi)$ for $0 < p \leqq \infty$.

Proof. Suppose that $0 < p < \infty$. Since it is clear that M is closed and $H^\infty(d\chi)\cdot M \subseteqq M$, we have only to show that $H_0^\infty(d\chi)\cdot M$ is dense in M. Take a nonzero $h \in H_0^\infty(R)$; then its boundary function $\hat{h}$ vanishes only on a negligible subset of Δ_1. So there exists a set $E \subseteqq \Delta_1$ with negligible complement such that $0 < |\hat{h}| < \infty$ on E. Set $u_n(b) = h(b)^{-1}$ if $b \in E$ with $|\hat{h}(b)| \geqq 1/n$ and $= 0$ otherwise on Δ_1. And for any given $f \in M$ we set $f_n = \hat{h}u_n f$. Since u_n is bounded,

$$u_n f \in u_n C_\Sigma L^p(d\sigma) \subseteqq C_\Sigma L^p(d\sigma) = M$$

and therefore $f_n \in H_0^\infty(d\sigma)\cdot M$. The definition of u_n implies that $|f - f_n| \leqq |f|$ and $f - f_n \to 0$ a.e. By Lebesgue's dominated convergence theorem we have $\|f - f_n\|_p \to 0$ as $n \to \infty$. Hence $H_0^\infty(d\chi)\cdot M$ is dense in M. The case $p = \infty$ is almost similar and the proof is omitted. □

3B. Here is our first main result in this section.

Theorem. Suppose that R is a regular PWS. Let M be a doubly invariant subspace of $L^p(d\chi)$ and let Σ be the support of M. Then $M = C_\Sigma L^p(d\chi)$ if one of the following holds:

(a) $1 \leqq p \leqq \infty$;

(b) $0 < p < 1$ and $\chi(\Sigma) < 1$.

Since $M \subseteqq C_\Sigma L^p(d\chi)$ is always true, the problem is to prove the reverse inclusion. By Lemma 1B $\{M\}_p$ is also doubly invariant and so, by Theorem 2C-(a), $\{M\}_p = C_S L^p(d\sigma)$ for some measurable subset S of $\mathbb{T}$. Since S is regarded as the support of $M \circ \hat{\Phi}$, $C_\Sigma \circ \hat{\Phi} = C_S$ a.e. by Lemma 1C. The proof is now divided into two cases (a) and (b).

3C. Suppose first that $1 \leqq p \leqq \infty$. Let $s^* \in L^{p'}(d\chi)$ with $p' = p/(p-1)$ be any element which is orthogonal to M. Take any nonzero $B \in S^{(0)}\mathcal{H}^\infty(R,(\xi^{(0)})^{-1})$ in the notation described in Ch. VII, 1A. If

u is any meromorphic function on R such that $|u|g^{(0)}$ is bounded, then $Bu \in H^\infty(R)$ and so $\hat{B}\hat{u}f^* \in M$ for any $f^* \in M$. Thus we have

$$(2) \qquad \int_{\Delta_1} \hat{B}\hat{u}f^*s^*d\chi = 0.$$

By the inverse Cauchy theorem (Theorem 1B, Ch. VII) there exists a function $k \in H^1(R)$ such that $\hat{k} = \hat{B}f^*s^*$ a.e. on Δ_1. Setting $u \equiv 1$ in (2), we have

$$(3) \qquad k(0) = \int_{\Delta_1} \hat{k}\, d\chi = 0.$$

On the other hand, the formula (12) in Ch. III shows that

$$H[(\hat{B}f^*s^*) \circ \hat{\phi}] = H[\hat{B}f^*s^*] \circ \phi = H[\hat{k}] \circ \phi = k \circ \phi \in H^1(\mathbb{D}).$$

For any $v \in H^\infty(\mathbb{D})$ we thus have, by (3) and Theorem 7B, Ch. III,

$$\int_{\mathbb{T}} v(e^{i\theta})((\hat{B}f^*s^*) \circ \hat{\phi})(e^{i\theta})d\sigma(\theta) = v(0)(k \circ \phi)(0)$$

$$= v(0)k(0) = 0.$$

By taking L^p-limits in $v\cdot(f^* \circ \hat{\phi})$, we see

$$\int_{\mathbb{T}} ((\hat{B}s^*) \circ \hat{\phi})(e^{i\theta})f_1(e^{i\theta})d\sigma(\theta) = 0$$

for any $f_1 \in \{M\}_p$. Since $\{M\}_p = C_S L^p(d\sigma)$, we conclude that $(\hat{B}s^*) \circ \hat{\phi}$ must vanish a.e. on S and consequently $\hat{B}s^* = 0$ a.e. on Σ. Since B is nonzero, $\hat{B}$ can vanish only on a set of χ-measure zero in Δ_1. Hence $s^* = 0$ a.e. on Σ. This implies that $s^* \perp C_\Sigma L^p(d\chi)$. We thus have $C_\Sigma L^p(d\chi) \subseteqq M$, so that the case (a) is verified.

3D. In order to deal with the case (b), we prove

<u>Lemma</u>. Let $f^* \in L^p(d\chi)$, $0 < p < \infty$. Then there exists a nonzero element $\hat{h} \in H^\infty(d\chi)$ such that $\hat{h}f^*$ is bounded.

<u>Proof</u>. (Lemma) Let $u^* = \log^+|f^*|$. Then u^* is positive and summable, for

$$0 \leqq u^* = \log|f^*| = p^{-1}\log|f^*|^p \leqq p^{-1}|f^*|^p$$

on the set $\{|f^*| > 1\}$. Let $u = H[u^*]$ be the harmonic extension to R of u^* and let v be the harmonic conjugate of u. Then u is a positive harmonic function on R and therefore e^{-u} is a bounded l.a. m. on R. If ξ denotes the line bundle associated to e^{-u}, then there exists an element $h_1 \in \mathcal{H}^\infty(R,\xi^{-1})$ with $\|h_1\|_\infty = 1$, for R is a PWS. Then $h_1 e^{-u-iv}$ determines a single-valued function, say h, belonging to $H^\infty(R)$ and we have $|\hat{h}f^*| \leq e^{-u^*}|f^*| \leq 1$, as desired. □

Let M be a doubly invariant subspace of $L^p(d\chi)$, $0 < p < 1$, whose support Σ has χ-measure < 1. Set $N = M \cap L^2(d\chi)$; then N is an invariant subspace of $L^2(d\chi)$. Since any nonzero element of $H^\infty(d\chi)$ vanishes only on a set of measure zero, the above lemma implies that M and N have the same support, namely Σ. So we have $N \subseteq C_\Sigma L^2(d\chi)$. On the other hand, the set S given in 3B is the support of $\{N\}_2$, so that $\{N\}_2 \subseteq C_S L^2(d\sigma)$. Since $\sigma(S) = \chi(\Sigma) < 1$, we see, in view of Theorem 2C, that $\{N\}_2$ must be doubly invariant. Hence $\{N\}_2 = C_S L^2(d\sigma)$. The argument in 3C then shows that $N = C_\Sigma L^2(d\chi)$ and therefore

$$C_\Sigma L^2(d\chi) = M \cap L^2(d\chi) \subseteq M \subseteq C_\Sigma L^p(d\chi).$$

Since $L^p(d\chi) = [L^2(d\chi)]_p$, we get the desired result $M = C_\Sigma L^p(d\chi)$. So the case (b) is proved. □

3E. <u>Problem</u>. Suppose that M is a doubly invariant subspace of $L^p(d\chi)$ with $0 < p < 1$ and that the support of M coincides with Δ_1. Can one show that $M = L^p(d\chi)$?

4. Simply Invariant Subspaces

4A. In what follows we work on a regular PWS R. Let us choose a point $a \in R$, which is held fixed. We denote by $F_a(R)$ the fundamental group of R, consisting of closed curves issuing from the point a, and consider a function Q on the product space $\Delta_1 \times F_a(R)$ with the following properties:

(A1) for each $\gamma \in F_a(R)$, $b \to Q(b;\gamma)$ is measurable on Δ_1;

(A2) there exists a line bundle ξ over R such that

$$Q(\cdot;\gamma_1) = \xi(\gamma_1)\xi(\gamma_2)^{-1}Q(\cdot;\gamma_2) \quad \text{a.e. on } \Delta_1$$

for any $\gamma_1, \gamma_2 \in F_a(R)$. Here ξ represents a line bundle and the corresponding character at the same time (Ch. II, 2B).

Such a function Q is called an <u>m-function</u> (m for "multiplicative") on $\Delta_1 \times F_a(R)$ of character ξ and will be used in order to describe the boundary values of a multiplicative analytic function. We say that two m-functions Q_1 and Q_2 are <u>equivalent</u> if they have the same bundle (or character), say ξ, and there exists a $\gamma_0 \in F_a(R)$ such that

$$Q_1(\cdot;\gamma) = \xi(\gamma_0)Q_2(\cdot;\gamma) \quad \text{a.e. on} \quad \Delta_1$$

for every $\gamma \in F_a(R)$. We write $Q_1 \equiv Q_2$ if Q_1 and Q_2 are equivalent.

To each $f \in \mathcal{H}^p(R,\xi)$ we associate an m-function $\check{f}$ on $\Delta_1 \times F_a(R)$ as follows: take a branch, say f_0, of f on the Green star region $\mathbb{G}'(a) = \mathbb{G}'(R,a)$ and denote it by $f_0(z;\mathbb{1})$, where $\mathbb{1}$ denotes the identity of the group $F_a(R)$. For each $\gamma \in F_a(R)$ we set

$$f_0(z;\gamma) = \xi(\gamma)f_0(z;\mathbb{1})$$

for $z \in \mathbb{G}'(a)$. Then, by Theorem 4A, Ch. VI, $\check{f}_0(\ell;\gamma)$ exists for almost every Green line $\ell \in \Lambda(a)$. We set

$$\check{f}_0(b;\gamma) = \check{f}_0(\ell;\gamma)$$

for $b = b_\ell$, $\ell \in \Lambda(a)$, if the right-hand side exists. So $\check{f}_0(b;\gamma)$ is determined a.e. on Δ_1 and

$$\check{f}_0(b;\gamma) = \xi(\gamma)\check{f}_0(b;\mathbb{1})$$

on Δ_1. If we choose another branch, say f_1, of f and define functions $f_1(z;\gamma)$ and $\check{f}_1(b;\gamma)$ as above, then there exists a $\gamma_0 \in F_a(R)$ such that $f_0(z;\gamma) = \xi(\gamma_0)f_1(z;\gamma)$ and $\check{f}_0(b;\gamma) = \xi(\gamma_0)\check{f}_1(b;\gamma)$; namely, $\check{f}_0$ and $\check{f}_1$ are equivalent. We call any one of $\check{f}$ the <u>boundary m-function</u> for f.

4B. In the following we consider the case $a = O$. In fact, as we shall see, results are independent of the choice of a. We first show the following

<u>Theorem</u>. Let Q be an m-function on $\Delta_1 \times F_O(R)$ of character ξ with $|Q(\cdot;\gamma)| = 1$ a.e. for each $\gamma \in F_O(R)$. Then $M = \{f^* \in L^p(d\chi): f^*/Q \equiv \check{h}$ for some $h \in \mathcal{H}^p(R,\xi^{-1})\}$, $1 \leqq p \leqq \infty$, is simply invariant.

<u>Proof</u>. Since R is a PWS, there exist $u_1 \in \mathcal{H}^\infty(R,\xi)$ and $u_2 \in \mathcal{H}^\infty(R,\xi^{-1})$ with $|u_1(O)| \neq 0$ and $|u_2(O)| \neq 0$ (Corollary 9A, Ch. V). Let $u_j(z;\gamma)$ and $\check{u}_j(b;\gamma)$, $j = 1, 2$, be constructed from u_j by the procedure described in 4A with $a = O$. Then $\check{u}_1(b;\gamma)/Q(b;\gamma)$ and $\check{u}_2(b;\gamma)Q(b;\gamma)$ are independent of $\gamma \in F_O(R)$ and define single-valued functions v_1^* and v_2^* in $L^\infty(d\chi)$, respectively. $u_1(z;\gamma)u_2(z;\gamma)$ is

also independent of γ and defines a function, say u_3, in $H^\infty(R)$.

We now take any $f^* \in M$, so that $f^*/Q \equiv \check{h}$ for an $h \in \mathcal{H}^p(R,\xi^{-1})$. If $\check{h}(b;\gamma)$ denotes the boundary m-function for h satisfying $\check{h}(b;\gamma) = f^*(b)/Q(b;\gamma)$ and if $h(z;\gamma)$ is the multiplicative analytic function corresponding to $\check{h}(b;\gamma)$, then $u_1(z;\gamma)h(z;\gamma)$ is independent of γ and defines an analytic function, say h_1, in $H^p(R)$. For any $k \in H_0^\infty(R)$, we have

$$\hat{k}(b)f^*(b)v_1^*(b) = \hat{k}(b)(f^*(b)/Q(b;\gamma))Q(b;\gamma)v_1^*(b)$$
$$= \hat{k}(b)\check{h}(b;\gamma)\check{u}_1(b;\gamma) = \hat{k}(b)\hat{h}_1(b)$$

a.e. on Δ_1 (cf. Theorem 6A, Ch. VI) and therefore

$$\int_{\Delta_1} \hat{k}(b)f^*(b)v_1^*(b)d\chi(b) = \int_{\Delta_1} \hat{k}(b)\hat{h}_1(b)d\chi(b) = k(\mathcal{O})h_1(\mathcal{O}) = 0.$$

Namely, $v_1^* \perp H_0^\infty(d\chi)\cdot M$. But, on the other hand, we have $v_2^* \in M$ and

$$\int_{\Delta_1} v_2^*(b)v_1^*(b)d\chi(b) = \int_{\Delta_1} \check{u}_2(b;\gamma)\check{u}_1(b;\gamma)d\chi(b)$$

$$= \int_{\Delta_1} \hat{u}_3(b)d\chi(b) = u_3(\mathcal{O}) \neq 0,$$

for $|u_3(\mathcal{O})| = |u_1(\mathcal{O})||u_2(\mathcal{O})| \neq 0$. Thus $H_0^\infty(d\chi)\cdot M$ cannot be dense in M, as was to be proved. □

4C. In order to proceed further, we need the direct Cauchy theorem (DCT_a) (Ch. VII, 3C), which we use here as a hypothesis.

<u>Theorem</u>. Suppose that ($DCT_\mathcal{O}$) holds. If M is a simply invariant subspace of $L^p(d\chi)$, $1 \leq p \leq \infty$, then there exists an m-function Q on $\Delta_1 \times F_\mathcal{O}(R)$ of some character ξ with $|Q(\cdot;\gamma)| = 1$ a.e. on Δ_1 such that

$$M = \{f^* \in L^p(d\chi)\colon f^*/Q \equiv \check{h} \text{ for some } h \in \mathcal{H}^p(R,\xi^{-1})\}.$$

The proof is divided into two parts. In the first part we define an m-function Q on $\Delta_1 \times F_\mathcal{O}(R)$ of modulus one a.e. in such a way that $M \subseteq \{f^* \in L^p(d\chi)\colon f^*/Q \equiv \check{h}$ for some $h \in \mathcal{H}^p(R,\xi^{-1})\}$, where ξ is the character of Q. And in the second part we verify the reverse inclusion.

4D. Let M be a simply invariant subspace of $L^p(d\chi)$ with $1 \leq p \leq \infty$. Then $\{M\}_p$ is a simply invariant subspace of $L^p(d\sigma)$. If it were doubly invariant, then M would turn out to be doubly invariant by our argument in 3C. So, by Theorem 2C, there exists a function $q \in L^\infty(d\sigma)$ with $|q| = 1$ a.e. on $\mathbb{T}$ such that

(4) $$\{M\}_p = qH^p(d\sigma).$$

Since $\{M\}_p$ is T-invariant, the uniqueness of q shows that $q \circ \tau = c(\tau)q$ a.e. on $\mathbb{T}$ for every $\tau \in T$, where $c(\tau)$ is a uniquely defined complex number of modulus one. The correspondence $\tau \to c(\tau)$ is easily seen to be a group homomorphism of T into the multiplicative group $\mathbb{T}$ of complex numbers of modulus one. By use of the canonical isomorphism $\tau \to \gamma_\tau$ of T onto $F_O(R)$ given by Theorem 7C, Ch. III, we can define a character $\xi \in F_O(R)^*$ of $F_O(R)$ by setting $\xi(\gamma_\tau) = c(\tau)$ for every $\tau \in T$. This ξ is also regarded as a line bundle over R according to Ch. II, 2B.

Take any nonzero member $N \in \mathcal{H}^\infty(R,\xi)$, which is held fixed. We take a single-valued branch of N on the Green star region $\mathbb{G}'(O)$ and denote it by $N(z;\mathbb{1})$, where $\mathbb{1}$ is the identity of the group $F_O(R)$. Then $N(z;\gamma)$ are defined as in 4A. Moreover, we define an analytic function $N_1(w)$ on $\mathbb{D}$ by the conditions

$$|N_1| = |N| \circ \phi \quad \text{and} \quad N_1(0) = N(O;\mathbb{1}).$$

Let $w \in \mathbb{D}$ satisfy $\phi(w) \in \mathbb{G}'(O)$; then, for any $\tau \in T$,

$$N_1 \circ \tau(w) = N(\phi(w);\gamma_\tau) = \xi(\gamma_\tau)N(\phi(w);\mathbb{1}) = c(\tau)N_1(w)$$

and therefore $N_1 \circ \tau = c(\tau)N_1$ everywhere on $\mathbb{D}$.

In order to define Q, we take any nonzero $f^* \in M$. Then $f^* \circ \hat{\phi} = q\hat{F}$ a.e. on $\mathbb{T}$ for some $F \in H^p(\mathbb{D})$. Multiplying $\hat{N}_1$ on both sides, we get

$$\hat{N}_1\bar{q}(f^* \circ \hat{\phi}) = \hat{N}_1\hat{F} \quad \text{a.e.}$$

Since q and $\hat{N}_1$ obey the same transformation rule with respect to T, N_1F is a T-invariant function in $H^1(\mathbb{D})$, so that

(5) $$N_1F = k \circ \phi$$

for some $k \in H^1(R)$. By the formula (12) in Ch. III we have

$$H[\hat{k} \circ \hat{\phi}] = H[\hat{k}] \circ \phi = k \circ \phi = N_1F = H[\hat{N}_1\hat{F}].$$

From this follows that $\hat{k} \circ \hat{\phi} = \hat{N}_1\hat{F}$ a.e. on $\mathbb{T}$ and therefore

(6) $$(f^*/\hat{k}) \circ \hat{\phi} = (f^* \circ \hat{\phi})/\hat{N}_1\hat{F} = q/\hat{N}_1 \quad \text{a.e.}$$

This means that $f^*/\hat{k}$ is independent of the choice of $f^* \in M$. Using the notation introduced in 4A, we set

$$Q(b;\gamma) = f^*(b)\check{N}(b;\gamma)/\hat{k}(b).$$

Then, for each $\gamma \in \mathcal{F}_0(R)$, $Q(b;\gamma)$ is defined a.e. on Δ_1 and is independent of the choice of $f^* \in M$. Moreover we have $Q(b;\gamma) = \xi(\gamma) \times Q(b;\mathbb{1})$.

Let $u = |N|$. Since u is a bounded l.a.m., Theorem 6B, Ch. VI, implies that $\check{u}(\ell)$ exists and is equal to $\hat{u}(b_\ell)$ for almost every $\ell \in \Lambda(\mathcal{O})$. We thus have, for every $\gamma \in \mathcal{F}_0(R)$,

$$\begin{aligned} |Q(b_\ell;\gamma)| &= |f^*(b_\ell)||\check{N}(\ell;\gamma)|/|\hat{k}(b_\ell)| \\ &= |f^*(b_\ell)|\check{u}(\ell)/|\hat{k}(b_\ell)| \\ &= |f^*(b_\ell)|\hat{u}(b_\ell)/|\hat{k}(b_\ell)| = 1 \end{aligned}$$

for almost every $\ell \in \Lambda(\mathcal{O})$ and so $|Q(b;\gamma)| = 1$ a.e. on Δ_1. We now recall the equation

(7) $$f^*(b)/Q(b;\gamma) = \hat{k}(b)/\check{N}(b;\gamma) \quad \text{a.e. on } \Delta_1,$$

where k is determined uniquely by (5). Since $|k(z)/N(z;\gamma)| \circ \phi = |N_1 F|/|N_1| = |F|$, we see that $k(z)/N(z;\gamma)$ is a section, say h, belonging to $\mathcal{H}^p(R,\xi^{-1})$. Hence

(8) $$M \subseteq \{f^* \in L^p(d\chi) : f^*/Q \equiv \check{h} \text{ for some } h \in \mathcal{H}^p(R,\xi^{-1})\}.$$

4E. To prove the reverse inclusion, we first note the following

Lemma. Let Q and ξ be as in 4D and let $J = \{h \in \mathcal{H}^p(R,\xi^{-1}) : \check{h} \equiv f^*/Q \text{ for some } f^* \in M\}$. Then J has no nonconstant common inner factors.

Proof. We set $u_h = \mathrm{pr}_I(\log|h|)$ for every $h \in J$. Since $I(R)$ is a band (Ch. II, 5A) and since $u_h \leqq 0$ for every $h \in J$ by Theorem 4A, Ch. IV, the family $\{u_h : h \in J\}$ has the least upper bound, say u, in $I(R)$, which is nonpositive. Let $q' \in H^\infty(\mathbb{D})$ be an inner function such that $|q'| = \exp(u \circ \phi)$. If $f^* \in M$, then, by (7), $f^*(b)/Q(b;\gamma) = \hat{k}(b)/\check{N}(b;\gamma)$ with $k/N \in \mathcal{H}^p(R,\xi^{-1})$. Thus we have $\mathrm{pr}_I(\log|k/N|) \leqq u$ on R. Since the map ϕ preserves the inner parts by Theorem 6B, Ch. III, $\mathrm{pr}_I(\log|(k\circ\phi)/N_1|) \leqq u \circ \phi$ and hence q' divides the inner factor of $(k\circ\phi)/N_1$. This means that $(k\circ\phi)/N_1 \in q'H^p(\mathbb{D})$. So, by use of (7), $f^* \circ \hat{\phi} = q(\hat{k}\circ\hat{\phi})/\hat{N}_1 \in qq'H^p(d\sigma)$. Since $f^* \in M$ is arbitrary and since $qq'H^p(d\sigma)$ is invariant, we have $\{M\}_p \subseteqq qq'H^p(d\sigma)$. In view of (4) we conclude that q' is a constant of modulus one and therefore that u vanishes identically on R. This proves the lemma as desired. □

4F. To finish the proof of Theorem 4C, let us take any $s^* \in L^{p'}(d\sigma)$, $p' = p/(p-1)$, orthogonal to M and show that it is orthogonal to the right-hand side of (8). We take a nonzero element h_0 in $\mathcal{H}^p(R,\xi^{-1})$, which is arbitrarily fixed. Clearly, $\check{h}_0(b;\gamma)Q(b;\gamma)$ is independent of $\gamma \in F_0(R)$.

Take any nonzero $f^* \in M$ and then choose $h \in J$ so that $\check{h}(b;\gamma) = f^*(b)/Q(b;\gamma)$ a.e. on Δ_1 for every $\gamma \in F_0(R)$. Since M is an $H^\infty(d\chi)$-module, s^*f^* is orthogonal to $H^\infty(d\chi)$. By Theorem 7A, Ch. VII, $s^*f^* \in \hat{K}(0) \subseteqq \hat{K}'(0)$. So there exists a meromorphic function $k_0 \in K'(0)$ such that $s^*f^* = \hat{k}_0$ a.e. on Δ_1 and thus

$$\begin{aligned}(9)\qquad s^*(b)Q(b;\gamma)\check{h}_0(b;\gamma) &= s^*(b)f^*(b)\check{h}(b;\gamma)^{-1}\check{h}_0(b;\gamma)\\ &= \hat{k}_0(b)\check{h}(b;\gamma)^{-1}\check{h}_0(b;\gamma).\end{aligned}$$

Since h_0 and h have the same character, $h_0(z;\gamma)/h(z;\gamma)$ is independent of γ and defines a single-valued meromorphic function. So the correspondence $z \to k_0(z)h_0(z;\gamma)/h(z;\gamma)$ is a meromorphic function of bounded characteristic, say u, which is independent of the choice of $h \in J$ in view of (9). Since $k_0 \in K'(0) = S'^{(0)}\mathcal{H}^1(R,\xi'^{-1})$, where ξ' is the character of $S'^{(0)}$, we have $k_0 = S'^{(0)}(z;\gamma)k_1(z;\gamma)$ for some $k_1 \in \mathcal{H}^1(R,\xi'^{-1})$. It follows that

$$\log(|u||S'^{(0)}|^{-1}) = \log|h_0| + \log|k_1| - \log|h|$$

and

$$\mathrm{pr}_I[\log(|u||S'^{(0)}|^{-1})] = \mathrm{pr}_I[\log(|h_0||k_1|)] - \mathrm{pr}_I[\log|h|].$$

Here we first note that $\mathrm{pr}_I[\log(|h_0||k_1|)] \leqq 0$. Since h can vary over J and since J has no nonconstant common inner factors by Lemma 4E, we conclude that

$$\mathrm{pr}_I[\log(|u||S'^{(0)}|^{-1})] \leqq 0.$$

Consequently, $|u||S'^{(0)}|^{-1}$ has a harmonic majorant, say v, on R; namely, we have

$$|u(z)|g^{(0)}(z) \leqq v(z)e^{-g(0,z)}.$$

In other words, u is a meromorphic function such that $u(O) = 0$ and $|u|g^{(O)}$ has a harmonic majorant. As we are assuming (DCT_O), we see that $\hat{u} \in L^1(d\chi)$ and

$$\int_{\Delta_1} s^*(b)Q(b;\gamma)\check{h}_0(b;\gamma)d\chi(b) = \int_{\Delta_1} \hat{u}(b)d\chi(b) = u(O) = 0.$$

Since $h_0 \in \mathcal{H}^p(R,\xi^{-1})$ is arbitrary, s^* is orthogonal to the right-hand side of (8). Hence the equality sign holds in (8). This finishes the proof of Theorem 4C. □

4G. As for the invariant subspaces of $H^p(R)$ the preceding theorem implies the following

<u>Theorem</u>. Suppose that (DCT_O) holds. Let $1 \leqq p \leqq \infty$ and let M be a closed (β-closed, if $p = \infty$) subspace of $H^p(R)$. Then, M is an $H^\infty(R)$-submodule of $H^p(R)$ if and only if there exists a bounded inner l.a.m. I such that, for $1 \leqq p < \infty$,

(10) $$M = \{f \in H^p(R)\colon (|f|/I)^p \text{ admits a harmonic majorant}\}$$

and, for $p = \infty$,

(11) $$M = \{f \in H^\infty(R)\colon |f|/I \text{ is bounded}\}.$$

<u>Proof</u>. (i) We show first that every subspace M of the form (10) or (11) is a closed (β-closed, if $p = \infty$) $H^\infty(R)$-submodule of $H^p(R)$. Since M is clearly an $H^\infty(R)$-module, we have only to show its closedness. Let $\phi\colon \mathbb{D} \to R$ be a universal covering map with $\phi(0) = O$ and let q be an inner function on $\mathbb{D}$ such that $I \circ \phi = |q|$ on $\mathbb{D}$. Then the set $M \circ \phi = \{f \circ \phi\colon f \in M\}$ is a subspace of $qH^p(\mathbb{D})$.

Suppose $1 \leqq p < \infty$. Let $\{f_n\colon n = 1, 2, \ldots\}$ be a sequence in M which converges to some $f \in H^p(R)$. Since $\hat{\phi}$ maps the measure $d\sigma$ on $\mathbb{T}$ to the measure $d\chi$ on Δ_1, the map $f \to f \circ \phi$ is an isometry from $H^p(R)$ into $H^p(\mathbb{D})$. So $\{f_n \circ \phi\}$ converges to $f \circ \phi$ in $H^p(\mathbb{D})$. As $qH^p(\mathbb{D})$ is closed in $H^p(\mathbb{D})$, so $f \circ \phi \in qH^p(\mathbb{D})$, which implies that $(|f|/I)^p$ has a harmonic majorant, as was to be proved.

The case $p = \infty$ can be proved similarly, if we note that $qH^\infty(\mathbb{D})$ is β-closed in $H^\infty(\mathbb{D})$ for any inner function q on $\mathbb{D}$.

(ii) To show the converse statement, let M be a nontrivial closed (β-closed, if $p = \infty$) $H^\infty(R)$-submodule of $H^p(R)$, $1 \leqq p \leqq \infty$. If we set $M_0 = \{\hat{f}\colon f \in M\}$, then M_0 is a closed (weakly* closed, if $p = \infty$) $H^\infty(d\chi)$-submodule of $H^p(d\chi)$ by Theorem 3B, Ch. IV (Corollary 5D, Ch. IV). Every nonzero function in M_0 cannot vanish identically on any nonnegligible subset of Δ_1 and so M_0 is not doubly invariant in view of Theorem 3B. M_0 is thus simply invariant. By Theorem 4C there exists an m-function Q on $\Delta_1 \times F_0(R)$ of character ξ such that $|Q(\cdot;\gamma)| = 1$ a.e. on Δ_1 and

$$M_0 = \{f^* \in L^p(d\chi): f^*/Q \equiv \check{h} \text{ for some } h \in \mathcal{H}^p(R,\xi^{-1})\}.$$

For any nonzero $f \in M$ there exists an element, say h_f, in $\mathcal{H}^p(R,\xi^{-1})$ such that $\hat{f}(b)/\check{h}_f(b;\gamma) = Q(b;\gamma)$ a.e. on Δ_1 for every $\gamma \in F_0(R)$. So $f(z)/h_f(z;\gamma)$ is a multiplicative meromorphic function of bounded characteristic, whose boundary values are independent of f. It follows therefore that $f(z)/h_f(z;\gamma)$ is also independent of f. Setting

$$Q_0(z;\gamma) = f(z)/h_f(z;\gamma), \tag{12}$$

we get a multiplicative meromorphic function Q_0 of bounded characteristic such that, for any fixed $\gamma \in F_0(R)$, $\check{Q}_0(\ell;\gamma) = Q(b_\ell;\gamma)$ for almost every $\ell \in \Lambda(O)$. Thus a function $f \in H^p(R)$ belongs to M if and only if $f/Q_0 \in \mathcal{H}^p(R,\xi^{-1})$. By setting $I(z) = |Q_0(z;\gamma)|$, we see easily that the latter occurs if and only if, for $1 \leqq p < \infty$, $(|f|/I)^p$ has a harmonic majorant and, for $p = \infty$, $|f|/I$ is bounded. As we shall see in the following lemma, Q_0 is a bounded inner function so that I is a bounded inner l.a.m. Hence M has the desired form. □

4H. <u>Lemma</u>. Q_0 is a (multiplicative) inner function.

<u>Proof</u>. We set $u_0(z) = |Q_0(z;\gamma)|$ on R. Then u_0 is an l.m.m. of bounded characteristic and, by Theorem 6B, Ch. VI,

$$\hat{u}_0(b_\ell) = \check{u}_0(\ell) = |\check{Q}_0(\ell;\gamma)| = |Q(b_\ell;\gamma)| = 1$$

for almost every $\ell \in \Lambda(O)$. We know that $\log u_0 \in SP'(R)$ and its quasibounded part vanishes identically, for

$$pr_Q[\log u_0] = H[\log \hat{u}_0] = H[0] = 0$$

by Theorem 5E, Ch. III. So $\log u_0 \in I(R)$. We set $v_1 = (-\log u_0) \vee 0$, $v_2 = (\log u_0) \vee 0$, $u_1 = \exp(-v_1)$ and $u_2 = \exp(-v_2)$. Then both u_1 and u_2 are bounded inner l.a.m. without nontrivial common inner factor and $u_0 = u_1/u_2$. Consequently, there exist multiplicative analytic functions $Q_1(z;\gamma)$ and $Q_2(z;\gamma)$ such that $u_1 = |Q_1|$, $u_2 = |Q_2|$ and

$$Q_0(z;\gamma) = Q_1(z;\gamma)/Q_2(z;\gamma).$$

We now show that Q_2 is a constant function. As we have seen, an element $f \in H^p(R)$ belongs to M if and only if, for $1 \leqq p < \infty$, $(|f|/u_0)^p$ has a harmonic majorant, say u_f, on R and, for $p = \infty$, f/u_0 is bounded on R. Suppose first that $1 \leqq p < \infty$. Since u_f can be chosen as quasibounded by Theorem 4A, Ch. IV, we obtain

$$pr_I(\log|f|) + v_1 - v_2 \leqq 0,$$

by applying the projection pr_I to the inequality $\log|f| + v_1 - v_2 \leqq p^{-1}u_f$. Since $pr_I(\log|f|) \leqq 0$ and $v_1 \wedge v_2 = 0$, we get

$$pr_I(\log|f|) + v_1 = v_1 \wedge (pr_I(\log|f|) + v_1) \leqq v_1 \wedge v_2 = 0.$$

Namely, the inner factor of f can be divided by Q_1. This in turn means that Q_2 is a common inner factor of

$$J = \{h \in \mathcal{H}^p(R,\xi^{-1}): h(z;\gamma) = f(z)/Q_0(z;\gamma) \text{ for some } f \in M\}.$$

In fact, if $h \in J$ and $f(z) = Q_0(z;\gamma)h(z;\gamma)$, then f is well-defined independently of γ and belongs to M. By what we have seen above,

$$|h(z;\gamma)/Q_2(z;\gamma)|^p = |f(z)/Q_1(z;\gamma)|^p \leqq u_f.$$

Thus Q_2 should divide the inner factor of h, as was to be proved. But, as shown in Lemma 4E, J has no nonconstant common inner factor and hence Q_2 must be a constant. The case $p = \infty$ can be treated in the same way by just changing u_f for a suitable constant. □

5. Equivalence of (DCT_a)

5A. Our characterization of invariant subspaces enables us to say more about the direct Cauchy theorem. Namely we have

Theorem. The following are equivalent for any regular PWS R:

(a) (DCT_a) holds for some $a \in R$;

(b) (DCT_a) holds for all $a \in R$;

(c) an ideal of $H^\infty(R)$ is β-closed if and only if it is of the form (11).

Proof. (a) ⇒ (c): This has been shown in Theorem 4G, for we have only to replace O by a.

(c) ⇒ (b): Take any $a \in R$, which is held fixed. Since $K(a)$ is included in $K'(a)$ and since the orthocomplement of $\hat{K}(a)$ with respect to the duality $(L^\infty(d\chi_a), L^1(d\chi_a))$ is $H^\infty(d\chi_a)$ by Theorem 7A, Ch. VII, the orthocomplement of $\hat{K}'(a)$ is included in $H^\infty(d\chi_a)$. We now set

$$M = \{h \in H^\infty(R): \int_{\Delta_1} \hat{h}(b)\hat{f}(b)d\chi_a(b) = 0 \quad \text{for any} \quad f \in K'(a)\}.$$

Since $\hat{K}'(a)^\perp$ is weakly* closed in $L^\infty(d\chi_a)$, Corollary 5D, Ch. IV, implies that M is a β-closed subspace of $H^\infty(R)$. Moreover M is an ideal of $H^\infty(R)$, for $K'(a)$ is closed under multiplication by functions in $H^\infty(R)$. By our hypothesis (c), there exists a bounded inner l.a.m. I such that $M = \{h \in H^\infty(R): h/I$ is bounded on $R\}$.

We claim that $J_n(a) \subseteqq M$ for every $n = 1, 2, \ldots$, where $J_n(a)$ was defined in Ch. VII, 6B. In fact, if $h \in J_n(a)$, $|h|/g_n^{(a)}$ (cf. Ch. VII, 1A) is bounded. So for any $f \in K'(a)$ we have

$$|fh|\exp(-\sum_{j=1}^{n} g(\cdot,z_j)) \leqq |f|g^{(a)}\cdot|h|/g_n^{(a)},$$

where the right-hand side has a harmonic majorant on R. By the weak Cauchy theorem (Theorem 4B, Ch. VII), $\hat{f}\hat{h}$ is χ_a-summable and

$$\int_{\Delta_1} \hat{f}\hat{h}\, d\chi_a = f(a)h(a) = 0.$$

Therefore $h \in M$, as desired.

Since $\cup_{n=1}^{\infty} J_n(a)$ has no nonconstant common inner factors by Lemma 6B, Ch. VII, we have $I = 1$ and thus $M = H^\infty(R)$. Namely, $K'(a)^\perp = H^\infty(d\chi_a)$. In particular,

$$\int_{\Delta_1} \hat{f}\, d\chi_a = 0$$

for any $f \in K'(a)$, which is clearly equivalent to (DCT_a).

(b) ⇒ (a): This is trivial. □

NOTES

There have appeared so many papers on invariant subspaces. In the case of Hardy classes, the origin is Beurling [2]. Detailed accounts with different flavor are found in Gamelin [10], Helson [33], Hoffman [34] and others. The case of compact bordered surfaces was studied by Forelli [9], Sarason [60], Voichick [66, 67] and Hasumi [16].

As for doubly invariant subspaces, an abstract general theorem is in Hasumi and Srinivasan [23]. The case of Riemann surface is in Hasumi [17]. As for simply invariant subspaces, the main theorem (Theorem 4C) was proved in Hasumi [17] under a stronger assumption. The present version is due to Hayashi [27], together with Theorem 5A.

CHAPTER IX. CHARACTERIZATION OF SURFACES OF PARREAU-WIDOM TYPE

In the preceding two chapters we have studied Cauchy theorems and invariant subspaces on PWS's. We are now going to push our previous investigation further and to exhibit intimate relations existing among these things.

Our objective in this chapter is twofold. First we look into the inverse Cauchy theorem. Here we find a remarkable fact that, roughly speaking, PWS's are characterized among hyperbolic Riemann surfaces as those for which the inverse Cauchy theorem holds. Next we investigate the direct Cauchy theorem (DCT). As we shall see in Chapter X, there exist PWS's for which (DCT) fails. Thus it is necessary for us to tell which PWS's satisfy (DCT). We have already had a criterion for this in Chapter VIII, 5A, so that the emphasis should now be placed on other conditions.

In §1 we show that the inverse Cauchy theorem essentially characterizes PWS's, the crucial point being a mean value theorem stated in 2D. And in §2 we provide a couple of conditions which make a given PWS R satisfy (DCT). There the principal role will be played by certain functionals on the character group $F_{\mathcal{O}}(R)^*$ corresponding to R.

In order to avoid unnecessary complication we assume through this chapter that R denotes a connected hyperbolic Riemann surface <u>carrying nonconstant bounded analytic functions</u>.

§1. THE INVERSE CAUCHY THEOREM AND SURFACES OF PARREAU-WIDOM TYPE

1. <u>Statement of the Main Result</u>

1A. We first fix some notations. Let $\mathcal{O}$ be a fixed point in R and $\phi = \phi_R$ a universal covering map $\mathbb{D} \to R$ with $\phi(0) = \mathcal{O}$. Then $\hat{\phi}$ is a measure-preserving map of the unit circumference $\mathbb{T}$ into the Martin boundary Δ of R with respect to the normalized Lebesgue measure $d\sigma(\theta) = d\theta/2\pi$ on $\mathbb{T}$ and the harmonic measure $d\chi$ on Δ_1 at the point $\mathcal{O}$ (Theorem 7B, Ch. III). Let T_R be the group of cover transformations for ϕ. By Theorem 7C, Ch. III, T_R is canonically identified with the fundamental group $F_0(R)$.

A measurable function Q on $\mathbb{T}$ is called an <u>m-function</u> with character $\xi \in T_R^*$ if $Q \circ \tau = \xi(\tau)Q$ a.e. on $\mathbb{T}$ for any $\tau \in T_R$. The character of Q is denoted by ξ_Q. An <u>i-function</u> with character ξ on $\mathbb{T}$ is an m-function Q with character ξ such that $|Q| = 1$ a.e. For $0 < p \leqq \infty$ and for any character ξ of T_R we set

$$H^p(\mathbb{D},\xi) = \{f \in H^p(\mathbb{D}) : f \circ \tau = \xi(\tau)f \text{ for any } \tau \in T_R\}$$

and

$$H^p(d\sigma,\xi) = \{f \in H^p(d\sigma) : f \circ \tau = \xi(\tau)f \text{ a.e. on } \mathbb{T} \text{ for any } \tau \in T_R\}.$$

For the sake of simplicity we use the same letter for a holomorphic function on $\mathbb{D}$ and for its boundary value function on $\mathbb{T}$. For instance, $F \in H^p(\mathbb{D},\xi)$ means a function on $\mathbb{D}$, while $F \in H^p(d\sigma,\xi)$ denotes the corresponding boundary function on $\mathbb{T}$. A character ξ of T_R is called <u>p-outer</u> if $H^p(\mathbb{D},\xi)$ (or, equivalently, $H^p(d\sigma,\xi)$) has no nonconstant common inner factors. It is called <u>outer</u> if it is p-outer for all p with $1 \leqq p \leqq \infty$. Given an i-function Q on $\mathbb{T}$ with character ξ_Q, we set

$$H^p(d\chi,Q) = \{f \in L^p(d\chi) : f \circ \hat{\phi} \in QH^p(d\sigma,\xi_Q^{-1})\}.$$

An expression $H^p(d\chi,Q)$ is called <u>standard</u> if the corresponding character ξ_Q^{-1} is p-outer. For any i-function Q on $\mathbb{T}$ the greatest common inner factor Q' of $H^p(d\sigma,\xi_Q^{-1})$ is again an i-function, so that $H^p(d\chi,QQ')$ is the standard expression for $H^p(d\chi,Q)$.

1B. We now briefly recall the inverse Cauchy theorem on PWS's. Let R be a (not necessarily regular) PWS and $R^\dagger$ the regularization of R in the sense of Theorem 3B, Ch. V. Let $g^\dagger(a,z)$ be the Green function for $R^\dagger$ and $Z(0;R^\dagger)$ the set of critical points in $R^\dagger$ of the function $z \to g^\dagger(0,z)$, which we repeat according to multiplicity. Set $q = g^{\dagger(0)}|R$, where

$$g^{\dagger(0)}(z) = \exp(-\sum \{g^\dagger(z,w) : w \in Z(0;R^\dagger)\})$$

for $z \in R^\dagger$. Then q is an inner l.a.m. on R and the inverse Cauchy theorem (Theorem 1C, Ch. VII) states the following:

(ICT) If $u^* \in L^1(d\chi)$ satisfies $\int_{\Delta_1} \hat{h}(b)u^*(b)d\chi(b) = 0$ for any h, meromorphic on R, such that $h(0) = 0$ and $|h|q$ is bounded on R, then there exists an $f \in H^1(R)$ such that $u^* = \hat{f}$ a.e. on Δ_1.

Let Q be the inner function on $\mathbb{D}$ such that $Q(0) \geq 0$ and $|Q| = q \circ \phi$. Then the character of Q is identified with that of q, i.e. $\xi_Q = \xi_q$ under the canonical isomorphism of T_R onto $\mathcal{F}_0(R)$. The condition on h says that $(h \circ \phi)Q \in H_0^\infty(\mathbb{D}) = \{F \in H^\infty(\mathbb{D}) : F(0) = 0\}$. Since $h \circ \phi(0) = 0$, we have $(h \circ \phi)(\tau(0)) = 0$ for any $\tau \in T_R$. Thus, $\{\tau(0) : \tau \in T_R\}$ is a subset of the zeros of the bounded analytic function $(h \circ \phi)Q$ on $\mathbb{D}$, so that we can find a Blaschke product B whose zeros are exactly $\{\tau(0) : \tau \in T_R\}$. Since $B \circ \tau$ is a Blaschke product with the same zeros as B, there exists a complex number $\eta(\tau)$ of modulus one such that $B \circ \tau = \eta(\tau)B$; namely, B has character η. We also see that Q and B have no common zeros, for $0 \notin Z(0;R^\dagger)$. So $h \circ \phi$ belongs to $(B/Q)H^\infty(\mathbb{D},\eta^{-1}\xi_Q)$; in other words, $\hat{h} \in H^\infty(d\chi,B/Q)$. It follows that the inverse Cauchy theorem is expressed as

$$H^\infty(d\chi,B/Q)^\perp \subseteqq H^1(d\chi), \tag{1}$$

where the orthocomplement is formed with respect to the dual pair $(L^1(d\chi),L^\infty(d\chi))$.

The inverse Cauchy theorem shows that the invariant subspace in $L^1(d\chi)$ given by $H^\infty(d\chi)^\perp$ has a special form. In fact, since we have $H^\infty(d\chi,B/Q) \subseteqq H^1(d\chi,B/Q)$, we infer from (1) that

$$H^1(d\chi,B/Q)^\perp \subseteqq H^\infty(d\chi,B/Q)^\perp \subseteqq H^1(d\chi)$$

and therefore that $H^1(d\chi,B/Q)^\perp \subseteqq H^1(d\chi) \cap L^\infty(d\chi) = H^\infty(d\chi)$. This means that $H^\infty(d\chi)^\perp \subseteqq H^1(d\chi,B/Q)$. Using the notations in Ch. VIII, we thus have $\{H^\infty(d\chi)^\perp\}_1 \subseteqq (B/Q)H^1(d\sigma)$; namely, $\{H^\infty(d\chi)^\perp\}_1$ is seen to be a simply invariant subspace of $L^1(d\sigma)$.

1C. Using the notations in 1A, we now state our main result of this section.

Theorem. Consider the following three conditions for a hyperbolic Riemann surface R carrying nonconstant bounded analytic functions:

(a) R is a PWS.

(b) There exists an inner l.a.m. q on R for which the inverse Cauchy theorem (ICT) is valid: namely, if $u^* \in L^1(d\chi)$ satisfies

$$\int_{\Delta_1} \hat{h}(b)u^*(b)d\chi(b) = 0$$

for any h, meromorphic on R, such that $h(0) = 0$ and $|h|q$ is bounded on R, then there exists an $f \in H^1(R)$ such that $u^* = \hat{f}$ a.e. on Δ_1.

(c) $H^\infty(d\chi)^\perp \subseteqq H^1(d\chi,Q)$ for some i-function Q on $\mathbb{T}$, where

$$H^\infty(d\chi)^\perp = \{f^* \in L^1(d\chi): \int_{\Delta_1} \hat{h}f^* d\chi = 0 \quad \text{for any} \quad h \in H^\infty(R)\}.$$

Then, we have (a) $\Rightarrow$ (b) $\Leftrightarrow$ (c).

If $H^\infty(R)$ separates the points of R, then all these conditions are equivalent.

1D. We have seen in 1B that (a) $\Rightarrow$ (b) $\Rightarrow$ (c). So we have only to show the remaining implication. As the proof is not so short, we first explain our plans for readers' convenience. The next subsection is devoted to preliminary observations, the main result being a mean value theorem given in 2D. The proof of (c) $\Rightarrow$ (b) is stated in 3A. The proof of (c) $\Rightarrow$ (a) under the additional assumption is given in 3B-3J. As we saw in Ch. V, a surface R is a PWS if (and only if) $m(a) > 0$ for some (and hence all) $a \in R$, where $m(a)$ was defined in Ch. V, 8A. We now write $m^1(a)$ instead of $m(a)$. More generally, we use the following notations:

$$m^p(\xi,a) = \sup\{|f(a)|: f \in \mathcal{H}^p(R,\xi),\ \|f\|_{p,a} \leq 1\}$$

for $a \in R$ and $\xi \in \mathcal{F}_0(R)^*$, and also

$$m^p(a) = \inf\{m^p(\xi,a): \xi \in \mathcal{F}_0(R)^*\}.$$

We shall finally prove in 3J that $m^\infty(0) > 0$ if (c) holds and if $H^\infty(R)$ separates the points of R. This will be done by showing, in three steps, that (i) $m^\infty(\xi,0) \geq K$ for any outer character ξ, where $K > 0$ is a fixed constant; (ii) the character of every finite Blaschke product on R is outer; and (iii) every character of $\mathcal{F}_0(R)$ can be approximated by characters of finite Blaschke products.

Before proceeding to the proof, we note that the condition (c) is independent of the choice of the origin (or, equivalently, the harmonic measure $d\chi$). To see this, we define

$$N_a = \{f^* \in L^1(d\chi_a): \int_{\Delta_1} \hat{h}f^* d\chi_a = 0 \quad \text{for any} \quad h \in H^\infty(R)\}$$

for $a \in R$, where $d\chi_a$ is the harmonic measure of R at the point a, i.e. $d\chi_a(b) = k_b(a)d\chi(b)$ (Ch. III, 2C). Then it is clear that $N_a \subseteq H^1(d\chi,Q)$ for some i-function Q on $\mathbb{T}$ if and only if $\{N_a\}_1$ is simply invariant. Now take any $a, a' \in R$ and set $s(b) = k_b(a')/k_b(a)$ for $b \in \Delta_1$. By use of Harnack's inequality we find a constant $c > 0$ such that $c^{-1} \leq s \leq c$ a.e. on Δ_1. This implies that $sN_{a'} = N_a$, because of $d\chi_{a'} = s\,d\chi_a$, and therefore that $\{N_a\}_1 = (s \circ \hat{\phi})\{N_{a'}\}_1$.

Since the multiplication by $s \circ \hat{\phi}$ defines a bicontinuous isomorphism of $L^1(d\sigma)$, we conclude that $e^{i\theta}\{N_a\}_1$ is L^1-dense in $\{N_a\}_1$ if and only if $e^{i\theta}\{N_{a'}\}_1$ is L^1-dense in $\{N_{a'}\}_1$, as was to be proved.

In what follows in §1, we assume that the condition (c), 1C, holds. As shown above, this is equivalent to the assumption that, for every $a \in R$, $H^\infty(d\chi_a)^\perp \subseteqq H^1(d\chi, Q_a)$ for some i-function Q_a on $\mathbb{T}$, where

$$H^\infty(d\chi_a)^\perp = \{f^* \in L^1(d\chi): \int_{\Delta_1} \hat{h}f^* d\chi_a = 0 \quad \text{for any} \quad h \in H^\infty(R)\}.$$

1E. Remark. We just look into the excluded case in which $H^\infty(R)$ contains only constant functions. Then $H^\infty(d\chi)^\perp$ consists of all functions $f^* \in L^1(d\chi)$ such that $\int_{\Delta_1} f^* d\chi = 0$. We also know that every function in $H^1(d\chi, Q)$ cannot vanish on a set of positive measure without vanishing identically. So if the condition (c), 1C, holds, then the Martin minimal boundary Δ_1 cannot be divided into three disjoint measurable subsets of positive measure. This means that R carries at most two linearly independent harmonic functions. We will not go into this matter further.

2. A Mean Value Theorem

2A. We now begin our preliminary observation for the proof of (c) $\Rightarrow$ (b) and (a) in Theorem 1C. By the condition (c) there exists a measurable function, say Q_A, on $\mathbb{T}$ with $|Q_A| = 1$ a.e. such that

$$\{H^\infty(d\chi)^\perp\}_1 = Q_A H^1(d\sigma).$$

Since $\{H^\infty(d\chi)^\perp\}_1$ is invariant under T_R, Q_A is seen to be an i-function and

$$H^\infty(d\chi)^\perp \subseteqq H^1(d\chi, Q_A),$$

the right-hand side being a standard expression. Let ξ_A be the character of Q_A^{-1}. Then ξ_A is 1-outer. To see this, let Q' be a common inner factor of $H^1(d\sigma, \xi_A)$. Then

$$H^\infty(d\chi)^\perp \circ \hat{\phi} \subseteqq Q_A H^1(d\sigma, \xi_A) \subseteqq Q_A Q' H^1(d\sigma)$$

and therefore $Q_A H^1(d\sigma) = \{H^\infty(d\chi)^\perp\}_1 \subseteqq Q_A Q' H^1(d\sigma)$. Hence, Q' should be a constant function, as desired.

We set $I_0 = H^1(d\chi, Q_A)^\perp$, which is seen to be a weakly* closed ideal of $H^\infty(d\chi)$. This ideal plays an important role in the following investigation. In the following, $[\cdot]_p$ denotes the closed (weakly* closed, if $p=\infty$) linear envelope in the space $L^p(d\chi)$. First we show:

Lemma. $H^\infty(d\chi)^\perp = [I_0 H^1(d\chi, Q_A)]_1$.

Proof. If $f^* \in I_0$, $h^* \in H^1(d\chi, Q_A)$ and $u^* \in H^\infty(d\chi)$, then h^*u^* belongs to $H^1(d\chi, Q_A)$ and thus $f^* \perp h^*u^*$. It follows that f^*h^* is orthogonal to $H^\infty(d\chi)$. Namely, $[I_0 H^1(d\chi, Q_A)]_1 \subseteqq H^\infty(d\chi)^\perp$. Next we take any $f^* \in L^\infty(d\chi)$ which is orthogonal to the set $I_0 H^1(d\chi, Q_A)$. Then, for any $h^* \in H^1(d\chi, Q_A)$, $f^*h^* \in I_0{}^\perp = H^1(d\chi, Q_A)^{\perp\perp} = H^1(d\chi, Q_A)$, which implies

$$(f^* \circ \hat{\phi})(H^1(d\chi, Q_A) \circ \hat{\phi}) \subseteqq H^1(d\chi, Q_A) \circ \hat{\phi}.$$

Since ξ_A is 1-outer, we have $(f^* \circ \hat{\phi}) Q_A H^1(d\sigma) \subseteqq Q_A H^1(d\sigma)$ and thus $f^* \in H^\infty(d\chi)$. So $[I_0 H^1(d\chi, Q_A)]_1{}^\perp \subseteqq H^\infty(d\chi)$ and the desired result is obtained. □

2B. Lemma. $I_0 \circ \hat{\phi}$ has no nonconstant common inner factors.

Proof. Let Q' be the greatest common inner factor of $I_0 \circ \hat{\phi}$. Then, $I_0 \circ \hat{\phi} \subseteqq Q' H^\infty(d\sigma)$ and therefore $I_0 \subseteqq H^\infty(d\chi, Q')$. So we have

$$I_0 H^1(d\chi, Q_A) \subseteqq H^\infty(d\chi, Q') H^1(d\chi, Q_A) \subseteqq H^1(d\chi, Q_A Q').$$

Using the preceding lemma, we get $H^\infty(d\chi)^\perp \subseteqq H^1(d\chi, Q_A Q')$ and so $Q_A H^1(d\sigma) = \{H^\infty(d\chi)^\perp\}_1 \subseteqq Q_A Q' H^1(d\sigma)$. Hence, Q' is a constant. □

2C. Here we determine $H^p(d\chi, Q)^\perp$.

Lemma. Let Q be an i-function such that $H^\infty(d\chi, Q) \neq \{0\}$. Let $1 \leq p \leq \infty$ and let Q' be an inner function on $\mathbb{D}$ such that $H^p(d\chi, Q) = H^p(d\chi, QQ')$, the right-hand side being a standard expression. Then,

$$H^p(d\chi, Q)^\perp = [I_0 H^{p'}(d\chi, Q_A/QQ')]_{p'} \tag{2}$$

and

$$[I_0 H^p(d\chi, Q)]_p{}^\perp = H^{p'}(d\chi, Q_A/QQ'), \tag{3}$$

where $p' = p/(p-1)$ and the orthocomplement is formed with respect to the dual pair $(L^p(d\chi), L^{p'}(d\chi))$.

Proof. In view of Lemma 2A we have

$$I_0 H^p(d\chi, Q) H^{p'}(d\chi, Q_A/QQ') \subseteqq I_0 H^1(d\chi, Q_A) \subseteqq H^\infty(d\chi)^\perp,$$

which in turn implies

(4) $$I_0 H^{p'}(d\chi, Q_A/QQ') \subseteqq H^p(d\chi, Q)^{\perp}$$

and

(5) $$H^{p'}(d\chi, Q_A/QQ') \subseteqq [I_0 H^p(d\chi, Q)]_p{}^{\perp}.$$

Next, let $s^* \in [I_0 H^p(d\chi,Q)]_p{}^{\perp}$. If $f^* \in H^p(d\chi,Q)$ and $h^* \in I_0$, then $f^*h^*s^* \in H^\infty(d\chi)^{\perp} \subseteqq H^1(d\chi,Q_A)$. We employ the fact $f^* \circ \hat{\phi} = QQ'F$ with $F \in H^p(d\sigma, \xi_{QQ'}{}^{-1})$ and get $(s^* \circ \hat{\phi})(h^* \circ \hat{\phi})QQ'F \in Q_A H^1(d\sigma)$. Since $\xi_{QQ'}{}^{-1}$ is p-outer, we have $(s^* \circ \hat{\phi})(h^* \circ \hat{\phi}) \in (Q_A/QQ')H^1(d\sigma)$. Moreover, by Lemma 2B, $I_0 \circ \hat{\phi}$ has no nonconstant common inner factors and therefore $s^* \circ \hat{\phi} \in ((Q_A/QQ')H^1(d\sigma)) \cap L^{p'}(d\sigma) = (Q_A/QQ')H^{p'}(d\sigma)$. Consequently, s^* belongs to $H^{p'}(d\chi, Q_A/QQ')$. Combining this with (5), we have the equality (3). On the other hand, suppose $s^* \in [I_0 H^{p'}(d\chi,Q_A/QQ')]_{p'}{}^{\perp}$. If $h^* \in I_0$, then $h^*s^* \in H^{p'}(d\chi,Q_A/QQ')^{\perp} = [I_0 H^p(d\chi,Q)]_p \subseteqq H^p(d\chi,Q)$. Since $I_0 \circ \hat{\phi}$ has no nonconstant common inner factors, we get $s^* \in H^p(d\chi,Q)$. Namely, $[I_0 H^{p'}(d\chi,Q_A/QQ')]_{p'}{}^{\perp} \subseteqq H^p(d\chi,Q)$, which, together with (4), implies the equality (2). □

2D. We are now able to prove a mean value theorem. Let Q_g be the inner function on $\mathbb{D}$ such that $Q_g(0) \geqq 0$ and $|Q_g| = \exp(-g_0 \circ \phi)$, where $g_0(z) = g(0,z)$ is the Green function for R with pole 0. Let $H_0^\infty(R)$ denote the ideal of $H^\infty(R)$ consisting of functions vanishing at the origin 0. Since R is assumed to carry nonconstant bounded analytic functions, we have

$$H_0^\infty(R) \neq \{0\}.$$

We see that $(\exp g_0)H_0^\infty(R)$ consists of bounded functions on R and therefore that $H_0^\infty(d\chi) = H^\infty(d\chi,Q_g)$. By Lemma 2C there exists an inner function Q_g' on $\mathbb{D}$ such that

$$H_0^\infty(d\chi)^{\perp} = H^\infty(d\chi,Q_g)^{\perp} = [I_0 H^1(d\chi,Q_A/Q_gQ_g')]_1,$$

where $H^\infty(d\chi,Q_gQ_g')$ is the standard expression for $H^\infty(d\chi,Q_g)$.

Let now Q_C be an i-function such that $H^1(d\chi,Q_C{}^{-1})$ is the standard expression for $H^1(d\chi,Q_A/Q_gQ_g')$ and let ξ_C be the character of Q_C. As we have $1 \in H^1(d\chi,Q_C{}^{-1})$, so we see that Q_C represents an inner function on $\mathbb{D}$. We may suppose, in what follows, that

$$Q_C(0) \geqq 0.$$

Thus every element f^* in $H_0^\infty(d\chi)^\perp$ is regarded as the boundary function of a meromorphic function f, of bounded characteristic, on R, so that its value at O is determined.

<u>Theorem</u>. For every $f^* \in H_0^\infty(d\chi)^\perp$ we have

$$f(O) = \int_{\Delta_1} f^*(b)d\chi(b).$$

<u>Proof</u>. Since $H_0^\infty(d\chi)^\perp = [I_O H^1(d\chi, Q_C^{-1})]_1 \subseteqq H^1(d\chi, Q_C^{-1})$, each f^* in $H_0^\infty(d\chi)^\perp$ determines an $F \in H^1(\mathbb{D})$ such that $f \circ \phi = Q_C^{-1}F$, where f is the meromorphic extension of f^* into R. We define a linear functional L on $H_0^\infty(d\chi)^\perp$ by setting $L(f^*) = F(0)$. Since we have

$$|L(f^*)| = |F(0)| \leqq \int_{\mathbb{T}} |F(e^{i\theta})|d\sigma(\theta) = \int_{\mathbb{T}} |Q_C(e^{i\theta})^{-1}F(e^{i\theta})|d\sigma(\theta)$$

$$= \int_{\Delta_1} |f^*(b)|d\chi(b) = \|f^*\|_1,$$

the Hahn-Banach theorem says that L has a norm-preserving extension to $L^1(d\chi)$. So there exists a function $u^* \in L^\infty(d\chi)$ such that $\|u^*\|_\infty = \|L\| \leqq 1$ and

$$F(0) = \int_{\Delta_1} f^*(b)u^*(b)d\chi(b)$$

for every $f^* \in H_0^\infty(d\chi)^\perp$. If $f^* \in H_0^\infty(d\chi)^\perp$ and $h \in H_0^\infty(R)$, then $\hat{h}f^*$ belongs to $H_0^\infty(d\chi)^\perp$ and $(\hat{h}f^*) \circ \hat{\phi} = (\hat{h} \circ \hat{\phi})(f^* \circ \hat{\phi}) = (\hat{h} \circ \hat{\phi})Q_C^{-1}F$, so that

$$\int_{\Delta_1} \hat{h}f^*u^*d\chi = L(\hat{h}f^*) = (h \circ \phi)(0)F(0) = 0.$$

As $h \in H_0^\infty(R)$ is arbitrary, $f^*u^* \in H_0^\infty(d\chi)^\perp \subseteqq H^1(d\chi, Q_C^{-1})$. Therefore, $Q_C^{-1}F\cdot(u^* \circ \hat{\phi}) \in Q_C^{-1}H^1(d\sigma)$, where F ranges over $(I_O \circ \hat{\phi})H^1(d\sigma, \xi_C)$. Since $(I_O \circ \hat{\phi})H^1(d\sigma, \xi_C)$ has no nonconstant common inner factors, we conclude that u^* belongs to $H^1(d\chi)$ and consequently to $H^\infty(d\chi)$. It thus follows that, for any $f^* \in H_0^\infty(d\chi)^\perp$,

$$(6) \qquad F(0) = \int_{\Delta_1} f^*u^*d\chi = \int_{\Delta_1} f^*(u^* - u(0))d\chi + u(0)\int_{\Delta_1} f^*d\chi$$

$$= u(0)\int_{\Delta_1} f^*d\chi.$$

Since $H^1(d\chi, \xi_C)$ has no common inner factors, it has no common zeros. Namely, there exists an $F_0 \in H^1(d\sigma,\xi_C)$ with $F_0(0) \neq 0$. Moreover, since $I_0 \circ \hat{\phi}$ has no nonconstant common inner factors, there is an $h \in I_0$ with $h(0) \neq 0$. Let $f_1^* \in L^1(d\chi)$ be defined by $f_1^* \circ \hat{\phi} = (\hat{h} \circ \hat{\phi}) Q_C^{-1} F_0$. Then, $f_1^* \in H_0^\infty(d\chi)^\perp$ and $f_1 \circ \phi = (h \circ \phi) Q_C^{-1} F_0$, where f_1 is the meromorphic extension of f_1^* into R. The function F_1 corresponding to this f_1^* is thus equal to $(h \circ \phi) F_0$. Consequently, $F_1(0) \neq 0$ and hence, by (6), $u(0) \neq 0$. In particular, for $f^* \equiv 1$ we have $F = Q_C$ and so, again by (6), $Q_C(0) = u(0) \neq 0$. Hence, for any $f^* \in H_0^\infty(d\chi)$,

$$\int_{\Delta_1} f^* d\chi = u(0)^{-1} F(0) = Q_C(0)^{-1} F(0)$$

$$= (f \circ \phi)(0) = f(0),$$

as was to be proved. □

In the proof we have shown the following

Corollary. $|Q_C(0)| \neq 0$.

3. Proof of the Main Theorem

3A. We are now in a position to prove the implication (c) ⇒ (b).

Theorem. Let q_C be the inner l.a.m. on R such that $q_C \circ \phi = |Q_C|$. Let $u^* \in L^1(d\chi)$ satisfy $\int_{\Delta_1} \hat{h}(b) u^*(b) d\chi(b) = 0$ for any h, meromorphic on R, such that $|h| q_C$ is bounded on R and $h(0) = 0$; then there exists an $f \in H^1(R)$ such that $u^* = \hat{f}$ a.e. on Δ_1.

Proof. Let $h^* \in H^1(d\chi)^\perp$. Since $H^1(d\chi)^\perp \subsetneqq H_0^\infty(d\chi)^\perp \subsetneqq H^1(d\chi, Q_C^{-1})$, we can find an $F \in H^1(d\sigma,\xi_C)$ such that $h^* \circ \hat{\phi} = Q_C^{-1} F$ a.e. on $\mathbb{T}$. The function h^* being bounded, F is also bounded, i.e. $F \in H^\infty(d\sigma,\xi_C)$. Let h be a meromorphic function on R with $h \circ \phi = Q_C^{-1} F$ on $\mathbb{D}$. Then $|h| q_C$ is bounded on R and $h^* = \hat{h}$ a.e. on Δ_1. On the other hand, since $h^* \in H^\infty(d\chi)^\perp$, Theorem 2D implies

$$h(0) = \int_{\Delta_1} \hat{h}(b) d\chi(b) = \int_{\Delta_1} h^*(b) d\chi(b) = 0,$$

the last equality sign being true because $h^* \perp 1$. So h satisfies the hypothesis in the theorem and thus $\int_{\Delta_1} u^* h^* d\chi = 0$. As $h^* \in H^1(d\chi)^\perp$ is arbitrary, we conclude that u^* belongs to $H^1(d\chi)$, as was to be proved. □

3B. We now begin the proof of the implication (c) ⇒ (a) under the assumption that $H^{\infty}(R)$ separates the points of R. For this purpose we have to look into $H^p(d\sigma,\xi)$ more closely.

<u>Lemma</u>. Let $\xi \in T_R^*$ with $H^{\infty}(d\sigma,\xi) \neq \{0\}$ and let Q_p be the greatest common inner factor of $H^p(d\sigma,\xi)$, $1 \leqq p \leqq \infty$. Then Q_p are the same up to constant factors of modulus one.

<u>Proof</u>. Take any nonzero $F \in H^{\infty}(d\sigma,\xi)$ and set $Q = |F|/F$. Then Q is an i-function with character ξ^{-1}. So $QH^p(d\sigma,\xi)$ consists of T_R-invariant functions and $H^p(d\chi,Q) \circ \hat{\phi} = QH^p(d\sigma,\xi)$. Our definition of Q_p shows that $H^p(d\chi,QQ_p)$ is the standard expression for $H^p(d\chi,Q)$, $1 \leqq p \leqq \infty$. Since Q_p divides Q_q for $p < q$, we have only to show that Q_∞ divides Q_1, i.e. Q_1/Q_∞ is an inner function. By Lemma 2C

$$\begin{aligned} H^1(d\chi,Q)^{\perp} &= [I_0H^{\infty}(d\chi,Q_A/QQ_1)]_{\infty} \\ H^{\infty}(d\chi,Q)^{\perp} &= [I_0H^1(d\chi,Q_A/QQ_{\infty})]_1. \end{aligned} \tag{7}$$

Since $H^{\infty}(d\chi,Q) = H^1(d\chi,Q) \cap L^{\infty}(d\chi)$, we see that $H^1(d\chi,Q)^{\perp}$ is L^1-dense in $H^{\infty}(d\chi,Q)^{\perp}$. Combined this with (7), we have

$$\begin{aligned} H^1(d\chi,Q_A/QQ_1) \supseteqq [I_0H^{\infty}(d\chi,Q_A/QQ_1)]_1 &= [H^1(d\chi,Q)^{\perp}]_1 \\ = H^{\infty}(d\chi,Q)^{\perp} &= [I_0H^1(d\chi,Q_A/QQ_{\infty})]_1. \end{aligned}$$

Since $I_0 \circ \hat{\phi}$ has no nonconstant common inner factors, we infer that $H^1(d\chi,Q_A/QQ_1) \supseteqq H^1(d\chi,Q_A/QQ_{\infty})$. Since Q_1 divides Q_{∞}, we thus have $H^1(d\chi,Q_A/QQ_1) = H^1(d\chi,Q_A/QQ_{\infty})$. Consequently,

$$\begin{aligned} H^{\infty}(d\chi,Q_A/QQ_1) &= H^1(d\chi,Q_A/QQ_1) \cap L^{\infty}(d\chi) \\ &= H^1(d\chi,Q_A/QQ_{\infty}) \cap L^{\infty}(d\chi) = H^{\infty}(d\chi,Q_A/QQ_{\infty}) \end{aligned}$$

and therefore

$$[I_0H^{\infty}(d\chi,Q_A/QQ_1)]_{\infty} = [I_0H^{\infty}(d\chi,Q_A/QQ_{\infty})]_{\infty}. \tag{8}$$

Since $H^{\infty}(d\chi,QQ_{\infty})$ is standard, so is $H^1(d\chi,QQ_{\infty})$. By (8) and Lemma 2C

$$\begin{aligned} H^1(d\chi,QQ_{\infty})^{\perp} &= [I_0H^{\infty}(d\chi,Q_A/QQ_{\infty})]_{\infty} \\ &= [I_0H^{\infty}(d\chi,Q_A/QQ_1)]_{\infty} = H^1(d\chi,QQ_1)^{\perp}. \end{aligned}$$

This implies that $H^1(d\chi,QQ_{\infty}) = H^1(d\chi,QQ_1)$ and hence that Q_{∞} divides Q_1. □

3C. Consider any $\xi \in T_R^*$ with $H^p(d\sigma,\xi) \neq \{0\}$ for some p with $1 \leqq p \leqq \infty$ and let Q_ξ be the greatest common inner factor of the space $H^p(d\sigma,\xi)$. Take any nonzero F_0 from $H^p(d\sigma,\xi)$ and set $Q = |F_0|/F_0$, so that Q is an i-function with character ξ^{-1}. We set $M = H^p(d\chi,Q)$ and define M^Δ as the orthocomplement, in $L^{p'}(d\chi)$, of the space $H_0^\infty(d\chi)M$, where $p' = p/(p-1)$.

<u>Lemma</u>. $M^\Delta \subseteqq H^{p'}(d\chi,1/QQ_\xi Q_C)$, where Q_C is the inner function defined in 2D.

<u>Proof</u>. Let $h^* \in M^\Delta$. Then, $h^*M \subseteqq H_0^\infty(d\chi)^\perp \subseteqq H^1(d\chi,1/Q_C)$ and so

$$(h^* \circ \hat{\phi})QQ_\xi H^p(d\sigma) = (h^* \circ \hat{\phi})\{M\}_p \subseteqq \{h^*M\}_1 \subseteqq (1/Q_C)H^1(d\sigma).$$

This means that

$$h^* \circ \hat{\phi} \in [(1/QQ_\xi Q_C H^1(d\sigma)] \cap L^{p'}(d\sigma) \subseteqq (1/QQ_\xi Q_C)H^{p'}(d\sigma).$$

Hence, $M^\Delta \subseteqq H^{p'}(d\chi,1/QQ_\xi Q_C)$. □

3D. <u>Lemma</u>. Let $\xi \in T_R^*$ and $1 \leqq p < \infty$ be given. If ξ is p-outer, then $H^{p'}(d\sigma,\xi^{-1}\xi_C) \neq \{0\}$ and

$$m^p(\xi,0)m^{p'}(\xi^{-1}\xi_C,0) \geqq |Q_C(0)|.$$

<u>Proof</u>. Let F_0 be any nonzero element in $H^p(d\sigma,\xi)$ and set $Q_0 = |F_0|/F_0$. We then define a linear functional L on $M = H^p(d\chi,Q_0)$ by setting $L(f^*) = F(0)$, where $f^* \circ \hat{\phi} = Q_0 F$ with $F \in H^p(d\sigma,\xi)$. Since $F \in H^p(d\sigma) \subseteqq H^1(d\sigma)$, we have

$$|L(f^*)| = |F(0)| \leq \int_{\mathbb{T}} |F|d\sigma = \int_{\mathbb{T}} |Q_0 F|d\sigma = \int_{\Delta_1} |f^*|d\chi$$
$$= \|f^*\|_1 \leq \|f^*\|_p.$$

By the Hahn-Banach theorem there exists a function $v^* \in L^{p'}(d\chi)$ such that $\|v^*\|_{p'} = \|L\|$ and $F(0) = L(f^*) = \int_{\Delta_1} f^*v^*d\chi$ for every $f^* \in M$. If $f^* \in M$ and $h \in H_0^\infty(R)$, then $\hat{h}f^* \in M$ and $(\hat{h}f^*) \circ \hat{\phi} = (\hat{h} \circ \hat{\phi})Q_0 F$, so that

$$\int_{\Delta_1} \hat{h}f^*v^*d\chi = L(\hat{h}f^*) = ((h \circ \phi)(0))F(0) = 0.$$

As $h \in H_0^\infty(R)$ is arbitrary, $f^*v^* \in H_0^\infty(d\chi)^\perp$. This implies that v^* belongs to M^Δ. Since $M^\Delta \subseteqq H^{p'}(d\chi,1/Q_0 Q_C)$ by Lemma 3C, there exists a $V \in H^{p'}(d\sigma)$ with $v^* \circ \hat{\phi} = (Q_0 Q_C)^{-1} V$ and therefore $(f^* v^*) \circ \hat{\phi} = Q_C^{-1} FV$ with $FV \in H^1(d\sigma, \xi_C)$. Namely, $Q_C^{-1} FV$ is the T_R-invariant meromorphic extension of $(f^* v^*) \circ \hat{\phi}$ into $\mathbb{D}$. By Theorem 2D we get

$$(9) \qquad Q_C(0)^{-1} F(0) V(0) = \int_{\Delta_1} f^* v^* d\chi = L(f^*) = F(0).$$

Since ξ has been assumed to be p-outer, $H^p(\mathbb{D}, \xi)$ has no common zeros and thus one can find an $f^* \in M$ such that $F(0) \neq 0$ with $f^* \circ \hat{\phi} = Q_0 F$. It follows from (9) that $V(0) = Q_C(0) \neq 0$, the last inequality sign being true in view of Corollary 2D.

Finally, our definition of m^p shows that $m^p(\xi, 0) = \|L\| = \|v^*\|_{p'}$. Since $V \in H^{p'}(d\sigma, \xi^{-1}\xi_C)$ and $\|V\|_{p'} = \|v^*\|_{p'}$, we have $m^{p'}(\xi^{-1}\xi_C, 0) \geqq |V(0)|/\|v^*\|_{p'}$. Hence,

$$m^p(\xi, 0) m^{p'}(\xi^{-1}\xi_C, 0) \geqq |V(0)| = |Q_C(0)| > 0. \quad \square$$

3E. We know that $H^1(d\chi, Q_C^{-1})$ is standard and therefore ξ_C is 1-outer. On the other hand, $H^\infty(d\sigma, \xi_C)$ contains the inner function Q_C so that $H^\infty(d\sigma, \xi_C) \neq \{0\}$. It then follows from Lemma 3B that ξ_C is outer.

Suppose now that $\xi \in T_R^*$ is 1-outer. Then, because of the inclusion relation $H^1(d\sigma, \xi) H^\infty(d\sigma, \xi_C) \subseteqq H^1(d\sigma, \xi\xi_C)$, $\xi\xi_C$ is also 1-outer. So, by Lemma 3D,

$$m^\infty(\xi^{-1}, 0) \geqq m^1(\xi\xi_C, 0) m^\infty((\xi\xi_C)^{-1}\xi_C, 0) \geqq |Q_C(0)| > 0.$$

This shows in particular that $H^\infty(d\sigma, \xi^{-1}) \neq \{0\}$. We further see the following

<u>Theorem</u>. If ξ is 1-outer, then ξ^{-1} is outer.

<u>Proof</u>. We have shown that $H^\infty(d\sigma, \xi^{-1}) \neq \{0\}$. Let P be the greatest common inner factor of $H^\infty(d\sigma, \xi^{-1})$. If ξ_P denotes the character of P, then $\xi^{-1}\xi_P^{-1}$ is outer. Since

$$H^\infty(d\sigma, \xi_P) H^\infty(d\sigma, \xi^{-1}\xi_P^{-1}) \subseteqq H^\infty(d\sigma, \xi^{-1}),$$

$H^\infty(d\sigma, \xi_P)$ should be divided by P. We thus conclude that $H^\infty(d\sigma, \xi_P) = PH^\infty(d\sigma, Id)$, where Id denotes the identity character of T_R. By use of induction we have

$$(10) \qquad H^\infty(d\sigma, \xi_P^n) = P^n H^\infty(d\sigma, Id)$$

for $n = 1, 2, \ldots$. On the other hand,

$$H^\infty(d\sigma,\xi_C)H^1(d\sigma,\xi)H^\infty(d\sigma,\xi^{-1}\xi_P{}^{-1}) \subseteqq H^1(d\sigma,\xi_P{}^{-1}\xi_C).$$

In view of our hypothesis, the left-hand side has no nonconstant common inner factors and therefore so does the right-hand side, i.e. $\xi_P{}^{-1}\xi_C$ is 1-outer. Again by induction we see that $\xi_P{}^{-n}\xi_C$ is 1-outer for $n = 1, 2,\ldots$. By Lemma 3D and (10) we get

$$|P(0)|^n = m^\infty(\xi_P{}^n,0) \geqq m^1(\xi_P{}^{-n}\xi_C,0)m^\infty((\xi_P{}^{-n}\xi_C)^{-1}\xi_C,0) \geqq |Q_C(0)| > 0.$$

As n is arbitrary, $|P(0)| = 1$ and hence P is a constant, as was to be proved. □

The theorem then implies almost immediately the following

Corollary. Suppose that ξ is p-outer for some p with $1 \leqq p < \infty$. Then both ξ and ξ^{-1} are outer.

3F. Let Q_0 be the greatest common inner factor of $H_0^\infty(R) \circ \phi$ so that $H_0^\infty(d\chi) = H^\infty(d\chi,Q_0)$. Using the notations in 2D, we have $Q_0 = Q_g Q_g'$. Namely, $H^\infty(d\chi,Q_0)$ is the standard expression for $H^\infty(d\chi,Q_g)$. Let ξ_0 be the character of Q_0. Then $\xi_0{}^{-1}$ is outer and, by the above corollary, ξ_0 is outer, too. So there exists a function $P_0 \in H^\infty(d\sigma,\xi_0)$ such that $P_0(0) \neq 0$ and $\|P_0\|_\infty \leqq 1/2$. For any $f \in H^\infty(R)$, $(P_0/Q_0)(f\circ\phi - f(0))$ is a T_R-invariant function in $H^\infty(\mathbb{D})$, so that we find a unique $h \in H^\infty(R)$ with $h\circ\phi = (P_0/Q_0)(f\circ\phi - f(0))$. We then define a linear operator T in $H^\infty(R)$ by setting $Tf = h$.

Lemma. (a) T is a nonzero bounded linear operator in $H^\infty(R)$ with $\|T\| \leqq 1$ and

$$T(fg) = (Tf)g + f(0)Tg \tag{11}$$

for any $f, g \in H^\infty(R)$.

(b) Define $T_0^n f = (T^n f)(0)$ for $f \in H^\infty(R)$ and $n = 0, 1,\ldots$. Then, $\|T_0^n\| \leqq 1$ for $n = 0, 1,\ldots$ and

$$T_0^n(fg) = \sum_{j=0}^{n} T_0^{n-j}(f)T_0^j(g) \tag{12}$$

for any $f, g \in H^\infty(R)$.

Proof. (a) T is obviously linear. For $f \in H^\infty(R)$ we write $f\circ\phi - f(0) = Q_0 F$ with $F \in H^\infty(d\sigma,\xi_0{}^{-1})$. Then

$$\|Tf\|_\infty = \|(P_0/Q_0)Q_0F\|_\infty = \|P_0F\|_\infty \leqq \|P_0\|_\infty\|F\|_\infty$$

$$= \|P_0\|_\infty \|f - f(0)\|_\infty \leq 2\|P_0\|_\infty \|f\|_\infty \leq \|f\|_\infty .$$

This means that $\|T\| \leq 1$. On the other hand, since ξ_0^{-1} is outer, there exists an $f_0 \in H_0^\infty(R)$ such that $f_0 \circ \phi = Q_0 F_0$ with $F_0(0) \neq 0$ and therefore $(Tf_0)(0) = P_0(0)F_0(0) \neq 0$. Thus T is nontrivial. The verification of (11) is easy and is omitted.

(b) For $f \in H^\infty(R)$ we have

$$|T_0^n f| = |(T^n f)(0)| = \left|\int_{\Delta_1} (T^n f)d\chi\right| \leq \|T^n f\|_\infty \leq \|f\|_\infty ;$$

so $\|T_0^n\| \leq 1$. The identity (12) follows immediately from the relation

$$T^n(fg) = (T^n f)g + \sum_{j=1}^{n} (T^{n-j}f)(0)T^j g,$$

which can be shown by an easy induction. □

3G. We now expand functions in $H^\infty(R)$ by means of the point derivation T. Since $\|T\| \leq 1$, $I - \zeta T$ is invertible for $|\zeta| < 1$. We set $S_\zeta = (I - \zeta T)^{-1}$ and $\Phi_\zeta f = (S_\zeta f)(0)$.

Lemma. (a) For every $f \in H^\infty(R)$ the function $\zeta \to \Phi_\zeta f$ is holomorphic in $\{|\zeta| < 1\}$ and

$$\Phi_\zeta f = \sum_{n=0}^{\infty} \zeta^n (T_0^n f). \tag{13}$$

(b) For every fixed ζ with $|\zeta| < 1$, the functional $f \to \Phi_\zeta f$ is a nontrivial complex homomorphism on $H^\infty(R)$.

(c) The map $\zeta \to \Phi_\zeta$, $|\zeta| < 1$, is injective.

(d) For every $h \in H_0^\infty(R)$ and $f \in H^\infty(R)$ we have

$$(h - \zeta Th)(TS_\zeta f) = (Th)(f - \Phi_\zeta f) \tag{14}$$

for any ζ with $|\zeta| < 1$.

Proof. (a) This is almost clear, for $S_\zeta = \sum_{n=0}^\infty \zeta^n T^n$.

(b) This follows easily from Lemma 3F and (13).

(c) Let $f_0 \in H_0^\infty(R)$ satisfy $T_0 f_0 \neq 0$, as in the proof of Lemma 3F. If $\Phi_\zeta = \Phi_{\zeta'}$ with $|\zeta|, |\zeta'| < 1$, then

$$\Phi_\zeta((I - \zeta T)(I - \zeta' T)f_0) = \Phi_{\zeta'}((I - \zeta T)(I - \zeta' T)f_0)$$

and thus $\zeta'(T_0 f_0) = \zeta(T_0 f_0)$. Since $T_0 f_0 \neq 0$, we have $\zeta = \zeta'$.

(d) In view of the identity (11), we have

$$T(hS_\zeta f) = (Th)(S_\zeta f) + h(0)(TS_\zeta f)$$
$$= (TS_\zeta f)h + (S_\zeta f)(0)Th,$$

so that $(TS_\zeta f)h = (Th)(S_\zeta f - \Phi_\zeta f)$. From this follows that

$$(h - \zeta Th)(TS_\zeta f) = h(TS_\zeta f) - \zeta(Th)(TS_\zeta f)$$
$$= (Th)(S_\zeta f - \Phi_\zeta f - \zeta(TS_\zeta f)) = (Th)(f - \Phi_\zeta f),$$

for $(I - \zeta T)S_\zeta = I$. □

3H. We now use the assumption that $H^\infty(R)$ separates the points of R. Choose an $f_0 \in H_0^\infty(R)$ with $T_0 f_0 = (Tf_0)(0) \neq 0$ and set

$$U_0 = \{z \in R: (Tf_0)(z) \neq 0\}.$$

Since $Tf_0 \in H^\infty(R)$ and $(Tf_0)(0) \neq 0$, U_0 is an open set containing the point 0 and $R \setminus U_0$ is discrete. We define a function $\psi: U_0 \to \mathbb{C}$ by setting $\psi(z) = f_0(z)/(Tf_0)(z)$. An application of (14) to $h = f_0$ then shows that

$$f(z) = \Phi_{\psi(z)} f$$

for any $f \in H^\infty(R)$ and $z \in U_0$. Namely, $\Phi_{\psi(z)}$ is exactly the evaluation functional ε_z at the point z. Choose any compact neighborhood U_0' of the origin 0 with $U_0' \subsetneqq U_0$. Since ψ is a nonconstant holomorphic function on U_0 and vanishes at 0, the image $\psi(U_0')$ includes a disk $\{|\zeta| < r_0\}$ for some $r_0 > 0$.

<u>Theorem</u>. Suppose that the condition (c), 1C, holds and that $H^\infty(R)$ separates the points of R. Let ξ_g be the character of Q_g. Then ξ_g^{-1} is outer.

<u>Proof</u>. We have only to show that Q_g' is a constant function. Since $H^\infty(R)$ separates the points of R, distinct points z in R give distinct evaluation functionals ε_z. Combined with the above remark, this means that $\Phi_{\psi(z)} \neq \Phi_{\psi(z')}$ for any distinct $z, z' \in U_0$. So, by Lemma 3G-(c), ψ is an injection. It follows that $(d\psi/dz)(0) \neq 0$ with respect to any local coordinate z at 0. Consequently, we have

$$(df_0/dz)(0) = (d\Phi_\zeta f_0/d\zeta)(0)\cdot(d\psi/dz)(0) \neq 0$$

for any $f_0 \in H_0^\infty(R)$ with $T_0 f_0 \neq 0$. So f_0 has a simple zero at the point 0. Since $Q_g Q_g'$ is an inner factor of f_0, we see that Q_g' cannot vanish at 0. Moreover Q_g' has no zeros in R. If this were not the case, then $Q_g'(a) = 0$ for some $a \neq 0$. Thus $f(0) = f(a) = 0$ for any $f \in H_0^\infty(R)$, so that $H^\infty(R)$ could not separate the points of R, contrary to our basic hypothesis.

To get the final conclusion, we suppose, on the contrary, that Q_g' be a singular inner function. Then, in the definition of the operator T, we could choose $P_0 \in H^\infty(d\sigma,\xi_0)$ in such a way that $P_0(0) \neq 0$ and P_0/Q_g' would be unbounded. Let h_0, k_0 be the l.a.m.'s with $h_0 \circ \phi = |P_0|$ and $k_0 \circ \phi = |Q_g'|$. Thus, $h_0(0) \neq 0$ and h_0/k_0 would be unbounded. Since k_0 is bounded away from zero on the compact set U_0', there would exist $z_n \in U_0 \setminus U_0'$ such that $(h_0/k_0)(z_n) \to \infty$. Then we would have

$$|\psi(z_n)| = |f_0(z_n)|/|(Tf_0)(z_n)| \leqq k_0(z_n)/h_0(z_n) \to 0.$$

So, for a sufficiently large n, we would have $|\psi(z_n)| < r_0$. The definition of r_0 shows that there exists a point $z^* \in U_0'$ with $\psi(z^*) = \psi(z_n)$. This is, however, impossible, for ψ is an injection. □

As we noticed in 1C, the condition (c) is independent of the choice of the origin and therefore we get the following

<u>Corollary</u>. Let $a_1,\ldots, a_n$ be a finite number of points in R and set $q(z) = \exp(-\sum_{j=1}^n g(a_j,z))$. Let ξ_q be the character of q. Then, under the same hypothesis on R as in the theorem, ξ_q^{-1} is outer.

3I. <u>Lemma</u>. Suppose that $H^\infty(R)$ separates the points of R. Let ξ be any character of $F_0(R)$. Then, for any canonical subregion G of R, there exists an l.a.m. q of the form $q(z) = \exp(-\sum_{j=1}^n g(\zeta_j,z))$ such that $\xi_q(\gamma) = \xi(\gamma)$ for any curve $\gamma \in F_0(R)$ contained in G, where ξ_q denotes the character of q and the points $\{\zeta_j\}$ can be taken from any fixed open subset of G.

<u>Proof</u>. The proof goes along the same lines as that of Lemma 8B, Ch. V. Let $R_0',\ldots, R_\ell'$ be an enumeration of the components of $R \setminus Cl(G)$ and let γ_j, $0 \leqq j \leqq \ell$, be the boundary contour of R_j', which is oriented positively with respect to G. So a homology basis for $Cl(G)$ consists of $\gamma_1,\ldots, \gamma_\ell$ and certain nondividing cycles $\gamma_{\ell+1},\ldots, \gamma_N$. Let V be a parametric disk such that $Cl(V) \subseteq G$ and the curves γ_k's lie off $Cl(V)$. Let $\sigma_k = \sigma(\gamma_k)$, $1 \leqq k \leqq N$, be the harmonic differentials in $\Gamma_{h0}(R)$ representing γ_k (Ch. I, 9B). If we write $\sigma_k = u_k dx + v_k dy$, $1 \leqq k \leqq N$, in V, then, as we saw in the proof of Lemma 8B, Ch. V, $\{u_k: 1 \leqq k \leqq N\}$ are linearly independent on V. So there exist points $a_1,\ldots, a_N$ in V such that

$$\Delta(a_1,\ldots,a_N) = \det(u_j(a_k))_{1\leq j,k\leq N} \neq 0.$$

By identifying V with the unit disk, we set

$$\Phi(x_1,\ldots,x_N) = (\sum_{k=1}^{N}\int_{a_k}^{a_k+x_k}\sigma_1,\ldots,\sum_{k=1}^{N}\int_{a_k}^{a_k+x_k}\sigma_N),$$

which is defined in $U = \{(x_1,\ldots,x_N): |x_k| < r\}$ with $r = \min\{1-|a_k|: k = 1,\ldots, N\}$. Since the Jacobian of Φ at the origin does not vanish, $\Phi(U)$ contains an open set in $\mathbb{R}^N$ and thus $(2\pi M)\Phi(U)$ contains an N-cube with sides of length 2π for a sufficiently large integer $M > 0$. By use of Lemma 7E, Ch. V, we have

$$M(\sum_{k=1}^{N}\int_{\gamma_1}{*dg}(\cdot,a_k+x_k),\ldots,\sum_{k=1}^{N}\int_{\gamma_N}{*dg}(\cdot,a_k+x_k))$$

$$= M(\sum_{k=1}^{N}\int_{\gamma_1}{*dg}(\cdot,a_k),\ldots,\sum_{k=1}^{N}\int_{\gamma_N}{*dg}(\cdot,a_k)) - 2\pi M\Phi(x_1,\ldots,x_N).$$

The definition of M implies that the right-hand side covers an N-cube with sides of length 2π when $(x_1,\ldots,x_N)$ ranges over U. So, for any $(b_1,\ldots,b_N) \in \mathbb{R}^N$ there exists an $(x_1,\ldots,x_N) \in U$ such that

$$M\sum_{k=1}^{N}\int_{\gamma_j}{*dg}(\cdot,a_k+x_k) \equiv b_j \pmod{2\pi}$$

for $j = 1,\ldots, N$. If $\zeta_1,\ldots,\zeta_n$ are the points a_k+x_k, $1 \leq k \leq N$, each repeated M times, then ζ_k's belong to V and

$$\sum_{k=1}^{n}\int_{\gamma_j}{*dg}(\cdot,\zeta_k) \equiv b_j \pmod{2\pi}$$

for $j = 1,\ldots, N$. Thus, given $\xi \in F_0(R)^*$, we find $\zeta_1,\ldots,\zeta_n \in V$ such that

$$\sum_{k=1}^{n}\int_{\gamma}{*dg}(\cdot,\zeta_k) \equiv -\arg(\xi(\gamma)) \pmod{2\pi}$$

for $\gamma = \gamma_j$, $1 \leq j \leq N$, and therefore for any $\gamma \in F_0(R)$ lying in G. We set $q(z) = \exp(-\sum_{k=1}^{n} g(z,\zeta_k))$ and let ξ_q be the character of the l.a.m. $q(z)$. Then $\xi_q(\gamma) = \xi(\gamma)$ for any $\gamma \in F_0(R)$ lying in G. □

3J. We are now able to prove our final objective easily.

<u>Theorem</u>. Suppose that (c), 1C, holds and that $H^\infty(R)$ separates the points of R. Then R is a PWS and $q_C = g^{\dagger(0)}|R$, where $g^{\dagger(0)}$ is the l.a.m. defined in 1B.

Proof. Let $\{G_n: n = 1, 2, \ldots\}$ be a canonical exhaustion of R (see Ch. I, 1C) with $0 \in G_1$. Take any character ξ of $F_0(R)$. By Lemma 3I and Corollary 3H there exists, for each n, an outer character ξ_n of $F_0(R)$ such that $\xi_n(\gamma) = \xi(\gamma)$ for $\gamma \in F_0(R)$ lying in G_n. Since ξ_n^{-1} is also outer, $\xi_n^{-1}\xi_C$ is outer and therefore, by Lemma 3D,

$$m^\infty(\xi_n) \geqq |Q_C(0)| > 0.$$

So we can choose, for each n, an $f_n \in H^\infty(R,\xi_n)$ such that $\|f_n\|_\infty = 1$ and $|f_n(0)| \geqq |Q_C(0)|/(1 + n^{-1})$. We regard ξ and ξ_n as characters of T_R and denote by F_n the function in $H^\infty(\mathbb{D},\xi_n)$ such that $F_n = (f_n)_0 \circ \phi$ in some neighborhood of the origin 0, where $(f_n)_0$ is the principal branch of f_n (Ch. II, 2D). Since $\{F_n\}$ is a bounded family, we may assume, passing to a subsequence if necessary, that $\{F_n\}$ converges to a function F almost uniformly on $\mathbb{D}$. Then, it is clear that F is a holomorphic function satisfying $\|F\|_\infty \leqq 1$ and $|F(0)| \geqq |Q_C(0)|$. Moreover, we have $F \in H^\infty(\mathbb{D},\xi)$. In fact, let $\tau \in T_R$ and let γ_τ be the image under ϕ of the radial segment $\tau(0)t$, $0 \leqq t \leqq 1$, so that γ_τ is the image of τ under the canonical isomorphism of T_R onto $F_0(R)$. Since $\{G_n\}$ is an exhaustion, there exists an $N > 0$ such that $\gamma_\tau \subseteqq G_n$ for all $n \geqq N$. Then the definition of ξ_n implies that $\xi_n(\gamma_\tau) = \xi(\gamma_\tau)$ for $n \geqq N$. It follows that

$$F_n \circ \tau = \xi_n(\tau)F_n = \xi_n(\gamma_\tau)F_n = \xi(\gamma_\tau)F_n = \xi(\tau)F_n.$$

Letting $n \to \infty$, we have $F \circ \tau = \xi(\tau)F$. As $\tau \in T_R$ is arbitrary, we conclude that $F \in H^\infty(\mathbb{D},\xi)$. If f denotes the multiplicative holomorphic function on R such that $f_0 \circ \phi = F$ in a neighborhood of 0, then it is clear that $f \in H^\infty(R,\xi)$, $\|f\|_\infty \leqq 1$ and $|f(0)| \geqq |Q_C(0)|$. It follows that $m^\infty(\xi,0) \geqq |Q_C(0)|$. As $\xi \in F_0(R)^*$ is arbitrary, we finally have $m^\infty(0) \geqq |Q_C(0)| > 0$. Hence, R is a PWS.

In order to show the second assertion of the theorem, let $R^\dagger$ be the regularization of R and use the notations in Ch. V, 3B. For each $a \in R^\dagger$ with $a \neq 0$ we set $q_a^\dagger(z) = \exp(-g^\dagger(a,z))$, $z \in R^\dagger$, and $q_a = q_a^\dagger|R$. Let $\xi_a^\dagger$ (resp. ξ_a) be the character of $F_0(R^\dagger)$ (resp. $F_0(R)$) defined by $q_a^\dagger$ (resp. q_a). Every bounded holomorphic function on R admits a unique holomorphic extension to the whole $R^\dagger$. So, if we set $H_a^\infty(R^\dagger) = \{f \in H^\infty(R^\dagger): f(a) = 0\}$, then $H_a^\infty(R^\dagger) = \{f \in H^\infty(R^\dagger): f/q_a^\dagger$ is bounded$\}$ and therefore the restrictions, to R, of functions in $H_a^\infty(R^\dagger)$ is exactly equal to

$$H_a^\infty(R) = \{f \in H^\infty(R): f/q_a \text{ is bounded}\}.$$

Let $S_a^\dagger(z) = \exp(-g^\dagger(a,z) - i\tilde{g}^\dagger(a,z))$ for $z \in R^\dagger$ and S_a the restriction of $S_a^\dagger$ to R, so that $S_a^\dagger$ (resp. S_a) has character $\xi_a^\dagger$ (resp. ξ_a). Then we have in an obvious sense $H_a^\infty(R^\dagger) = S_a^\dagger \mathcal{H}^\infty(R^\dagger, \xi_a^{\dagger -1})$ and $H_a^\infty(R) = S_a \mathcal{H}^\infty(R, \xi_a^{-1})$. Therefore, $\mathcal{H}^\infty(R,\xi_a^{-1})$ is regarded as the restriction of $\mathcal{H}^\infty(R^\dagger, \xi_a^{\dagger -1})$ to R. Since $\xi_a^{\dagger -1}$ is outer by Theorem 3H, $\mathcal{H}^\infty(R^\dagger, \xi_a^{\dagger -1})$ has no nonconstant common inner factors and so does the space $\mathcal{H}^\infty(R,\xi_a^{-1})$. Hence ξ_a^{-1} is outer for any $a \in R^\dagger$.

Let now $\{z_1, z_2, \ldots\}$ be an enumeration of $Z(0;R^\dagger)$. For every $n = 1, 2, \ldots$ let

$$q_n^\dagger = \exp(-\sum_{j=1}^{n} g^\dagger(z_j, z))$$

for $z \in R^\dagger$, $q_n = q_n^\dagger | R$, and let ξ_n be the character of $F_0(R)$ determined by q_n. Then the above observation shows that each ξ_n is an outer character. Let Q_n be the inner function on $\mathbb{D}$ such that $Q_n(0) \geqq 0$ and $|Q_n| = q_n \circ \phi$. Let f be meromorphic on R such that $|f|q_n$ has a harmonic majorant on R. As is seen easily, f has a unique meromorphic extension to $R^\dagger$, which we denote by the same letter. Take any $h \in H_0^\infty(R)$, which is also regarded as an element in $H_0^\infty(R^\dagger)$. Then $|fh|q_n^\dagger$ has a harmonic majorant on $R^\dagger$. It follows from Theorem 4B, Ch. VII, and Theorem 3B, Ch. V, that

$$\int_{\Delta_1} \hat{f}\hat{h}\,d\chi = \int_{\Delta_1(R^\dagger)} \hat{f}\hat{h}\,d\chi^\dagger = f(0)h(0) = 0.$$

This means that $H^1(d\chi, Q_n^{-1}) \subseteqq H_0^\infty(d\chi)^\perp \subseteqq H^1(d\chi, Q_C^{-1})$. Since ξ_n is outer, we see that Q_C/Q_n is inner. Namely, q_C/q_n is an inner l.a.m. on R; or, equivalently,

$$q_C(z) \leqq \exp(-\sum_{j=1}^{n} g^\dagger(z_j, z))$$

on R for any $n = 1, 2, \ldots$. Letting $n \to \infty$, we conclude that q_C is divided by $q = g^{\dagger(0)}|R$, where

$$g^{\dagger(0)}(z) = \exp(-\sum_{j=1}^{\infty} g^\dagger(z_j, z)).$$

In other words, if $Q^{\dagger(0)}$ is the inner function on $\mathbb{D}$ such that $Q^{\dagger(0)}(0) \geqq 0$ and $|Q^{\dagger(0)}| = q \circ \phi$, then $Q^{\dagger(0)}$ is an inner factor of Q_C.

On the other hand, we claim that $H^\infty(d\chi, 1/Q^{\dagger(0)})^\perp \subseteqq H_0^1(d\chi)$. To see this, let $f^* \in H^\infty(d\chi, 1/Q^{\dagger(0)})^\perp$. This means that $f^* \in L^1(d\chi)$ and

$\int_{\Delta_1} f^*\hat{h}d\chi = 0$ for any h, meromorphic on R, such that $|h|q$ is bounded. By the inverse Cauchy theorem (Theorem 1C, Ch. VII) there is a unique $f \in H^1(R)$ with $f^* = \hat{f}$ a.e. on Δ_1. Since $H^\infty(d\chi, 1/Q^{\dagger(0)})$ contains constant functions, we have

$$f(0) = \int_{\Delta_1} \hat{f}d\chi = \int_{\Delta_1} f^*d\chi = 0.$$

Hence, $f \in H_0^1(R)$, as claimed. We now repeat our argument in 1B and get

$$H_0^\infty(d\chi)^\perp \subseteq H^1(d\chi, 1/Q^{\dagger(0)}).$$

So, in view of the definition of Q_C (see 2D), $[I_0 H^1(d\chi, Q_C^{-1})]_1 \subseteq H^1(d\chi, 1/Q^{\dagger(0)})$. Since I_0 has no nonconstant common inner factors and ξ_C is an outer character, this implies that $Q^{\dagger(0)}/Q_C$ is also an inner function. Summing up, we have shown that Q_C and $Q^{\dagger(0)}$ differ only by a constant factor of modulus one. (Indeed, $Q_C = Q^{\dagger(0)}$ because we assumed that $Q_C(0) > 0$ and $Q^{\dagger(0)}(0) > 0$.) Hence, $q_C = g^{\dagger(0)}|R$. □

§2. CONDITIONS EQUIVALENT TO THE DIRECT CAUCHY THEOREM

4. General Discussion

4A. The direct Cauchy theorem was studied in the preceding two chapters. We are now able to deepen our discussion by taking into account of the results in §1. Our present objective is to seek for a "direct Cauchy theorem" which contains Theorem 4B, Ch. VII, for every possible choice of $\{z_j\}$. We need a definition valid for general hyperbolic Riemann surfaces.

Definition. Let q be an inner l.a.m. on R. We say that the direct Cauchy theorem (abbreviated to (DCT)) holds for q, if we have

$$f(0) = \int_{\Delta_1} \hat{f}(b)d\chi(b)$$

for any meromorphic function f on R such that $|f|q$ has a harmonic majorant.

There are two types of (DCT), which are valid for any R.

(i) It is trivial that the constant function $q \equiv 1$ satisfies (DCT), for we have $f \in H^1(R)$ in this case.

(ii) If $\{z_1, z_2, \dots\}$ is an enumeration of $Z(O;R)$, then Theorem 4B, Ch. VII, shows that any inner l.a.m. q of the form

$$q(z) = \exp(-\sum_{j=1}^{n} g(z_j, z))$$

satisfies (DCT) for $n = 1, 2, \dots$.

Suppose now that R carries an inner l.a.m. q such that q satisfies a (DCT) which includes the case (ii) for all $n = 1, 2, \dots$. Then q can be divided by $\exp(-\sum_{j=1}^{n} g(z_j, z))$ for $n = 1, 2, \dots$ in the set of inner l.a.m.'s. It follows that q should be divided by $\exp(-\sum_{j=1}^{\infty} g(z_j, z))$ and therefore that $\sum_{j=1}^{\infty} g(z_j, z)$ should converge for some z. If, moreover, R is assumed regular in the sense of potential theory, then our result implies that R is a PWS.

4B. We thus assume, in what follows in this section, that R is a regular PWS. Then the direct Cauchy theorem we wish to have is just the (DCT_a) given in Ch. VII, 3C. By Theorem 5A, Ch. VIII, (DCT_a)'s with $a \in R$ are mutually equivalent, so that these conditions are denoted collectively by (DCT).

Lemma. Let R be a regular PWS for which (DCT) holds. Let $Q^{(O)}$ be the inner function on $\mathbb{D}$ such that $g^{(O)} \circ \phi = |Q^{(O)}|$ and $Q^{(O)}(0) > 0$, where $g^{(O)}$ was defined in Ch. VII, 1A. Then we have

$$H_0^\infty(d\chi)^\perp = H^1(d\chi, 1/Q^{(O)}).$$

Proof. Let $f^* \in H^1(d\chi, 1/Q^{(O)})$. Then there exists an $F \in H^1(d\sigma)$ such that $f^* \circ \hat{\phi} = (1/Q^{(O)})F$. Since $(1/Q^{(O)})F$ is T_R-invariant on $\mathbb{T}$, it is T_R-invariant on $\mathbb{D}$. Thus there exists a meromorphic function f on R such that $f \circ \phi = (1/Q^{(O)})F$ on $\mathbb{D}$ and $f^* = \hat{f}$ a.e. on Δ_1. Take any $h \in H_0^\infty(R)$. Then $|fh|g^{(O)}$ has a harmonic majorant on R, so that

$$\int_{\Delta_1} \hat{f}\hat{h}\, d\chi = f(O)h(O) = 0$$

by (DCT). Namely, $f^* \in H_0^\infty(d\chi)^\perp$ and therefore $H^1(d\chi, 1/Q^{(O)})$ is included in $H_0^\infty(d\chi)^\perp$. On the other hand, we have shown in the proof of Theorem 3J that $H_0^\infty(d\chi)^\perp \subseteqq H^1(d\chi, 1/Q^{(O)})$. Hence we are done. □

4C. It is now easy to prove our first characterization of (DCT).

Theorem. Let R be a regular PWS with origin O. Then the following conditions are equivalent:

(a) (DCT) holds.

(b) $H_0^\infty(d\chi)^\perp = H^1(d\chi, 1/Q^{(0)})$.

(c) Every simply invariant subspace of $L^p(d\chi)$, $1 \leqq p \leqq \infty$, is of the form $H^p(d\chi, Q)$ for an i-function Q on $\mathbb{T}$.

Proof. By Lemma 4B, (a) implies (b). Conversely, suppose that (b) is satisfied. Let f be a meromorphic function on R such that $|f|g^{(0)}$ has a harmonic majorant on R. Then, $|(f \circ \phi)Q^{(0)}|$ has a harmonic majorant on $\mathbb{D}$ and therefore $\hat{f} \in H^1(d\chi, 1/Q^{(0)}) = H_0^\infty(d\chi)^\perp$. So, by Theorem 2D, $f(0) = \int_{\Delta_1} \hat{f} d\chi$. Hence, (DCT) holds. On the other hand, Theorem 4C, Ch. VIII, states that (a) implies (c). Finally, suppose that (c) holds. Since $H_0^\infty(d\chi)^\perp$ is easily seen to be a simply invariant subspace of $L^1(d\chi)$, there exists an i-function Q on $\mathbb{T}$ such that $H_0^\infty(d\chi)^\perp = H^1(d\chi, Q)$. Then

$$[I_0 H^1(d\chi, 1/Q^{(0)})]_1 = H^1(d\chi, Q),$$

by use of Theorems 2D and 3J. Lifting to $\mathbb{T}$, we see that

$$(I_0 \circ \hat{\phi})(1/Q^{(0)}) H^1(d\sigma, \xi_C) \subseteqq Q H^1(d\sigma, \xi_Q^{-1}).$$

Since $I_0 \circ \hat{\phi}$ has no nonconstant common inner factors and since ξ_C is outer, we see that $(QQ^{(0)})^{-1}$ is an inner function, say P. Thus

$$H^1(d\chi, 1/Q^{(0)}) \subseteqq H^1(d\chi, 1/PQ^{(0)}) = [I_0 H^1(d\chi, 1/Q^{(0)})]_1 \subseteqq H^1(d\chi, 1/Q^{(0)}),$$

which shows the statement (b). This completes the proof. □

5. Functions $m^p(\xi, a)$ and (DCT)

5A. We will give our second characterization of (DCT), in which the functions $m^p(\xi, a)$ (see 1D) play a principal role. By use of the notations given in Ch. VII, 1A, we first get the following

Theorem. Let R be a regular PWS. Then it satisfies (DCT) if and only if

$$m^1(\xi^{(a)}, a) = g^{(a)}(a) \tag{15}$$

for some (and hence all) $a \in R$.

Proof. Suppose that (DCT) holds and choose any $a \in R$. Let h be in $\mathcal{H}^1(R, \xi^{(a)})$ with $\|h\|_{1,a} \leqq 1$ and set $f = h/S^{(a)}$. Then f is a meromorphic function on R such that $|f|g^{(a)}$ has a harmonic majorant on

R. Setting $u(z) = |h(z)|$, we get an l.a.m. u such that $LHM(u) = H[\hat{u}]$ and $|\hat{f}| = \hat{u}/\hat{g}^{(a)} = \hat{u}$ a.e. on Δ_1. As we are assuming (DCT), we have

$$f(a) = \int_{\Delta_1} \hat{f}(b)d\chi_a(b),$$

with $d\chi_a(b) = k_b(a)d\chi(b)$, and therefore

$$|f(a)| \leq \int_{\Delta_1} |\hat{f}(b)|d\chi_a(b) = \int_{\Delta_1} \hat{u}(b)d\chi_a(b)$$
$$= \|h\|_{1,a} \leq 1.$$

So $|h(a)| = |f(a)|\, g^{(a)}(a) \leq g^{(a)}(a)$. As h is arbitrary, we have

$$g^{(a)}(a) \geq \sup\{|h(a)|: h \in \mathcal{H}^1(R,\xi^{(a)}),\ \|h\|_{1,a} \leq 1\}$$
$$= m^1(\xi^{(a)},a).$$

On the other hand, the function $S^{(a)}$ belongs to $\mathcal{H}^1(R,\xi^{(a)})$ with the properties: $\|S^{(a)}\|_{1,a} \leq \|S^{(a)}\|_\infty \leq 1$ and $|S^{(a)}(a)| = g^{(a)}(a)$. Thus $m^1(\xi^{(a)},a) \geq g^{(a)}(a)$. Hence the equation (15) holds for all $a \in R$.

(b) Suppose conversely that the equation (15) holds for some $a \in R$. We denote by B the set of functions h defined on $R \setminus Z(a;R)$ such that (i) $|h|$ has a harmonic majorant on R and (ii) $h/g^{(a)}$ extends to a meromorphic function on R. In other words,

$$B = (\exp(i\tilde{s}^{(a)}))\mathcal{H}^1(R,\xi^{(a)}).$$

Take any $h \in B$. Then $h/g^{(a)}$ is a meromorphic function on R, which, by Lemma 3B, Ch. VII, admits fine boundary values a.e. on Δ_1. As we have $\hat{g}^{(a)} = 1$ a.e. on Δ_1 (Lemma 2A, Ch. VII), so $\hat{h}$ exists a.e. and, in fact, belongs to the space $L^1(d\chi)$. We set

$$\|h\|_B = \int_{\Delta_1} |\hat{h}(b)|d\chi_a(b),$$

which defines a norm in the linear space B, so that B is identified with a subspace of $L^1(d\chi_a)$. Define a linear functional L on B by setting $L(h) = h(a)/g^{(a)}(a)$ for $h \in B$. If $h \in B$, then we have $h\cdot\exp(-i\tilde{s}^{(a)}) \in \mathcal{H}^1(R,\xi^{(a)})$ and so

$$|h(a)| = |h(a)\exp(-i\tilde{s}^{(a)}(a)|$$
$$\leq \|h\|_B\cdot m^1(\xi^{(a)},a)$$

$$= \|h\|_B \cdot g^{(a)}(a)$$

by use of (15). This means that $|L(h)| \leqq \|h\|_B$ for every $h \in B$. So we see via the Hahn-Banach theorem that L has a norm-preserving extension to $L^1(d\chi_a)$ and that there exists a function $w \in L^\infty(d\chi_a)$ with $\|w\|_\infty = \|L\| \leqq 1$ and

$$L(h) = \int_{\Delta_1} \hat{h}(b)w(b)d\chi_a(b)$$

for every $h \in B$. In particular, for $h = g^{(a)}$ we have

$$1 = L(g^{(a)}) = \int_{\Delta_1} w(b)d\chi_a(b) \leqq \int_{\Delta_1} |w(b)|d\chi_a(b) \leqq 1.$$

Thus $w = 1$ a.e. on Δ_1 and therefore

$$L(h) = \int_{\Delta_1} \hat{h}(b)d\chi_a(b)$$

for every $h \in B$.

To finish the proof, let us take any meromorphic function f on R such that $|f|g^{(a)}$ has a harmonic majorant. Then $h = fg^{(a)} \in B$ and $\hat{f} = \hat{h}$ a.e. on Δ_1. From what we have seen above follows that

$$f(a) = h(a)/g^{(a)}(a) = L(h) = \int_{\Delta_1} \hat{h}(b)d\chi_a(b) = \int_{\Delta_1} \hat{f}(b)d\chi_a(b).$$

Hence, (DCT_a) holds, as was to be proved. □

5B. Let $\{\alpha_n: n = 1, 2, \ldots\}$ be a strictly decreasing sequence of positive numbers tending to 0 such that each $\{z \in R: g(\mathit{0},z) = \alpha_n\}$ is disjoint from $Z(\mathit{0};R)$. We set $R_n = R(\alpha_n,\mathit{0})$ and denote by $g_n(a,z)$ the Green function for R_n. Let η_n be the character of the l.a.m. $u_n^{(\mathit{0})}(z) = \exp(-\sum\{g_n(w,z): w \in Z(\mathit{0};R_n)\})$. Since $g_n(\mathit{0},z) = g(\mathit{0},z) - \alpha_n$, we have $Z(\mathit{0};R_n) = Z(\mathit{0};R) \cap R_n$ for $n = 1, 2, \ldots$. We define a multiplicative holomorphic function $B \in \mathcal{H}^\infty(R,\xi^{(\mathit{0})})$ (resp. $B_n \in \mathcal{H}^\infty(R_n,\eta_n)$) by the conditions $B_{\mathit{0}}(\mathit{0}) > 0$ and $|B(z)| = g^{(\mathit{0})}(z)$ (resp. $(B_n)_{\mathit{0}}(\mathit{0}) > 0$ and $|B_n(z)| = u_n^{(\mathit{0})}(z)$), where the suffix $\mathit{0}$ denotes the principal branch at $\mathit{0}$. For any fixed $a \in R$, $g_n(a,z)$ converge to $g(a,z)$ almost uniformly on $R \setminus \{a\}$. Passing to a subsequence if necessary, we may assume that $B_n(z)$ converge to $B(z)$ almost uniformly on R. So, for each fixed $\gamma \in F_{\mathit{0}}(R)$, we have $\eta_n(\gamma) \to \xi^{(\mathit{0})}(\gamma)$ as $n \to \infty$.

The following lemma gives a result which is an adaptation of the equation (15) to an arbitrary PWS R.

Lemma. Let $1 \leq p < \infty$ and $p' = p/(p-1)$. Then

$$g^{(0)}(0) \leq m^{p'}(\xi,0)m^p(\xi^{-1}\xi^{(0)},0) \leq m^1(\xi^{(0)},0)$$

for any $\xi \in F_0(R)^*$.

Proof. By the normal family argument we easily see that

$$m^{p'}(\xi,0) = \lim_{n\to\infty} m^{p'}(\xi|_n,0), \tag{16}$$

where $\xi|_n$ denotes the restriction of ξ to R_n. Next we recall that the formula (7) in Ch. V, 4C, asserts

$$\inf\{\|f\|_{p'}: f \in \mathcal{H}^\infty(R_n,\xi|_n), |f(0)| = 1\}$$
$$= \sup\{|h(0)|: h \in \mathcal{H}^p(R_n,\xi|_n{}^{-1}\eta_n), \|h\|_{p,n} = r_n{}^{-1}\},$$

where $\|h\|_{p,n} = [-(1/2\pi)\int_{\partial R_n}|h|^p d\tilde{g}_n(\cdot,0)]^{1/p}$ and $r_n = u_n^{(0)}(0)$. This then implies

$$m^{p'}(\xi|_n,0)m^p(\xi|_n{}^{-1}\eta_n,0) = r_n. \tag{17}$$

We now estimate the limit of $m^p(\xi|_n{}^{-1}\eta_n,0)$ as $n \to \infty$. For this aim we choose $h_n \in \mathcal{H}^p(R_n,\xi|_n{}^{-1}\eta_n)$, for each n, such that $\|h\|_{p,n} = 1$ and $|h_n(0)| \geq m^1(\xi|_n{}^{-1}\eta_n,0)(n-1)/n$. Since R is a PWS and since every character of $F_0(R_n)$ is the restriction of some character of $F_0(R)$ by Theorem 3B, Ch. I and Theorem 1B, Ch. II, we have

$$m^p(\xi|_n{}^{-1}\eta_n,0) \geq \inf\{m^p(\xi',0): \xi' \in F_0(R)^*\} = m^p(0) > 0$$

and therefore $|h_n(0)| \geq (n-1)m^p(0)/n$ for $n = 1, 2, \ldots$. As we have seen in the proof of Theorem 5A, Ch. V, $\{|h_n(z)|\}$ is uniformly bounded on any compact subset of R and, by passing to a subsequence if necessary, we may assume that the sequence $\{h_n(z)\}$ converges almost uniformly to some multiplicative analytic function, say h, on R. It is immediate that $\|h\|_p \leq 1$ and $h \in \mathcal{H}^p(R,\xi^{-1}\xi^{(0)})$. So

$$m^p(\xi^{-1}\xi^{(0)},0) \geq \limsup_{n\to\infty} m^p(\xi|_n{}^{-1}\eta_n,0). \tag{18}$$

Combining (16), (17) and (18), we get

$$m^{p'}(\xi,0)m^p(\xi^{-1}\xi^{(0)},0) \geq \limsup_{n\to\infty} r_n. \tag{19}$$

In order to get the desired inequalities, we first show that the right-hand side of (19) is not smaller than $g^{(0)}(0)$. Since $g_n(0,z) = g(0,z) - \alpha_n$, we have

$$-\log r_n = \sum\{g(0,w) - \alpha_n : w \in Z(0;R_n)\}$$

$$\leqq \sum\{g(0,w) - \alpha_{n+1} : w \in Z(0;R_n)\} + \sum\{g(0,w) - \alpha_{n+1} : w \in Z(0;R_{n+1}) \setminus R_n\}$$

$$= -\log r_{n+1} \leqq -\log g^{(0)}(0).$$

So, $\lim \log r_n$ exists and, for any fixed $m > 0$,

$$\sum\{g(0,w) - \alpha_n : w \in Z(0;R_m)\} \leqq -\lim_{k\to\infty} \log r_k \leqq -\log g^{(0)}(0)$$

for all $n \geqq m$. Letting $n \to \infty$ and then $m \to \infty$, we see that r_k tend to $g^{(0)}(0)$, which shows our claim.

Finally we show that the left-hand side of (19) is not larger than $m^1(\xi^{(0)},0)$. But this easily follows from the inclusion relation

$$\mathcal{H}^{p'}(R,\xi)\mathcal{H}^p(R,\xi^{-1}\xi^{(0)}) \subseteqq \mathcal{H}^1(R,\xi^{(0)}).$$

This finishes the proof. □

5C. We will now study the continuity of functionals $\xi \to m^p(\xi,0)$, $1 \leqq p \leqq \infty$, on the character group $F_0(R)^*$. For this purpose we equip the group $F_0(R)$ with the discrete topology and therefore its character group $F_0(R)^*$ with the compact topology; namely, a sequence $\{\xi_n\}$ of characters converges to a character ξ if and only if $\xi_n(\gamma) \to \xi(\gamma)$ for each fixed $\gamma \in F_0(R)$. Since $F_0(R)$ is countable, $F_0(R)^*$ is a compact metric space. We begin with remarking that, for any regular PWS R, (DCT) implies

$$m^{p'}(\xi,0)m^p(\xi^{-1}\xi^{(0)},0) = m^1(\xi^{(0)},0) \tag{20}$$

for any $\xi \in F_0(R)^*$ and any $1 \leqq p \leqq \infty$, where $p' = p/(p-1)$. This is an immediate consequence of Theorem 5A and Lemma 5B. We next prove the following

<u>Lemma</u>. If (20) holds for any $\xi \in F_0(R)^*$, then the functional $\xi \to m^p(\xi,0)$ is continuous on $F_0(R)^*$ for every p with $1 \leqq p \leqq \infty$.

<u>Proof</u>. Take any sequence $\{\xi_n : n = 1, 2, \ldots\} \subseteqq F_0(R)^*$ which converges to the identity character Id. Since $m^\infty(\xi_n,0) \leqq 1$, the equation (20) implies that $m^1(\xi_n{}^{-1}\xi^{(0)},0) \geqq m^1(\xi^{(0)},0)$ for $n = 1, 2, \ldots$ and thus

$$\liminf_{n\to\infty} m^1(\xi_n{}^{-1}\xi^{(0)},0) \geqq m^1(\xi^{(0)},0).$$

We now claim the reverse inequality:

$$\limsup_{n\to\infty} m^1(\xi_n{}^{-1}\xi^{(0)},0) \leqq m^1(\xi^{(0)},0).$$

If this were not the case, then we could assume, passing to a subsequence if necessary, that $m^1(\xi_n{}^{-1}\xi^{(0)},0) > m^1(\xi^{(0)},0) + \varepsilon$ for $n = 1, 2,\ldots$ and for some fixed $\varepsilon > 0$. Then, choose an $f_n \in \mathcal{H}^1(R,\xi_n{}^{-1}\xi^{(0)})$ with $\|f_n\|_1 = 1$ and $|f_n(0)| > m^1(\xi^{(0)},0) + \varepsilon/2$ for each $n = 1, 2,\ldots$. By passing to a further subsequence if necessary, we could also assume that $\{f_n\}$ would converge almost unformly on R to a multiplicative function $f \in \mathcal{H}^1(R,\xi^{(0)})$ with $\|f\|_1 \leqq 1$ and $|f(0)| \geqq m^1(\xi^{(0)},0) + \varepsilon/2$. This contradiction shows our claim. Hence we have

$$\lim_{\xi\to Id} m^1(\xi^{-1}\xi^{(0)},0) = m^1(\xi^{(0)},0).$$

Combining this with (20), we have shown

(21) $$\lim_{\xi\to Id} m^\infty(\xi,0) = 1 = m^\infty(Id,0).$$

Now take any p with $1 \leqq p \leqq \infty$. Since $\mathcal{H}^p(R,\eta)\mathcal{H}^\infty(R,\eta^{-1}\xi) \subseteqq \mathcal{H}^p(R,\xi)$ for any ξ, $\eta \in F_0(R)^*$, we have

$$m^p(\eta,0)m^\infty(\eta^{-1}\xi,0) \leqq m^p(\xi,0).$$

By letting $\xi \to \eta$ and using (21), we have

$$\liminf_{\xi\to\eta} m^p(\xi,0) \geqq m^p(\eta,0).$$

Changing the role of ξ and η, we get the reverse inequality:

$$\limsup_{\xi\to\eta} m^p(\xi,0) \leqq m^p(\xi,0).$$

These two inequalities together show that $m^p(\xi,0)$ is continuous in ξ on the character group $F_0(R)^*$. □

5D. Let $\{z_j: j = 1, 2,\ldots\}$ be an enumeration of $Z(0;R)$. For every $n = 1, 2,\ldots$ let θ_n be the character of the l.a.m.

$$\exp(-\sum_{j=1}^{n} g(z_j,z))$$

and let $C_n(z) \in \mathcal{H}^\infty(R,\theta_n)$ be defined by $|C_n(z)| = \exp(-\sum_{j=1}^n g(z_j,z))$ and $(C_n)_0(0) > 0$. Since $\|C_n\|_p \leqq \|C_n\|_\infty = 1$, we see first that

$$\exp(-\sum_{j=1}^{n} g(z_j,0)) = |C_n(0)| \leqq m^p(\theta_n,0)$$

for $1 \leqq p \leqq \infty$.

Let p, $1 \leqq p \leqq \infty$, be fixed and take any $f \in \mathcal{H}^p(R,\theta_n)$. The quotient $f_0/(C_n)_0$ of the principal branches then extends uniquely to a meromorphic function, say h_f, on R. Since $|h_f|\exp(-\sum_{j=1}^{n} g(z_j,\cdot))$, being equal to $|f|$, has a harmonic majorant, the weak Cauchy theorem (Theorem 4B, Ch. VII) shows that $\hat{h}_f \in L^1(d\chi)$ and $h_f(0) = \int_{\Delta_1} \hat{h}_f d\chi$. Since $|h_f(0)| = |f(0)|/|C_n(0)|$ and $\|\hat{h}_f\|_1 = \|f\|_1 \leqq \|f\|_p$, we have

$$|f(0)| \leqq |C_n(0)| \int_{\Delta_1} |\hat{h}_f| d\chi = |C_n(0)| \, \|\hat{h}_f\|_1$$

$$= |C_n(0)| \, \|f\|_1 \leqq |C_n(0)| \, \|f\|_p .$$

As $f \in \mathcal{H}^p(R,\theta_n)$ is arbitrary, $m^p(\theta_n,0) \leqq |C_n(0)| = \exp(-\sum_{j=1}^{n} g(z_j,0))$ and thus

$$m^p(\theta_n,0) = \exp(-\sum_{j=1}^{n} g(z_j,0))$$

for $1 \leqq p \leqq \infty$.

Since $\sum_{j=1}^{\infty} g(z_j,z)$ is uniformly convergent on any compact subset of $R \setminus Z(0;R)$, we see that $\{C_n\}$ converges almost uniformly on R to the function B, which was defined in 5B, and therefore that $\{\theta_n\}$ converges to $\xi^{(0)}$. If $m^1(\xi;0)$ is continuous in ξ, then

$$m^1(\xi^{(0)},0) = \lim_{n\to\infty} m^1(\theta_n,0) = \exp(-\sum_{j=1}^{\infty} g(z_j,0)) = g^{(0)}(0),$$

which implies, by Theorem 5A, that R satisfies (DCT).

Summing up our considerations in §2, we finally get the following characterization of (DCT):

Theorem. Let R be a regular PWS with origin 0. Then the following conditions are equivalent:

(a) (DCT) holds.

(b) $H_0^\infty(d\chi)^\perp = H^1(d\chi, 1/Q^{(0)})$.

(c) Every simply invariant subspace of $L^p(d\chi)$, $1 \leqq p \leqq \infty$, is of the form $H^p(d\chi,Q)$ for some i-function Q on $\mathbb{T}$.

(d) $m^1(\xi^{(0)},0) = g^{(0)}(0)$, where $\xi^{(0)}$ is the character of the l.a.m. $g^{(0)}(z) = \exp(-\sum\{g(w,z): w \in Z(0;R)\})$.

(e) $m^\infty(\xi,0)m^1(\xi^{-1}\xi^{(0)},0) = m^1(\xi^{(0)},0)$ for any $\xi \in F_0(R)^*$.

(f) $\xi \to m^1(\xi,0)$ is continuous on $F_0(R)^*$.

(g) $\xi \to m^p(\xi,0)$ is continuous on $F_0(R)^*$ for every p with $1 \leqq p \leqq \infty$.

NOTES

Most of the observations stated above are based on the idea of Hayashi [28] with a couple of modification by the author. A discussion involving the function $m^{\infty}(\xi;O)$ is also in Pranger [55].

CHAPTER X. EXAMPLES OF SURFACES OF PARREAU-WIDOM TYPE

The objective of this chapter is to provide some examples of surfaces of Parreau-Widom type in order to show that our foregoing discussion is by no means a labor in vain. Although every compact bordered Riemann surface is evidently a PWS, such can never be typical of PWS's. The emphasis should be placed on PWS's of infinite connectivity. Here we are going to give three different types of construction: (i) PWS's of infinite genus for which (DCT) holds; (ii) planar PWS's for which (DCT) fails but the corona theorem is true; and (iii) PWS's for which the corona theorem is false. A remaining problem is to find out more examples aiming at a reasonable classification of PWS's.

In §1 we give a necessary and sufficient condition for a Riemann surface of Myrberg type to be a PWS and then show that such a PWS always satisfies (DCT). Next, we construct in §2 planar regions of Parreau-Widom type for which (DCT) fails. As a preliminary for discussing the corona problem for PWS's, we show in §3 that every PWS R can be embedded in the maximal ideal space of $H^{\infty}(R)$ as an open subset. It is also shown here that $H^{\infty}(R)$ is dense in the space of all holomorphic functions in the topology of almost uniform convergence. Finally in §4, positive as well as negative examples are constructed concerning the corona problem. We show, on the one hand, that our examples in §2 can be modified so as to satisfy the corona theorem but not (DCT) and, on the other, that there exist PWS's for which the corona theorem fails.

§1. PWS OF INFINITE GENUS FOR WHICH (DCT) HOLDS

1. PWS's of Myrberg Type

1A. Our objective in this section is to construct PWS's of infinite genus, for which (DCT) holds. A Riemann surface R is said to be of Myrberg type over the unit disk $\mathbb{D}$ if there exists an analytic function $\psi: R \to \mathbb{D}$ which makes R an n-sheeted, branched, full covering surface of $\mathbb{D}$, i.e. each point of $\mathbb{D}$ has exactly n pre-images, counting multiplicities. In §1 R denotes a Riemann surface of this type and $\{\zeta_j\} = \{\zeta_j : j = 1, 2, \ldots\}$ is the sequence of points in $\mathbb{D}$ over which ψ is branched. Then the following holds.

Theorem. Let R be a Riemann surface of Myrberg type over $\mathbb{D}$ as defined above. Then the following conditions are equivalent:

(a) R is a PWS.

(b) The set $H^\infty(R)$ separates the points of R.

(c) The sequence $\{\zeta_j\}$ satisfies the Blaschke condition, i.e.

$$\sum_{j=1}^{\infty} (1 - |\zeta_j|) < \infty.$$

The proof will be given in 1G below, after some preliminary observations.

1B. Lemma. Suppose that a bounded analytic function f on R separates the points in $\psi^{-1}(\zeta^*)$ for some $\zeta^* \in \mathbb{D} \setminus \{\zeta_j\}$. Then $\{\zeta_j\}$ satisfies the Blaschke condition.

Proof. We define, for every $\zeta \in \mathbb{D} \setminus \{\zeta_j\}$,

$$A_k(\zeta) = (-1)^k \sigma_k(f(a_1), \ldots, f(a_n)),$$

where $\psi^{-1}(\zeta) = \{a_1, \ldots, a_n\}$ and σ_k denotes the k-th elementary symmetric function of n variables. Clearly, A_k is a bounded analytic function in $\mathbb{D} \setminus \{\zeta_j\}$ and thus can be continued analytically to the whole $\mathbb{D}$. The definition of A_k's implies that

$$f^n + (A_1 \circ \psi) f^{n-1} + \cdots + A_n \circ \psi = 0$$

holds on $\psi^{-1}(\mathbb{D} \setminus \{\zeta_j\})$ and thus everywhere on R. Let $D(\zeta)$ be the discriminant of the equation $X^n + A_1(\zeta) X^{n-1} + \cdots + A_n(\zeta) = 0$ in X. We see that the function $D(\zeta)$ belongs to $H^\infty(\mathbb{D})$ and that $D(\zeta) \neq 0$ if and only if the equation in X has n distinct roots. Since f separates the points in $\psi^{-1}(\zeta^*)$, $D(\zeta)$ does not vanish identically. At the points ζ_j we have $D(\zeta_j) = 0$. Namely, ζ_j's are among the zeros of the bounded analytic function $D(\zeta)$ on $\mathbb{D}$. So $\{\zeta_j\}$ satisfies the Blaschke condition. □

1C. Next we want to see when Widom's integral $\int_0^\infty B(\alpha, a)\, d\alpha$ converges, where $B(\alpha, a)$ is the first Betti number of the region $R(\alpha, a) = \{z \in R : g(a, z) > \alpha\}$. Since the convergence does not depend on the choice of a, we may assume that the set $\psi^{-1}(\psi(a))$ consists of n distinct points. We may also assume, by applying a conformal transformation of the disk if necessary, that $\psi(a) = 0$. Namely, we assume that $\psi^{-1}(0)$ contains n distinct points $a_1, \ldots, a_n$ and $a = a_1$. We set

$$h(z) = \sum_{k=1}^{n} g(z, a_k).$$

Then we have $h(z) = -\log|\psi(z)|$. In fact, the local form of the Green function (see Theorem 6A, Ch. I) shows that $u(z) = h(z) + \log|\psi(z)|$ is harmonic everywhere on R. Take any ρ with $0 < \rho < 1$. Then, since $h(z)$ is positive, $u(z) \geqq \log\rho$ on the curve $|\psi(z)| = \rho$. By use of the minimum principle for harmonic functions we have $u(z) \geqq \log\rho$ on the closed region $\{z \in R: |\psi(z)| \leqq \rho\}$. Letting $\rho \to 1$, we see that $u(z)$ is nonnegative on R. On the other hand, we have $\log|\psi(z)| < 0$ and therefore $0 \leqq u(z) \leqq h(z)$ on R. Since h is a potential, it follows that $u(z) \equiv 0$, as claimed.

For each $\alpha > 0$ let $S_\alpha = \{z \in R: h(z) > \alpha\}$ and $B'(\alpha, a_1)$ the first Betti number of S_α. We fix a positive number r, $0 < r < 1$, such that $\{z \in R: |\psi(z)| < r\}$ is connected. By Harnack's inequality we find a constant $A > 1$ such that

$$(1) \qquad A^{-1} g(z, a_1) \leqq h(z) \leqq A g(z, a_1)$$

for $|\psi(z)| \geqq r$. Moreover, given any $a' \in R$ with $|\psi(a')| < r < 1$, we find a constant $A_{a'} > 1$ such that

$$A_{a'}^{-1} g(z, a') \leqq -\log\left|\frac{\psi(z) - \psi(a')}{1 - \overline{\psi(a')}\psi(z)}\right| \leqq A_{a'} g(z, a')$$

for any $z \in R$ with $|\psi(z)| \geqq r$.

<u>Lemma</u>. Let $0 < \alpha < -\log r$. Then S_α is connected and

$$R(\alpha A, a_1) \subseteqq S_\alpha \subseteqq R(\alpha/A, a_1).$$

<u>Proof</u>. We have $S_\alpha = \{h(z) > \alpha\} = \{|\psi(z)| < e^{-\alpha}\}$, which includes the connected set $\{|\psi(z)| < r\}$ and so is connected.

Let $z \in R(\alpha A, a_1)$; then $|\psi(z)| < e^{-\alpha}$, i.e. $z \in S_\alpha$. So $R(\alpha A, a_1)$ is included in S_α. Conversely, take any $z \in R$ with $r < |\psi(z)| < e^{-\alpha}$. Then, by (1), $\alpha < h(z) \leqq A g(z, a_1)$ and thus $z \in R(\alpha/A, a_1)$. As $R(\alpha/A, a_1)$ cannot have any compact complementary component, it includes all of S_α. □

1D. <u>Lemma</u>. Let $0 < \alpha < -\log r$. Then

$$B(\alpha A, a_1) \leqq B'(\alpha, a_1) \leqq B(\alpha/A, a_1).$$

Proof. We claim that $S_\alpha \setminus R(\alpha A, a_1)$ has no components U, which are relatively compact in S_α. If U were such a component, then we would have $\partial U \subseteq \partial R(\alpha A, a_1)$. It would follow that $g(z,a_1) = \alpha A$ on ∂U and therefore $g(z,a_1)$ would be constant on U, a contradiction. So, if a cycle γ in $R(\alpha A, a_1)$ bounds in S_α, then it should bound in the region $R(\alpha A, a_1)$. This shows the first inequality. The other half can be shown similarly. □

We thus have for any $0 < \varepsilon < -\log r$

$$\int_\varepsilon^{-\log r} B(\alpha A, a_1)\,d\alpha \leq \int_\varepsilon^{-\log r} B'(\alpha, a_1)\,d\alpha \leq \int_\varepsilon^{-\log r} B(\alpha/A, a_1)\,d\alpha,$$

which implies

$$A^{-1}\int_{A\varepsilon}^{-\log r} B(\alpha, a_1)\,d\alpha \leq \int_\varepsilon^{-\log r} B'(\alpha, a_1)\,d\alpha \leq A\int_{\varepsilon/A}^{-\log r} B(\alpha, a_1)\,d\alpha.$$

Hence, $\int_0^\infty B(\alpha,a_1)\,d\alpha$ converges if and only if $\int_0^\infty B'(\alpha,a_1)\,d\alpha$ does.

1E. We now estimate $\int_0^\infty B'(\alpha,a_1)\,d\alpha$. Let $0 < \alpha < -\log r$ and suppose that the circle $\{|\zeta| = \alpha\}$ contains no ζ_j's. We then form the double $\hat{S}_\alpha$ of S_α so that the function $\psi(z)$ can be extended to a meromorphic function, say $\hat{\psi}$, of $\hat{S}_\alpha$ onto the extended complex plane $\overline{\mathbb{C}}$. We see that $\hat{\psi}$ makes $\hat{S}_\alpha$ an n-sheeted, branched, full covering of $\overline{\mathbb{C}}$. It is then clear that $B'(\alpha,a_1)$ is exactly the genus of $\hat{S}_\alpha$. In order to obtain a further information about $B'(\alpha,a_1)$, we need the Riemann-Hurwitz relation: if a compact Riemann surface F is an n-sheeted, branched, full covering of another compact surface F_0, then

(2) $$\chi = n\chi_0 + V,$$

where χ (resp. χ_0) is the Euler characteristic of F (resp. F_0) and V is the sum of the orders of all the branch points of F (see Nevanlinna [44], p. 324).

Lemma.

(3) $$B'(\alpha,a_1) = b(e^{-\alpha}) + 1 - n,$$

where $b(e^{-\alpha})$ is the total order of branching over the disk given by $|\zeta| < e^{-\alpha}$.

Proof. Let $\hat{\chi}$ and χ be the Euler characteristic of $\hat{S}_\alpha$ and $\overline{\mathbb{C}}$, respectively; so $\chi = -2$. Since $\hat{S}_\alpha$ is an n-sheeted, full covering of $\overline{\mathbb{C}}$, the Riemann-Hurwitz relation (2) gives us that $\hat{\chi} = -2n + 2b(e^{-\alpha})$.

Since $\hat{S}_\alpha$ is a compact surface, $\hat{\chi}+2$ is equal to the first Betti number of $\hat{S}_\alpha$, i.e. $2B'(\alpha,a_1)$ (cf. Ahlfors and Sario [AS], p. 55). The formula (3) then follows at once. □

The meaning of $B'(\alpha,a_1)$ and $b(e^{-\alpha})$ shows that the formula (3) still holds when ∂S_α contains some branching points.

1F. Suppose that $0 < \varepsilon < -\log r$ and there is no branch point z with $|\psi(z)| = \varepsilon$. Then we have

$$\int_\varepsilon^\infty B'(\alpha,a_1)d\alpha = \alpha B'(\alpha,a_1)\Big|_\varepsilon^\infty - \int_\varepsilon^\infty \alpha dB'(\alpha,a_1)$$

$$= -\varepsilon B'(\varepsilon,a_1) - \int_\varepsilon^\infty \alpha dB'(\alpha,a_1).$$

Let r' be fixed with $r < r' < 1$. Then $dB'(\alpha,a_1) = db(e^{-\alpha})$ for $0 < \alpha < -\log r'$. So, if n_j denotes the total order of branching over the point ζ_j, then we have, for $0 < \varepsilon < -\log r'$,

$$-\int_\varepsilon^{-\log r'} \alpha dB'(\alpha,a_1) = -\int_\varepsilon^{-\log r'} \alpha db(e^{-\alpha})$$

$$= \int_{r'}^{\exp(-\varepsilon)} \log r \, db(r)$$

$$= -\sum_{r'<|\zeta_j|<\exp(-\varepsilon)} n_j \log|\zeta_j|.$$

When $\sum_j n_j \log|\zeta_j|$ is convergent, $-\int_0^{-\log r'} \alpha dB'(\alpha,a_1)$ converges and therefore $\int_0^\infty B'(\alpha,a_1)d\alpha$ converges. Conversely, suppose that the last integral converges. Since $B'(\alpha,a_1)$ is decreasing in α, we see that $\varepsilon B'(\varepsilon,a_1) \leqq \int_0^\varepsilon B'(\alpha,a_1)d\alpha \to 0$ as $\varepsilon \to 0$ and so $\int_0^\infty \alpha dB'(\alpha,a_1)$ converges. It follows that the series

$$-\sum_{r'<|\zeta_j|<1} n_j \log|\zeta_j|$$

is convergent. Hence we have the following

<u>Lemma</u>. The integral $\int_0^\infty B'(\alpha,a_1)d\alpha$ converges if and only if the series $\sum\{n_j \log|\zeta_j| : 0 < |\zeta_j| < 1\}$ converges.

Since $1 \leqq n_j \leqq n-1$ for all j, $\sum_j n_j \log|\zeta_j|$ converges if and only if $\sum_j \log|\zeta_j|$ converges or, equivalently, $\sum_j (1-|\zeta_j|) < \infty$.

1G. Proof of Theorem 1A. The implication (a) $\Rightarrow$ (b) was already shown in Theorem 9B, Ch. V. The fact (b) $\Rightarrow$ (c) is contained in Lemma 1B. Namely, if $H^\infty(R)$ separates the points in R, then, for any $\zeta^* \in \mathbb{D}\setminus\{\zeta_j\}$, there exists a function $f \in H^\infty(R)$ which takes n distinct values at the points of $\psi^{-1}(\zeta^*)$. Thus Lemma 1B shows that $\{\zeta_j\}$ satisfies the Blaschke condition. Finally, (c) $\Rightarrow$ (a) follows from Lemma 1F and the the remark given at the end of 1D. □

2. Verification of (DCT)

2A. We begin with constructing a class of examples of PWS's of Myrberg type. Let $\{x_j: j = 1, 2, \dots\}$ be a strictly increasing sequence of positive real numbers converging to 1. Let R_1 and R_2 be two copies of the open unit disk each slit along the closed segments $[x_{2j-1}, x_{2j}]$, $j = 1, 2, \dots$. We define a Riemann surface R by joining R_1 and R_2 crosswise along these segments in the usual fashion. Let ψ be the natural projection of R onto the unit disk $\mathbb{D}$, which identifies each leaf R_j with $\mathbb{D}$. Then R is a surface of Myrberg type with a branch point of order two over each of the points x_j, $j = 1, 2, \dots$. By Theorem 1A, R is a PWS if and only if $\{x_j\}$ satisfies the Blaschke condition $\sum_{j=1}^\infty (1 - x_j) < \infty$. We note that such a surface is always of infinite genus.

2B. Theorem. Let R be a PWS of Myrberg type over $\mathbb{D}$ with covering map ψ. Then the direct Cauchy theorem (DCT) holds for R.

Proof. Let $a \in R$ be fixed once and for all. As we saw in 1C, there exist positive numbers r and A_a with $0 < r < 1$ and $A_a > 1$ such that $\{z \in R: |\psi(z)| < r\}$ is connected, $|\psi(a)| < r$ and

$$(4) \qquad -\log\left|\frac{\psi(z) - \psi(a)}{1 - \overline{\psi(a)}\psi(z)}\right| \leqq A_a g(z,a)$$

for any $z \in R$ with $|\psi(z)| \geqq r$. Let $\{z_j: j = 1, 2, \dots\}$ be an enumeration of all critical points of $g(z,a)$, counting multiplicities. We then take an integer $N > 0$ so large that $|\psi(z_j)| \geqq r$ for every $j \geqq N$. From (4) follows that

$$-\log\left|\frac{\psi(z_j) - \psi(a)}{1 - \overline{\psi(a)}\psi(z_j)}\right| \leqq A_a g(z_j,a)$$

for $j \geqq N$. Since R is a PWS, $\sum_j g(z_j,a) < \infty$ and therefore

$$\left\{ \frac{\psi(z_j) - \psi(a)}{1 - \overline{\psi(a)}\psi(z_j)} : j = 1, 2, \ldots \right\}$$

satisfies the Blaschke condition. We further infer that $\{\psi(z_j): j = 1, 2, \ldots\}$ also satisfies the Blaschke condition. Put

$$B_n(\zeta) = \prod_{j=n+1}^{\infty} \frac{|\psi(z_j)|}{\psi(z_j)} \frac{\psi(z_j) - \zeta}{1 - \overline{\psi(z_j)}\zeta}$$

for $n = 1, 2, \ldots$. Then

$$|B_n(\psi(z))| = \prod_{j=n+1}^{\infty} \left| \frac{\psi(z_j) - \psi(z)}{1 - \overline{\psi(z_j)}\psi(z)} \right| \leqq \exp(-\sum_{j=n+1}^{\infty} g(z,z_j))$$
$$= g_n^{(a)}(z).$$

Since we have

$$\int_{\mathbb{T}} |1 - \hat{B}_n(e^{it})|^2 d\sigma(t) = \int_{\mathbb{T}} (2 - 2\mathrm{Re}(\hat{B}_n(e^{it})))d\sigma(t)$$
$$= 2 - 2\mathrm{Re}\left(\int_{\mathbb{T}} \hat{B}_n(e^{it})d\sigma(t)\right) = 2 - 2\mathrm{Re}(B_n(0))$$
$$= 2(1 - \prod_{j=n+1}^{\infty} |\psi(z_j)|) \to 0$$

as $n \to \infty$, a suitable subsequence $\{\hat{B}_{n(k)}: k = 1, 2, \ldots\}$ converges to 1 a.e. on $\mathbb{T}$. We also note that $B_n \to 1$ almost uniformly in $\mathbb{D}$.

Now let f be a meromorphic function on R such that $|f|g^{(a)}$ has a harmonic majorant, say u, on R. If $h_n = B_n f$, then h_n is meromorphic on R and

$$|h_n(z)|\exp(-\sum_{j=1}^{n} g(z,z_j)) = |B_n(z)f(z)|\exp(-\sum_{j=1}^{n} g(z,z_j))$$
$$\leqq |f(z)|g^{(a)}(z) \leqq u(z).$$

By the weak Cauchy theorem (Theorem 4B, Ch. VII) we have

$$f(a)B_n(\psi(a)) = \int_{\Delta_1} \hat{f}(b)\hat{B}_n(\psi(b))d\chi_a(b).$$

Taking the limit along the subsequence $\{B_{n(k)}\}$, we get

$$f(a) = \lim_{k\to\infty} f(a)B_{n(k)}(\psi(a))$$

$$= \int_{\Delta_1} f(b)[\lim_{k\to\infty} \hat{B}_{n(k)}(\psi(b))]d\chi_a(b)$$

$$= \int_{\Delta_1} \hat{f}(b)d\chi_a(b).$$

This shows that (DCT_a) holds. By Theorem 5A, Ch. VIII, (DCT_a) holds for all $a \in R$, i.e. (DCT) holds for this R. □

As each surface given in 2A has infinite genus, we have shown that there exist PWS's of infinite genus for which (DCT) is valid.

§2. PLANE REGIONS OF PARREAU-WIDOM TYPE FOR WHICH (DCT) FAILS

We are going to construct plane regions of Parreau-Widom type for which (DCT) fails. It will be seen later that the corona problem may have an affirmative solution for some regions of this kind.

3. Some Simple Lemmas

3A. In what follows we shall use various disks with centers on the real axis. For every pair of real numbers a, b with $a < b$ we denote by $\Delta(a,b)$ the closed disk with center at $(a+b)/2$ and radius $(b-a)/2$, so that the interval $[a, b]$ is a diameter of $\Delta(a,b)$. Moreover, $\Delta(c,\infty)$ denotes the infinite "disk" $\{z \in \overline{\mathbb{C}}: \mathrm{Re}(z) \geqq c\}$. We begin with proving some simple facts.

Lemma. Let $K = \Delta(a,b)$ be the closed disk with $0 < a < b < \infty$ and F an arbitrary closed subset of $-K = \Delta(-b,-a)$. Suppose that F has a connected complement in $\overline{\mathbb{C}}$ and let u be the harmonic measure of the boundary ∂K of K with respect to the region $\overline{\mathbb{C}} \setminus (K \cup F)$, i.e. the solution of the Dirichlet problem for the region $\overline{\mathbb{C}} \setminus (K \cup F)$ with the boundary data equal to 1 on ∂K and to 0 on ∂F. Then $u(0) \geqq 1/2$.

Proof. If u_0 denotes the harmonic measure of ∂K with respect to the region $\overline{\mathbb{C}} \setminus (K \cup (-K))$, then we have $u_0(0) = 1/2$. In fact, we see that $u_0(-z)$ is the harmonic measure of $-\partial K$ with respect to the same region. Since $u_0(z) + u_0(-z)$ is a bounded harmonic function on $\overline{\mathbb{C}} \setminus (K \cup (-K))$ with boundary values identically equal to 1, it is identically equal to 1 on $\overline{\mathbb{C}} \setminus (K \cup (-K))$. Thus $u_0(0) = 1/2$. Since $\overline{\mathbb{C}} \setminus (K \cup F)$ includes $\overline{\mathbb{C}} \setminus (K \cup (-K))$, we have $u(z) \geqq u_0(z)$ in the domain of u_0. In particular, $u(0) \geqq u_0(0) = 1/2$. □

3B. <u>Lemma</u>. Let $0 < a < b < c < \infty$. Then the disks $\Delta(a,b)$ and $\Delta(c,\infty)$ are symmetric with respect to the origin in the sense of the conformal structure of $\overline{\mathbb{C}}$, provided that $a^{-1} - b^{-1} = c^{-1}$. If this is the case, we have $b - a = a^2/(c - a)$.

<u>Proof</u>. We have only to find a fractional linear transformation $L(z)$ which preserves the origin and carries the disks $\Delta(a,b)$ and $\Delta(c,\infty)$ to disks symmetric about the origin. So let $b < d < c$ and set $L(z) = (z - x)^{-1} + d^{-1}$. Trivially, L maps the real axis onto itself and $L(0) = 0$. We also see that $L([a,b])$ and $L([c,\infty])$ are diameters of the transformed disks $L(\Delta(a,b))$ and $L(\Delta(c,\infty))$, respectively. So the disks $L(\Delta(a,b))$ and $L(\Delta(c,\infty))$ are symmetric about the origin if and only if so are the intervals $L([a,b])$ and $L([c,\infty])$. This happens if $-L(a) = L(\infty)$ and $-L(b) = L(c)$. The desired result follows by a simple computation. □

3C. <u>Lemma</u>. Let $K_i = \Delta(a_i,b_i)$, $i = 0, 1,\dots, n$, be closed disks with $0 < a_n < b_n < a_{n-1} < b_{n-1} < \cdots < a_0 < b_0 < \infty$ and set

$$D_n = \overline{\mathbb{C}} \setminus (\bigcup_{i=0}^{n} K_i). \tag{5}$$

We denote by $g_n(\infty,z)$ the Green function for the region D_n with pole at the point at infinity ∞. Then $g_n(\infty,z)$ has exactly n critical points $x_1,\dots, x_n$, for which we have $b_i < x_i < a_{i-1}$, $i = 1, 2,\dots, n$.

<u>Proof</u>. Since the region D_n is symmetric with respect to the real axis, so is the Green function $g_n(\infty,z)$. So $[\partial g_n(\infty,x + iy)/\partial y]_{y=0} = 0$ on $\mathbb{R} \setminus (\cup_{i=0}^{n} K_i)$. Every boundary point of D_n is regular, so that $g_n(\infty,b_i) = g_n(\infty,a_{i-1}) = 0$ for $i = 1, 2,\dots, n$. The Rolle theorem thus implies that, for each $i = 1, 2,\dots, n$, there exists a point x_i in the interval (b_i, a_{i-1}) with $\partial g_n(\infty,x_i)/\partial x = 0$. So these x_i's are critical points of $g_n(\infty,z)$. On the other hand, since D_n has $n + 1$ contours, $g_n(\infty,z)$ has exactly n critical points, as the Riemann-Roch theorem shows (see the proof of Theorem 1C, Ch. V). This establishes the lemma. □

As a consequence, we get the following

<u>Corollary</u>. $g_n(\infty,x)$ is a strictly decreasing function on the interval $(-\infty, a_n)$.

3D. As a trivial consequence of the definition of the Green functions, we get the following

Lemma. Using the notations in 3C, we have $g_n(\infty,z) \geqq g_{n+1}(\infty,z)$ in the region D_{n+1}.

4. Existence Theorem

4A. We are in a position to prove the second main result of the present chapter.

Theorem. There exists a regular region D in $\overline{\mathbb{C}}$ of Parreau-Widom type such that $H^\infty(D)$ possesses a β-closed maximal ideal which is not of the form (11) in Ch. VIII, 4G.

In view of Theorem 5A, Ch. VIII, the direct Cauchy Theorem (DCT) fails for such a region D.

Proof. Let $K_i = \Delta(a_i, b_i)$ be closed disks with

$$0 < \cdots < a_n < b_n < a_{n-1} < \cdots < a_0 < b_0 < \infty,$$

D_n the region defined by (5), and $g_n(\infty,z)$ the Green function for the region D_n with pole ∞. Suppose that $a_n \to 0$ and set

$$D = \overline{C} \setminus (\bigcup_{i=1}^{\infty} K_i \cup \{0\}).$$

Finally, let $g(\infty,z)$ be the Green function for D with pole ∞.

We construct our disks K_i in the following way. Let K_0 be an arbitrary fixed closed disk with $0 < a_0 < b_0 < \infty$ and set $\varepsilon_0 = g_0(\infty,0)$. By Corollary 3C the function $g_0(\infty,x)$ is strictly decreasing on the interval $(-\infty, a_0)$, we have $0 < g_0(\infty,z) \leqq \varepsilon_0$ for $0 \leqq x < a_0$. Suppose that we have chosen disks $K_0, \ldots, K_n$, $n \geqq 0$, with $\varepsilon_i = g_i(\infty,0)$. Then we choose a_{n+1}, b_{n+1} with $0 < a_{n+1} < b_{n+1} < a_n$ satisfying the following conditions:

(a) $a_{n+1} = c_n a_n$ with $0 < c_n < 1/2$,

(b) $b_{n+1} - a_{n+1} = a_{n+1}^2/(a_n - a_{n+1}) = \{c_n/(1-c_n)\}a_{n+1}$,

(c) $\inf\{g_n(\infty,z)\colon z \in \partial K_{n+1}\} \geqq \alpha\varepsilon_n$ with $K_{n+1} = \Delta(a_{n+1}, b_{n+1})$,

where α is a fixed constant with $0 < \alpha < 1$ and the sequence $\{c_n : n = 0, 1, \ldots\}$ should satisfy an additional condition $\sum_{n=0}^{\infty} c_n < \infty$. This construction is always possible by taking c_n sufficiently small.

Let u_n (resp. v_n) be the solution of the Dirichlet problem for the region D_{n+1} with the boundary data equal to $g_n(\infty,z)$ (resp. to 1) on ∂K_{n+1} and to 0 on $\partial K_0 \cup \cdots \cup \partial K_n$. It follows from the condition (c) that $u_n(z) \geqq \alpha\varepsilon_n v_n(z)$ on D_{n+1}. On the other hand, Lemma 3B implies that the disk K_{n+1} and the infinite disk $\Delta(a_n,\infty)$ are symmetric with respect to the origin. Since $K_0 \cup \cdots \cup K_n$ is included in $\Delta(a_n,\infty)$, Lemma 3A shows that $v_n(0) \geqq 1/2$. Since $g_{n+1}(\infty,z) = g_n(\infty,z) - u_n(z)$ in D_{n+1}, we have in particular

$$g_{n+1}(\infty,0) = g_n(\infty,0) - u_n(0) \leqq g_n(\infty,0) - \alpha\varepsilon_n v_n(0)$$
$$\leqq (1-\alpha/2)\varepsilon_n.$$

Thus we have by induction

$$\varepsilon_n \leqq (1-\alpha/2)^n \varepsilon_0. \tag{6}$$

By Lemma 3D, $g_n(\infty,z) \geqq g_{n+1}(\infty,z) \geqq g(\infty,z)$ on the region D for all n. So $\{g_n(\infty,z) - g(\infty,z): n = 0, 1,\ldots\}$ is a monotonically decreasing sequence of bounded positive harmonic functions on D. By the Harnack theorem this sequence converges to a bounded nonnegative harmonic function, say u, on D. The function u is easily seen to have boundary value 0 everywhere on ∂D and therefore vanishes identically on D. Namely, $g_n(\infty,z)$ converge decreasingly to $g(\infty,z)$ on D. The argument principle then shows that $g(\infty,z)$ has a single critical point, say z_n, in each of the intervals (b_{n+1}, a_n), $n = 0, 1,\ldots,$ and there exist no other critical points. The relation (6) implies that $g_n(\infty,z)$ can be made arbitrarily small in a neighborhood of the origin by letting n large enough. Since $g(\infty,z) \leqq g_n(\infty,z)$ for every n, we conclude that $g(\infty,z)$ tends to 0 as z tends to the origin. Namely, D is a regular region. Moreover, we have

$$\sum_{n=0}^{\infty} g(\infty,z_n) \leqq \sum_{n=0}^{\infty} g_n(\infty,z_n) \leqq \sum_{n=0}^{\infty} g_n(\infty,0)$$
$$= \sum_{n=0}^{\infty} \varepsilon_n \leqq \sum_{n=0}^{\infty} (1-\alpha/2)^n \varepsilon_0 < \infty.$$

Hence, D is a regular PWS.

Next take any $f \in H^\infty(D)$. We see that f has the non-tangential boundary values at almost every point on the boundary ∂D. Since the total length of ∂D is finite,

$$f(z) = f(\infty) + \sum_{n=0}^{\infty} \frac{1}{2\pi i} \int_{\partial K_n} \frac{f(\zeta)}{\zeta - z}\, d\zeta \tag{7}$$

for z in D, where each path ∂K_n is directed clockwise. Since we see that $|\partial K_n|$ = the length of $\partial K_n = \pi c_{n-1} a_n/(1-c_{n-1}) \leqq 2\pi c_{n-1} a_n$ for $n = 1, 2,\ldots,$ we have, for every z in the negative real axis,

$$\left|\int_{\partial K_n} \frac{f(\zeta)}{\zeta - z}\, d\zeta\right| \leq \|f\|_\infty |\partial K_n|/a_n \leq 2\pi c_{n-1}\|f\|_\infty .$$

So the right-hand side of (7) converges even for $z = 0$, for the series $\sum_{n=0}^\infty c_n$ is convergent. We define $f(0)$ by the formula (7) with $z = 0$. Then $f(z)$ tends to $f(0)$ as z tends to 0 along the negative real axis.

4B. We now interrupt the proof and show the following

Lemma. If D is as above and if $\Phi(f) = f(0)$ for $f \in H^\infty(D)$, then Φ is a β-continuous complex homomorphism of $H^\infty(D)$.

Proof. As we have remarked above,

$$\Phi(f) = f(0) = \lim_{x\to -0} f(x).$$

It is then clear that Φ is an algebraic homomorphism of $H^\infty(D)$ into $\mathbb{C}$. We then surround each disk K_n by a circle γ_n in D such that (i) γ_n, $n = 0, 1, \ldots$, are mutually exclusive; (ii) each γ_n is oriented clockwise; and (iii) $|\gamma_n|$ (= the length of γ_n) $\leq 3|\partial K_n|/2$ for each n. Then it follows from the usual Cauchy integral formula that

$$\frac{1}{2\pi i}\int_{\partial K_n} \frac{f(\zeta)}{\zeta - z}\, d\zeta = \frac{1}{2\pi i}\int_{\gamma_n} \frac{f(\zeta)}{\zeta - z}\, d\zeta$$

for any z in $\mathrm{Cl}(D)$ lying in the exterior of γ_n. In particular, this is true for $z = 0$. The property (iii) of γ_n implies that for each ζ on γ_n

$$|\zeta| \geq a_n - (b_n - a_n)/2 = a_n - \frac{c_{n-1}}{2(1 - c_{n-1})}\cdot a_n$$
$$= \frac{1 - 3c_{n-1}/2}{1 - c_{n-1}}\cdot a_n > a_n/2.$$

It follows that

$$(8) \qquad \left|\int_{\gamma_n} \frac{f(\zeta)}{\zeta}\, d\zeta\right| \leq \|f\|_\infty\cdot\frac{|\gamma_n|}{a_n/2} \leq 3\|f\|_\infty\cdot\frac{|\partial K_n|}{a_n} \leq 6\pi\|f\|_\infty c_{n-1}.$$

For each $n = 0, 1, \ldots$ we define a measure μ_n on D by the formula $d\mu_n(\zeta) = (2\pi i\zeta)^{-1} d\zeta|_{\gamma_n}$. Then the inequality (8) implies that $\|\mu_n\| \leq 3c_{n-1}$. Since $\sum_{n=0}^\infty c_n$ is convergent, $\mu = \sum_{n=0}^\infty \mu_n + \mu_\infty$ is a well-defined finite Borel measure on D, where μ_∞ denotes the Dirac measure at the point ∞. Since the formula (7) with $z = 0$ shows that $\Phi(f) = \int_D f(z)d\mu(z)$ for $f \in H^\infty(D)$, it follows from Theorem 5C, Ch. IV, that

Φ is β-continuous on $H^\infty(D)$. □

4C. In order to finish the proof of Theorem 4A, we consider D and Φ constructed above and set $J = \{f \in H^\infty(D): \Phi(f) = 0\}$. Since Φ is a β-continuous complex homomorphism of $H^\infty(D)$, J is a β-closed maximal ideal of $H^\infty(D)$. We claim that J has no nonconstant common inner factors. If this is the case, then it is easy to see that J cannot be written in the form (11) in Ch. VIII, 4G.

To see our claim concerning J, let ω be the harmonic measure, with respect to D, of the boundary $\partial D = \bigcup_{n=0}^{\infty} \partial K_n \cup \{0\}$ at the point ∞ (see Ch. I, 5D for the definition of harmonic measures). Since the single point $\{0\}$ forms a polar set, we have $\omega(\{0\}) = 0$ and therefore $\sum_{n=0}^{\infty} \omega(\partial K_n) = 1$. Take any nonnegative continuous function u_0^* on $\bigcup_{n=0}^{\infty} \partial K_n$ such that $u_0^*(z) \to \infty$ as $z \to 0$ through $\bigcup_{n=0}^{\infty} \partial K_n$ and

$$\int_{\partial D} u_0^*(\zeta)d\omega(\zeta) < \infty.$$

Let u_0 be the solution of the Dirichlet problem for the region D with the boundary data equal to u_0^* on $\bigcup_{n=0}^{\infty} \partial K_n$. Then it is easy to see that u_0 is quasibounded on D. As shown in 4A, $g(\infty,z)$ tends to 0 as z tends to 0 through D. This means that the origin 0 is a regular boundary point for the Dirichlet problem in the sense of Ch. I, 5B (cf. Tsuji [65], p. 82, Theorem III.36). We thus see that $u_0(z) \to \infty$ as z tends to 0 through D. Look at the function $z \to \exp(-u_0(z))$ on D. Clearly, it is a bounded l.a.m. and tends to 0 as $z \to 0$ in D. Let ξ be the line bundle determined by this l.a.m. in the sense of Ch. II, 2C. So there exists a bounded holomorphic section, say h_0, of the line bundle ξ on D (i.e. $h_0 \in \mathcal{H}^\infty(D,\xi)$) such that $|h_0| = \exp(-u_0)$. Since $\log|h_0| = -u_0$ is quasibounded, h_0 is an outer l.a.m. (cf. Ch. II, 5D). This clearly means that $\mathcal{H}^\infty(D,\xi)$ has no nonconstant common inner factors. By Lemma 6A, Ch. VII, we see that $\mathcal{H}^\infty(D,\xi^{-1})$ has no nonconstant common inner factors.

We now suppose, on the contrary, that J has a nonconstant common inner factor h. Since $\mathcal{H}^\infty(D,\xi^{-1})$ has no common inner factors, it contains a function, say h_1, whose inner factor is not divisible by h. We define $k \in H^\infty(D)$ by setting $k = h_0h_1$. Then,

$$|k(z)| = |h_0(z)||h_1(z)| \leq \|h_1\|_\infty \cdot \exp(-u_0(z)),$$

so that $k(z) \to 0$ as $z \to 0$ through D. This means in particular that $k \in J$. But our construction shows that the inner factor of k is not divisible by h, which is a contradiction. This proves Theorem 4A. □

4D. As we easily see, our proof of Theorem 4A does not depend on the fact that each K_n is a disk. The same proof goes if we replace disks by any closed sets which are symmetric with respect to the real axis; e.g. closed intervals on the real axis.

§3. FURTHER PROPERTIES OF PWS

5. Embedding into the Maximal Ideal Space

5A. Our objective is to investigate the Banach algebra of bounded holomorphic functions on a PWS. Let R be a PWS, which is held fixed. The set $H^\infty(R)$ of all bounded holomorphic functions on R forms a Banach algebra under the pointwise algebraic operations on functions and the supremum norm $\|f\|_\infty = \sup\{|f(z)|: z \in R\}$. We refer the reader to Hoffman [34] or Garnett [12] for a detailed exposition of Banach algebras and especially of the algebra $H^\infty(\mathbb{D})$.

Let $M(H^\infty(R))$ be the maximal ideal space of $H^\infty(R)$, i.e. the set of all homomorphisms ω of the algebra $H^\infty(R)$ into $\mathbb{C}$ with $\omega(1) = 1$. The set $M(H^\infty(R))$ is equipped with the weak topology relative to $H^\infty(R)$; namely, a net $\{\omega_\lambda\}$ in $M(H^\infty(R))$ converges to an element $\omega \in M(H^\infty(R))$ if and only if $\omega_\lambda(f) \to \omega(f)$ for each $f \in H^\infty(R)$. The topological space $M(H^\infty(R))$ thus defined is called the maximal ideal space of the Banach algebra $H^\infty(R)$. Since $H^\infty(R)$ has the identity 1, we know by the general theory of Banach algebras that $M(H^\infty(R))$ is a compact Hausdorff space. With each element f in $H^\infty(R)$ we associate a function, say $\hat{f}$, on $M(H^\infty(R))$ which is defined by setting $\hat{f}(\omega) = \omega(f)$ for every $\omega \in M(H^\infty(R))$. The function $\hat{f}$ is easily seen to be continuous and is called the Gelfand transform of f. (Namely, the meaning of $\hat{f}$ here is different from that in the preceding, where $\hat{f}$ meant the fine boundary function.)

For every point $z \in R$ the evaluation functional

$$\varepsilon_z: f \to f(z)$$

is a nonzero homomorphism of the algebra $H^\infty(R)$ into $\mathbb{C}$ and so determines an element in $M(H^\infty(R))$. Since every f in $H^\infty(R)$ is continuous on R, the map $z \to \varepsilon_z$ is seen to be a continuous function from R into $M(H^\infty(R))$. At this point we use our hypothesis that R is a PWS. Then the set $H^\infty(R)$ separates the points of R by Theorem 9B, Ch. V. This means in turn that distinct points z in R determine distinct evaluation functionals of $H^\infty(R)$. So, by identification of each point

z with the corresponding functional ε_z, we regard R as a subset of $M(H^\infty(R))$. In view of the existing theory of $H^\infty(\mathbb{D})$, we naturally come to the following questions:

(i) Is the embedding $z \to \varepsilon_z$ of R into $M(H^\infty(R))$ a homeomorphism onto an open subset?

(ii) Is R dense in the space $M(H^\infty(R))$?

The first question has been answered affirmatively. The result is explained right below. As for the second, which is usually called the corona problem, the answer is not definite. We will discuss it in the next section.

5B. We now settle the question (i).

Theorem. If R is a PWS, then every open subset of R is open in the maximal ideal space $M(H^\infty(R))$ of $H^\infty(R)$.

Proof. Let R be a PWS. Take an arbitrary point in R, which we denote by $\mathcal{O}$. What we need to see is that each neighborhood of $\mathcal{O}$ in R is again a neighborhood of $\mathcal{O}$ in the space $M(H^\infty(R))$. To show this, we first note that all the discussions in §1 of Ch. IX can be used here without change.

Let ϕ be the universal covering map $\mathbb{D} \to R$ with $\phi(0) = \mathcal{O}$ and let T_R be the group of cover transformations for ϕ. We repeat the arguments in 3F-3H of Ch. IX. Set $H_0^\infty(R) = \{f \in H^\infty(R): f(\mathcal{O}) = 0\}$. Let Q_0 be the greatest common inner factor of $H_0^\infty(R) \circ \phi$ and ξ_0 the character of Q_0. We then take a function P_0 in $H^\infty(\mathbb{D}, \xi_0)$ such that $P_0(0) \neq 0$ and $\|P_0\|_\infty \leq 1/2$. For each $f \in H^\infty(R)$ there exists a unique element $h \in H^\infty(R)$ such that

$$h \circ \phi = (P_0/Q_0)(f \circ \phi - f(\mathcal{O})).$$

We define a linear operator T in $H^\infty(R)$ by setting $Tf = h$. For each fixed ζ in $\mathbb{C}$ with $|\zeta| < 1$ we set $S_\zeta f = (I - \zeta T)^{-1} f$ and $\Phi_\zeta f = (S_\zeta f)(\mathcal{O})$. As shown in Ch. IX, 3G, S_ζ (resp. Φ_ζ) is a bounded linear operator (resp. a complex homomorphism) in $H^\infty(R)$.

Choose an element $f_0 \in H_0^\infty(R)$ with $(Tf_0)(\mathcal{O}) \neq 0$, which is held fixed, and set

$$V = \{\omega \in M(H^\infty(R)): |\omega(f_0)| < |\omega(Tf_0)|\}.$$

Since $Tf_0 \in H^\infty(R)$, $f_0(\mathcal{O}) = 0$ and $(Tf_0)(\mathcal{O}) \neq 0$, V is an open subset of $M(H^\infty(R))$ containing the point $\mathcal{O}$. Define $\Psi: V \to \mathbb{C}$ by the formula $\Psi(\omega) = \omega(f_0)/\omega(Tf_0)$. Clearly, Ψ is continuous and has modulus less than 1 on V. Take any ω in V and apply it to both sides of the equation (14) in Ch. IX, 3G after setting $h = f_0$ and $\zeta = \Psi(\omega)$. We get

$$\omega(f) = \Phi_{\Psi(\omega)} f \tag{9}$$

for every $f \in H^\infty(R)$. Since $H^\infty(R)$ separates the points of $M(H^\infty(R))$, the equation (9) implies that Ψ is an injection of V into $\mathbb{C}$. We set $U = V \cap R$. Then, $U = \{z \in R: |f_0(z)| < |(Tf_0)(z)|\}$, so that it is an open subset of R containing 0. Let ψ be the restriction of Ψ to the set U. By identifying z with ε_z, we write $\psi(z) = \Psi(\varepsilon_z)$. Since $\psi(z) = f_0(z)/(Tf_0)(z)$ for $z \in U$, it is holomorphic in U with modulus less than 1. Since Ψ is injective on V, ψ should be injective on U, too. In other words, ψ is a univalent holomorphic function on U with $\psi(0) = 0$. So, $\psi(U)$ includes a disk $\{|\zeta| < r_0\}$, $r_0 > 0$. For $0 < r < r_0$ we set $U_r = \{z \in U: |\psi(z)| < r\}$ and $V_r = \{\omega \in V: |\Psi(\omega)| < r\}$. Since ψ is a continuous function on the open subset U of R, each U_r is open in R. Similarly, each V_r is open in $M(H^\infty(R))$. The definition of r_0 implies that ψ (resp. Ψ) maps U_r (resp. V_r) univalently onto the disk $\{|\zeta| < r\}$ for each $0 < r < r_0$. Since ψ is the restriction of Ψ, we see that $U_r = V_r$ for $0 < r < r_0$. The set U_r, $0 < r < r_0$, is thus open as a subset of the space $M(H^\infty(R))$. Thus every neighborhood of 0 in the space R is a neighborhood of 0 in the space $M(H^\infty(R))$. As 0 is arbitrary in R, we conclude that every open subset of R is open in $M(H^\infty(R))$. □

6. Density of $H^\infty(R)$

6A. We prove here that every regular PWS R is convex with respect to the set $H^\infty(R)$ of bounded holomorphic functions and also that $H^\infty(R)$ is dense in the space of all holomorphic functions on R. More precisely, we have the following

Theorem. Let R be a regular PWS. Then the following hold:

(a) If R' is a connected Riemann surface which includes R as a subregion and if each bounded holomorphic function on R extends analytically into R', then $R' = R$.

(b) For every line bundle ξ over R and every compact subset K of R, the $\mathcal{H}^\infty(R,\xi)$-convex hull of K, namely

$$\{z \in R: |f(z)| \leq \sup\{|f(\zeta)|: \zeta \in K\} \text{ for each } f \in \mathcal{H}^\infty(R,\xi)\},$$

is compact.

(c) $H^{\infty}(R)$ is dense in the space of all holomorphic functions on R in the topology of uniform convergence on compact subsets of R.

The statement (a) is easy and is proved in the next subsection. To prove (b) and (c), we use some general results due to E. Bishop. The proof of (b) and (c) will be described in 6D.

6B. Proof of (a). Suppose on the contrary that $R' \neq R$. Since R' is connected, the topological boundary of R in the space R' is nonempty. Let b be a boundary point of R within R' and choose a sequence $\{a_n: n = 1, 2, \ldots\}$ in R such that $a_n \to b$ in R' and $\sum_{n=1}^{\infty} g(a_n, z) < \infty$ for some point z in R, where $g(a,z)$ denotes the Green function for R. Then $\sum_{n=1}^{\infty} g(a_n, z) < \infty$ for every $z \in R \setminus \{a_n: n = 1, 2, \ldots\}$ and, by the argument in Ch. VII, 2A, the function $z \to \exp(-\sum_{n=1}^{\infty} g(a_n, z))$ is seen to be an inner l.a.m., which determines a line bundle, say ξ. Since R is a PWS, we find, by Widom's theorem (Theorem 2B, Ch. V), a nonzero section $h \in \mathcal{H}^{\infty}(R, \xi^{-1})$ and therefore an $f \in H^{\infty}(R)$ such that

$$|f(z)| = |h(z)| \exp(-\sum_{n=1}^{\infty} g(a_n, z)).$$

Thus, f is a nonzero bounded holomorphic function on R vanishing at a_n, $n = 1, 2, \ldots$. By the hypothesis in (a) f extends holomorphically into R'. Thus f should be holomorphic at the point b. This, however, is impossible because the zeros of any nonzero holomorphic function cannot have a cluster point in its domain. The statement (a) is proved. □

6C. In order to prove the properties (b) and (c), we use two general results of Bishop. To make the description somewhat simpler, we use the following notations: let X be a set and $\mathcal{B}$ a set of complex-valued functions on X. For an arbitrary subset F of X we denote by $\tilde{F}(X, \mathcal{B})$ the following set:

$$\{x \in X: |f(x)| \leq \sup\{|f(\zeta)|: \zeta \in F\} \text{ for all } f \in \mathcal{B}\},$$

which may be called the $\mathcal{B}$-convex hull of F. Now, let R be a connected hyperbolic Riemann surface and let A be an algebra of holomorphic functions on R such that A contains the constant function 1 and separates the points of R. The theorems of Bishop we need are the following (B1) and (B2).

<u>Theorem</u>. (B1) If F is a compact subset of R with $F = \tilde{F}(R;A)$, then every function, continuous on R and holomorphic on the interior of F, can be approximated uniformly on F by elements in A.

(B2) Suppose moreover that A is closed in the space $H(R)$ of all holomorphic functions on R in the topology τ_c of uniform convergence on compact subsets of R. Then there exist a connected Riemann surface R' and an algebra A' of holomorphic functions on R' such that the following hold:

(i) R is embedded in R' as a subregion.

(ii) Every element in A is extended holomorphically to an element in A' and A' consists precisely of these extensions.

(iii) The set

$$T = \{(z,w) \in R' \times R' : z \neq w,\ f(z) = f(w) \text{ for all } f \in A'\}$$

is a countable subset of $R' \times R'$ which has no cluster points in $R' \times R'$.

(iv) For each compact subset F' of R' there exists a compact subset F of R such that $F' \subseteq \tilde{F}(R';A')$.

(v) For each compact subset F' of R' the set $\tilde{F}'(R';A')$ is the union of a compact set L and all points $z \in R'$ for which there exists $w \in L$ with $(z,w) \in T$. The set L can be taken to be the union of F' and all relatively compact components of $R' \setminus F'$.

See Bishop [3] for the proof of (B1) and Bishop [4] for the proof of (B2).

6D. We are going to prove (b) and (c). We begin with some general comments. Let $H(R)$ be, as above, the space of all holomorphic functions on R, which is equipped with the topology τ_c of uniform convergence on compact subsets of R. The space $H(R)$ is then a complete topological algebra; namely, the addition and the multiplication in $H(R)$ are continuous and every Cauchy sequence in $H(R)$ is convergent in $H(R)$. Since $H^\infty(R)$ is a subalgebra of $H(R)$, its closure, say A, in $H(R)$ is also a complete topological algebra in the topology τ_c. We now assume that R is a PWS. Then, by Theorem 9B, Ch. V, $H^\infty(R)$ separates the points of R and a fortiori so does the algebra A. So, by Bishop's theorem (B2), there exist a connected Riemann surface R' and an algebra A' of holomorphic functions on R' which satisfy the properties (i) - (v) of (B2). Since $H^\infty(R) \subseteq A$, every bounded holomorphic function on R extends holomorphically into R'. The statement (a) then implies that $R' = R$ and therefore $A' = A$. Since A separates the points of R, the set T in (iii) of (B2) is void.

Proof of (b). Let F be an arbitrary compact subset of R. By (v) of (B2) the set $\tilde{F}(R;A)$ is compact, for T is void. We claim

$$(10) \qquad \tilde{F}(R;\mathcal{H}^\infty(R,\xi)) \subseteqq \tilde{F}(R;H^\infty(R)) \subseteqq \tilde{F}(R;A)$$

for every line bundle ξ over R. Once this is shown, the statement (b) will be established, because $\tilde{F}(R;\mathcal{H}^\infty(R,\xi))$ is a closed subset of a compact set $\tilde{F}(R;A)$. First we prove the second inclusion relation in (10). Choose a point $z \in \tilde{F}(R;H^\infty(R))$ and an element $f \in A$. Since $H^\infty(R)$ is dense in A in the topology τ_c, there exists a sequence $\{f_n: n = 1, 2,\ldots\}$ in $H^\infty(R)$ such that $\{f_n\}$ tends to f uniformly on the compact set $F \cup \{z\}$. Since z belongs to $\tilde{F}(R;H^\infty(R))$, we have $|f_n(z)| \leqq \sup\{|f_n(\zeta)|: \zeta \in F\}$, $n = 1, 2,\ldots$. By taking the limits in n, we have $|f(z)| \leqq \sup\{|f(\zeta)|: \zeta \in F\}$. As f is arbitrary in A, we see that $z \in \tilde{F}(R;A)$. This being true for every $z \in \tilde{F}(R;H^\infty(R))$, we find that $\tilde{F}(R;H^\infty(R)) \subseteqq \tilde{F}(R;A)$. Next, we show the first inclusion in (10). Suppose on the contrary that this is not the case. Then there exists a point $z_0 \in \tilde{F}(R;\mathcal{H}^\infty(R,\xi))$ lying outside $\tilde{F}(R;H^\infty(R))$. So we can find a bounded holomorphic function h on R such that

$$\sup\{|h(\zeta)|: \zeta \in F\} < c < 1 = |h(z_0)|,$$

where c is a positive constant. Take any $f \in \mathcal{H}^\infty(R,\xi)$. As we have $h^n f \in \mathcal{H}^\infty(R,\xi)$ for all $n = 1, 2,\ldots$ and $z_0 \in \tilde{F}(R;\mathcal{H}^\infty(R,\xi))$, so we get

$$\begin{aligned}|f(z_0)| = |h(z_0)^n f(z_0)| &\leqq \sup\{|h(\zeta)^n f(\zeta)|: \zeta \in F\} \\ &\leqq c^n \cdot \sup\{|f(\zeta)|: \zeta \in F\}.\end{aligned}$$

As n is arbitrary and $0 < c < 1$, so we should have $f(z_0) = 0$. The point z_0 should thus be a common zero of elements in $\mathcal{H}^\infty(R,\xi)$, contradicting to Corollary 9A, Ch. V. Hence, the first inclusion in (10) is established. □

Proof of (c). Let f be any holomorphic function on R. Take any compact subset F of R and set $F' = \tilde{F}(R;H^\infty(R))$. Then, by the property (b) just proved, F' is compact. Moreover the definition of convex hulls clearly implies that $F' = \tilde{F}'(R;H^\infty(R))$. By Bishop's theorem (B1) with $A = H^\infty(R)$, the restriction $f|_{F'}$ of f to the set F' can be approximated uniformly on F' by elements in $H^\infty(R)$. Since $F \subseteqq F'$, f can be approximated uniformly on F by elements in $H^\infty(R)$. As F is arbitrary, we have shown that f can be approximated uniformly on each compact subset of R by bounded holomorphic functions on R, as desired. □

§4. THE CORONA PROBLEM FOR PWS

7. (DCT) and the Corona Theorem: Positive Examples

7A. Let R be a PWS. Then, by Theorem 5B, R is identified with an open subset of the maximal ideal space $M(H^\infty(R))$ of the Banach algebra $H^\infty(R)$. We will now set about studying the corona problem: Is R dense in $M(H^\infty(R))$? We say that the corona conjecture holds for R if R is dense in $M(H^\infty(R))$. Our first result is the following

Theorem. There exists a plane region of Parreau-Widom type for which (DCT) fails but the corona conjecture holds.

This will be proved in 7C by making use of our construction in 4A.

7B. Let V be a connected subregion of the extended complex plane $\overline{\mathbb{C}}$ such that $H^\infty(V)$ separates the points of V. This is the case if, for instance, the complement $\overline{\mathbb{C}} \setminus V$ includes a continuum. If $M(H^\infty(V))$ denotes the maximal ideal space of $H^\infty(V)$, then V is homeomorphically embedded into $M(H^\infty(V))$ as an open subset, although V is not necessarily of Parreau-Widom type. By the corona conjecture we again mean the statement that V is dense in $M(H^\infty(V))$. L. Carleson's famous result [7] says that the conjecture is true for the open unit disk. Later M. Behrens [1] devised a method of constructing a variety of regions for which the corona conjecture holds. This is expressed as follows. Consider a sequence $\{K_n : n = 1, 2, \dots\}$ of disjoint closed disks included in V with centers α_n which cluster only on ∂V. We denote by $\mathrm{rad}(K_n)$ the radius of the disk K_n. We say that the sequence $\{K_n\}$ is hyperbolically rare in the region V if there exist disjoint closed disks K_n' with centers α_n such that $K_n \subseteq K_n' \subseteq V$ for each n and such that

$$\sum_{n=1}^{\infty} \mathrm{rad}(K_n)/\mathrm{rad}(K_n') < \infty.$$

Theorem. Suppose that the corona conjecture holds for V and a region U is obtained from V by deleting a hyperbolically rare sequence of closed disks. Then the corona conjecture holds for U, too.

See Behrens [1] for the proof.

7C. Proof of Theorem 7A. We are going to show that the region D in 4A does the required job after a slight modification in the construction. We recall that $D = \overline{\mathbb{C}} \setminus (\cup_{n=0}^{\infty} \Delta(a_n,b_n) \cup \{0\})$ with the following requirements: first, let α, a_0 and b_0 be arbitrarily fixed with $0 < \alpha < 1$ and $0 < a_0 < b_0 < \infty$; then determine three sequences $\{c_n: n = 0, 1,\ldots\}$, $\{a_n: n = 1, 2,\ldots\}$ and $\{b_n: n = 1, 2,\ldots\}$ recurrently by the conditions:

(C1) $0 < c_n < 1/2$ and $\sum_{n=0}^{\infty} c_n < \infty$,

(C2) $a_{n+1} = c_n a_n$ for $n = 0, 1,\ldots,$

(C3) $b_{n+1} - a_{n+1} = a_{n+1}^2/(a_n - a_{n+1}) = \{c_n/(1 - c_n)\}a_{n+1}$,

(C4) $\inf\{g_n(\infty,z): z \in \partial\Delta(a_{n+1},b_{n+1})\} \geqq \alpha g_n(\infty,0)$ for $n = 0, 1,\ldots,$

where $g_n(\infty,z)$ is the Green function with pole ∞ for the region $D_n = \overline{\mathbb{C}} \setminus (\cup_{i=0}^{n} \Delta(a_i,b_i))$.

Take a sequence $\{\lambda_n: n = 1, 2,\ldots\}$ of positive numbers $\lambda_n > 1$ such that $\sum_{n=1}^{\infty} \lambda_n^{-1} < \infty$. We then set

$$a_n' = \frac{1 - (\lambda_n+1)c_{n-1}/2}{1 - c_{n-1}} \cdot a_n \quad \text{and} \quad b_n' = \frac{1 + (\lambda_n-1)c_{n-1}/2}{1 - c_{n-1}} \cdot a_n$$

for $n = 1, 2,\ldots$. By a simple computation we see that, for every n, the disk $\Delta(a_n',b_n')$ is concentric with $\Delta(a_n,b_n)$ and $\mathrm{rad}(\Delta(a_n',b_n')) = \lambda_n \mathrm{rad}(\Delta(a_n,b_n))$. Moreover, it is easy to see that the disks $\Delta(a_n',b_n')$ with $n = 1, 2,\ldots$ are mutually disjoint and are included in the region $\overline{\mathbb{C}} \setminus (\Delta(a_0,b_0) \cup \{0\})$ if we have

(C5) $2c_n(1 + \lambda_{n+1}c_n) < 1 - \lambda_n c_{n-1}$ for $n = 1, 2,\ldots$.

Since the last condition can be fulfilled by taking c_n's sufficiently small, we thus get a plane region D satisfying all the conditions. Then, by the proof of Theorem 4A, D is a region of Parreau-Widom type, for which (DCT) fails. So we have only to show that the corona conjecture holds for D. Since the region $\overline{\mathbb{C}} \setminus (\Delta(a_0,b_0) \cup \{0\})$ is conformal to the punctured disk, Carleson's theorem implies that the corona conjecture holds for $\overline{\mathbb{C}} \setminus (\Delta(a_0,b_0) \cup \{0\})$, for the algebra H^{∞} for the punctured disk is essentially the same as that for the disk. We see also that the sequence $\{\Delta(a_n,b_n): n = 1, 2,\ldots\}$ of disks is hyperbolically rare in the domain $\overline{\mathbb{C}} \setminus (\Delta(a_0,b_0) \cup \{0\})$ because the centers of $\Delta(a_n,b_n)$ converge to 0; $\Delta(a_n',b_n')$ are mutually disjoint with

$$\Delta(a_n,b_n) \subseteqq \Delta(a_n',b_n') \subseteqq \overline{\mathbb{C}} \setminus (\Delta(a_0,b_0) \cup \{0\})$$

for $n = 1, 2,\ldots$; and

$$\sum_{n=1}^{\infty} \frac{\mathrm{rad}(\Delta(a_n,b_n))}{\mathrm{rad}(\Delta(a_n',b_n'))} = \sum_{n=1}^{\infty} 1/\lambda_n < \infty.$$

Thus Theorem 7B implies that the corona conjecture holds for the region D, as was to be proved. □

8. Negative Examples

8A. Our objective here is to present PWS's for which the corona problem has a negative solution. As shown by Nakai [43], the construction of famous examples of B. Cole can be modified in a simple way to yield such PWS's. We begin with Nakai's very simple trick for defining PWS's of infinite genus. All the Riemann surfaces considered here are supposed to be connected.

Let $\{S_n: n = 1, 2, \ldots\}$ be a sequence of interiors S_n of compact bordered Riemann surfaces $\overline{S}_n$ with analytic borders ∂S_n. We suppose that $\overline{S}_n$'s are mutually disjoint. Let $\{X_n: n = 1, 2, \ldots\}$ be a sequence of disjoint copies of the rectangular strip

$$\{z: 0 \leq \mathrm{Re}(z) \leq 2,\ 0 < \mathrm{Im}(z) < 1\}.$$

We denote by α_n' (resp. β_n') the left (resp. right) vertical side of X_n. We then form a Riemann surface $R' = R(\{S_n\},\{X_n\})$, out of the disjoint union $(\cup_{n=1}^{\infty} S_n) \cup (\cup_{n=1}^{\infty} X_n)$, by identifying for each n the side α_n' with an open arc α_n in ∂S_n and the side β_n' with an open arc β_n in ∂S_{n+1}. Here the arcs α_n and β_{n-1} are assumed to be disjoint. Since R' can be embedded in a larger Riemann surface, say R'', in which R' is not dense, it is a hyperbolic Riemann surface. Although we do not yet know whether R' is a PWS or not, it is possible to make R' into a PWS after a suitable simple modification.

Take a sequence $\{\eta_n: n = 1, 2, \ldots\}$ of positive numbers with $0 < \eta_n < 1$ and define a vertical slit, say σ_n, in each X_n by setting

$$\sigma_n = \{z \in X_n: \mathrm{Re}(z) = 1,\ \eta_n \leq \mathrm{Im}(z) < 1\}.$$

We set

$$R(\{S_n\};\{X_n\};\{\eta_n\}) = R' \setminus (\bigcup_{n=1}^{\infty} \sigma_n);$$

so R' may be denoted by $R(\{S_n\};\{X_n\};\{1\})$.

<u>Theorem</u>. If the sequence $\{\eta_n\}$ converges to zero sufficiently rapidly, then the surface $R(\{S_n\};\{X_n\};\{\eta_n\})$ is a regular PWS.

<u>Proof</u>. We write $W = R(\{S_n\};\{X_n\};\{\eta_n\})$ and denote by W_n that part of W corresponding to the finite union

$$\left(\bigcup_{i=1}^{n} S_i\right) \cup \left(\bigcup_{i=1}^{n-1} X_i \setminus \sigma_i\right) \cup \{z \in X_n : 0 \leq \mathrm{Re}(z) < 1\}.$$

Each W_n is easily seen to be the interior of a compact bordered surface and therefore has a finite first Betti number, which we call B_n. We fix a sequence of positive numbers $\{\varepsilon_n : n = 1, 2, \ldots\}$ strictly decreasing to zero such that $\sum_{n=1}^{\infty} B_{n+1}(\varepsilon_n - \varepsilon_{n+1}) < \infty$. Take a fixed point $a \in S_1$ and let $g(a,z)$ (resp. $g'(a,z)$) be the Green function for the surface W (resp. R') with pole a. Since W is a subregion of R', $g(a,z) \leq g'(a,z)$ for every $z \in W$.

In order to determine η_n's, take an arbitrary n and consider the local coordinate $z = x + iy$ in X_n which we used at the beginning, so that $X_n = \{0 \leq x \leq 2,\ 0 < y < 1\}$. Since the lower horizontal side $\{0 \leq x \leq 2,\ y = 0\}$ may be regarded as a boundary arc of R', each point in it is regarded as a regular boundary point of R'. In particular, we have

$$g'(a, 1 + iy) \to 0$$

as $y > 0$ tends to zero. So there exists a positive number $\eta_n > 0$ such that $g'(a, 1 + iy) < \varepsilon_n$ for all $1 + iy \in X_n$ with $0 < y < \eta_n$. We claim that the sequence $\{\eta_n\}$ thus obtained has the desired property. First we see that $g(a, 1 + iy) \leq g'(a, 1 + iy) < \varepsilon_n$ for all $1 + iy \in X_n$ with $0 < y < \eta_n$. By the maximum principle for the Green function we have $g(a,z) < \varepsilon_n$ for all $z \in W \setminus \overline{W}_n$. Now choose any α with $\varepsilon_n > \alpha > \varepsilon_{n+1}$. Then the region $W(\alpha,a) = \{z \in W : g(a,z) > \alpha\}$ is included in the region W_{n+1} as a regular subregion (Ch. I, 1A). It follows that the first Betti number $B(\alpha,a)$ of $R(\alpha,a)$ is not larger than that of W_{n+1}, i.e. $B(\alpha,a) \leq B_{n+1}$. Thus

$$\int_{\varepsilon_{n+1}}^{\varepsilon_n} B(\alpha,a)\,d\alpha \leq B_{n+1}(\varepsilon_n - \varepsilon_{n+1}).$$

Our assumption on the sequence $\{\varepsilon_n\}$ then implies that

$$\int_0^{\infty} B(\alpha,a)\,d\alpha < \infty;$$

namely, W is a PWS. On the other hand, W_{n+1} is the interior of a compact bordered Riemann surface, every boundary points of which is regular. Since $g(a,z)$ is continuous on $\overline{W}_{n+1}$ and is less than ε_n at every boundary point, we see that $\{g(a,z) \geq \alpha\}$ is compact whenever $\alpha > \varepsilon_n$. As ε_n tend to zero, the surface is thus seen to be regular. This completes the proof. □

8B. A very useful reformulation of the corona problem is given by the following

Theorem. A PWS R is dense in the space $M(H^{\infty}(R))$ if and only if, for every finite sequence of elements $\{f_i: i = 1, 2, \ldots, n\}$ in $H^{\infty}(R)$ such that $|f_1(z)| + \cdots + |f_n(z)| \geqq \delta > 0$ for every $z \in R$, there exists a sequence $\{g_i: i = 1, 2, \ldots, n\}$ in $H^{\infty}(R)$ such that $f_1 g_1 + \cdots + f_n g_n = 1$ on R.

Hoffman [34; p. 163] proves this in the case of the unit disk. The proof continues to be valid in the present case as well.

8C. In view of the preceding theorem it is sufficient to prove the following in order to obtain a negative example for the corona problem.

Theorem. There exists a regular PWS R with the following property: there exist two elements $f, g \in H^{\infty}(R)$ such that

$$|f| + |g| \geqq \delta > 0$$

everywhere on R but the equation

$$fh + gk = 1$$

has no solutions h, k in $H^{\infty}(R)$.

Proof. The construction follows almost the same lines as that of Cole's examples. Let $0 < \delta < 1$ be fixed. For each $m = 1, 2, \ldots$ W_m denotes Cole's malformed finite Riemann surface: namely, W_m is the interior of a compact bordered Riemann surface $\overline{W}_m$ with analytic border having the following property: there exists two functions f_m, g_m, holomorphic in W_m and continuous on $\overline{W}_m$, such that

$$|f_m| \leqq 1, \quad |g_m| \leqq 1, \quad |f_m| + |g_m| \geqq \delta \quad \text{on} \quad W_m;$$

and such that $f_m h + g_m k = 1$ with $h, k \in H^{\infty}(W_m)$ implies

$$\sup_{W_m} |h| + \sup_{W_m} |k| \geqq m.$$

For each $m = 1, 2, \ldots$ let S_m be a finite Riemann surface obtained from W_m by attaching an annulus to each boundary component of $\overline{W}_m$. We then form a Riemann surface $R(\{S_m\};\{X_m\};\{\eta_m\})$ as in 8A, where the sequence $\{\eta_m\}$ is fixed in such a way that the resulting surface, say W, is a regular PWS. Then let γ_m be a simple analytic arc that starts at ∂W_m, passes directly through the slit rectangle $X_m \setminus \sigma_m$ joining S_m to S_{m+1}, and terminates at ∂W_{m+1}. We assume that γ_m's are mutually disjoint and that the union

$$F = \left(\bigcup_{m=1}^{\infty} \overline{W}_m\right) \cup \left(\bigcup_{m=1}^{\infty} \gamma_m\right)$$

has no relatively compact complementary components in W. This is always possible as is easily seen. Define next continuous functions f_0 and g_0 on the set F by the following conditions:

$$f_0 = f_m, \quad g_0 = g_m \quad \text{on } \overline{W}_m;$$

$$|f_0| \leqq 1, \quad |g_0| \leqq 1 \quad \text{and} \quad |f_0| + |g_0| \geqq \delta \quad \text{on } F.$$

Then, using Carleman type approximation based on Bishop's theorems (B1) and (B2) in 6C, we find holomorphic functions f, g on W such that

(11) $$\sup_{\overline{W}_m \cup \gamma_m} |f - f_0| + \sup_{\overline{W}_m \cup \gamma_m} |g - g_0| \leqq \delta/4m$$

for $m = 1, 2, \ldots$. (See Kaplan [36], for instance, for the Carleman approximation.) Let R be a connected neighborhood of F in W on which $|f| \leqq 2$, $|g| \leqq 2$ and $|f| + |g| \geqq \delta/2$. Restricting in an obvious way, if necessary, we may assume that $W \setminus R$ has no compact components. So, using the same notations as in the proof of Theorem 8A and denoting by $g_R(a,z)$ the Green function for R with pole a, we see that $R \cap W_n$ and W_n have the same first Betti number B_n and that $g_R(a,z) \leqq g(a,z) < \varepsilon_n$ for all $z \in R \setminus W_n$. It follows, as we did in 8A, that R is a regular PWS.

Finally we show that the equation $fh + gk = 1$ has no solution in $H^\infty(R)$. We suppose on the contrary that there exist elements $h, k \in H^\infty(R)$ such that $fh + gk = 1$. Define the function λ_m on W_m by setting $\lambda_m = f_m h + g_m k$. It follows from (11) that

$$\begin{aligned} \sup_{W_m} |1 - \lambda_m| &= \sup_{W_m} |h(f - f_m) + k(g - g_m)| \\ &= \sup_{W_m} |h(f - f_0) + k(g - g_0)| \\ &\leqq (\sup_{W_m} |h| + \sup_{W_m} |k|) \cdot \frac{\delta}{4m} \\ &\leqq (\|h\|_\infty + \|k\|_\infty)/4m. \end{aligned}$$

If $m \geqq \|h\|_\infty + \|k\|_\infty$, then $|\lambda_m| \geqq 3/4$ on W_m. Thus h/λ_m and k/λ_m belong to $H^\infty(W_m)$ and satisfy $f_m(h/\lambda_m) + g_m(k/\lambda_m) = 1$ on W_m. It then follows from the property of f_m, g_m that

$$\sup_{W_m} |h/\lambda_m| + \sup_{W_m} |k/\lambda_m| \geq m$$

and therefore

$$\sup_{W_m} |h| + \sup_{W_m} |k| \geq 3m/4.$$

As m can be made arbitrarily large, this contradicts the boundedness of h and k. Hence the equation $fh + gk = 1$ has no solutions in the space $H^\infty(R)$, as was to be proved. □

NOTES

Theorem 1A together with its proof is due to Stanton [64]. That such a PWS satisfies (DCT) was found by Hayashi [27].

Theorem 4A is due to Hayashi [30], but a simple argument described here has been devised by the author. In a sense our example is similar to that of Rudin [58].

The embedding theorem (Theorem 5B) is taken from Stanton [64] and the density theorem (Theorem 6A) is adapted from Pranger [56].

The corona problem for the unit disk was solved by Carleson in his very famous paper [7]. A simple new proof was recently furnished by Wolff. A detailed account may be found either in Garnett [12] or in Koosis [40]. The positive example (Theorem 7A) is new, while the negative example (Theorem 8C) is due to Nakai [43]. See Gamelin [11] for Cole's malformed finite Riemann surfaces. Hara [14] constructed PWS's of infinite genus for which the corona conjecture holds.

Neville [47] contains further examples of PWS's satisfying (DCT).

CHAPTER XI. CLASSIFICATION OF PLANE REGIONS

This chapter deals with the classification problem of plane regions in terms of Hardy classes or, more precisely, in terms of Hardy-Orlicz classes. The result extends Heins' classification table for Riemann surfaces given in bis book [31]. Using notations to be explained below, our result can be expressed roughly as follows:

$$\mathcal{O}_G = \mathcal{O}_{AB^*} = \mathcal{O}_{AS} < \cap\{\mathcal{O}_q : 0 < q < \infty\} < \mathcal{O}_p^- < \mathcal{O}_p < \mathcal{O}_p^+$$

$$< \cup\{\mathcal{O}_q : 0 < q < \infty\} < \mathcal{O}_{AB}.$$

The first equality is long known (cf. Sario and Nakai [62], p. 280 and p. 332). The second one concerns the quasibounded majoration and is due to Segawa [63]. Except these equalities, all that happen in the category of Riemann surfaces can already be realized in the category of plane regions. The crux of our discussion lies in constructing a certain kind of exceptional sets in the plane. After having done this, we shall see little difficulty in solving the problem.

In §1 we describe some basic properties of Hardy-Orlicz classes of analytic functions. §2 is devoted to the construction of exceptional sets, belonging to a preassigned class, of positive logarithmic capacity. A merit of our construction is that the size and position of such sets can be at our disposal. By making use of exceptional sets we prove in §3 our classification result, which implies the inclusion relation mentioned above as a special case.

§1. HARDY-ORLICZ CLASSES

1. Definitions

1A. Let Φ be a <u>convex function</u>, by which we mean, through this chapter, any nonconstant, nondecreasing, convex function defined on $[0, \infty)$ with $\Phi(0) = 0$. For any region D in the extended complex plane $\overline{\mathbb{C}}$ we denote by $H^\Phi(D)$ the set of holomorphic functions f on D such that $\Phi(\log^+|f|)$ has a harmonic majorant on D. If this is the case, then $\Phi(\log^+|f|)$ has the LHM, for it is subharmonic on D.

We say that a set $E \subseteq \overline{\mathbb{C}}$ is a null set of class N_Φ if E is a totally disconnected compact subset of $\overline{\mathbb{C}}$ and if $H^\Phi(V \setminus E) = H^\Phi(V)$ for any region V in $\overline{\mathbb{C}}$ including E. Moreover, let us denote by $\mathcal{O}_\Phi$ the collection of nonvoid regions D in $\overline{\mathbb{C}}$ for which $H^\Phi(D)$ contains only constant functions.

If $0 < p < \infty$, then the function $\Phi(t) = e^{pt} - 1$, $0 \leqq t < \infty$, is a convex function and the corresponding class $H^\Phi(D)$ coincides with the Hardy class $H^p(D)$ defined in Ch. IV. In this case we write N_p and $\mathcal{O}_p$ in place of N_Φ and $\mathcal{O}_\Phi$, respectively. We denote by $AB^*(D)$ (resp. N_{AB^*}, $\mathcal{O}_{AB^*}$) the class $H^\Phi(D)$ (resp. N_Φ, $\mathcal{O}_\Phi$) with $\Phi(t) = t$. We also denote by $\mathcal{O}_G$ (resp. $\mathcal{O}_{AB}$) the class of subregions D of $\overline{\mathbb{C}}$ on which there exist no Green functions (resp. no nonconstant bounded holomorphic functions).

In what follows we often use the condition:

$$\lim_{t\to\infty} \frac{\Phi(t)}{t} = \infty. \tag{1}$$

1B. Every $f \in H^\Phi(D)$ is of bounded characteristic on D in the sense that $\log|f| \in SP'(D)$, i.e. $\log^+|f|$ has a harmonic majorant on D. In fact, there exist a positive numbers c_0 and t_0 such that $\Phi(t) \geqq c_0 t$ for $t \geqq t_0$. So, if $f \in H^\Phi(D)$ and u is the LHM of the function $\Phi(\log^+|f|)$ on D, then we have $\log^+|f| \leqq c_0^{-1}u + t_0$ on D. Hence, f is of bounded characteristic. On the other hand, every bounded holomorphic function on D belongs to any $H^\Phi(D)$. Summing up,

$$AB(D) \subseteqq H^\Phi(D) \subseteqq AB^*(D),$$

where $AB(D)$ is the same as $H^\infty(D)$. Concerning the classes $\mathcal{O}_\Phi$, we have

$$\mathcal{O}_G \subseteqq \mathcal{O}_{AB^*} \subseteqq \mathcal{O}_\Phi \subseteqq \mathcal{O}_{AB}.$$

A precise result will be given later.

2. Some Basic Properties

2A. We are going to state some basic properties of Hardy-Orlicz classes following after Parreau [51]. Although the properties can be stated for general Riemann surfaces, we restrict our attention to plane regions.

Let S be a proper subregion of $\overline{\mathbb{C}}$ and let G be a region in S such that the relative boundary $S \cap \partial G$ ($= C_0$, say) is nonvoid and consists of analytic curves. We suppose, moreover, that G lies on one side of C_0.

Lemma. Let u be a harmonic function on $S \cap \mathrm{Cl}(G)$ such that $u = 0$ on C_0 and $\Phi(|u|)$ has a harmonic majorant on G. Then both $\Phi(u^+)$ and $\Phi(u^-)$ have harmonic majorants on G.

Proof. As was shown in 1B, u is of bounded characteristic, so that u^+ and u^- exist (Ch. II, 5A).

Let $\{S_n: n = 1, 2, \ldots\}$ be a regular exhaustion of S (Ch. I, 1A) such that $S_1 \cap C_0$ is nonvoid. We choose a component, say G_1, of $G \cap S_1$ such that $S_1 \cap \partial G_1 \neq \emptyset$. Moreover, let G_n be the component of $G \cap S_n$ which includes G_1. We set $C_{0n} = S_n \cap \partial G_n$ and $C_{1n} = \partial G_n \setminus C_{0n}$. It is then clear that C_{0n} is a part of C_0.

Now let u_n^+ be the solution of the Dirichlet problem for G_n with the boundary data equal to 0 on C_{0n} and to $\max\{u,0\}$ on C_{1n}. As $\max\{u,0\}$ is subharmonic, we see that u_n^+ is the LHM of $\max\{u,0\}$ on G_n. If we denote by U the LHM of $\Phi(|u|)$ on G, then

$$\Phi(u_n^+) = \Phi(\max\{u,0\}) \leqq \Phi(|u|) \leqq U$$

on ∂G_n. Since $\Phi(u_n^+)$ is subharmonic on G_n, we have $\Phi(u_n^+) \leqq U$ on G_n. The definition of u_n^+ implies that $u_1^+ \leqq u_2^+ \leqq \cdots$ and so

$$\Phi(u_n^+) \leqq \Phi(u_{n+1}^+) \leqq \cdots \leqq U$$

on G_n. Thus $\{u_n^+(z): n = 1, 2, \ldots\}$ is bounded at each $z \in G$. By the Harnack theorem the sequence $\{u_n^+\}$ converges to a nonnegative harmonic function, say v, on G. We see in fact that $v = \mathrm{LHM}(\max\{u,0\}) = u^+$. Since $\Phi(v) = \lim_{n\to\infty} \Phi(u_n^+) \leqq U$, we conclude that $\Phi(u^+) \leqq U$.

Similarly, we can show that $\Phi(u^-) \leqq U$. □

2B. Theorem. Let G be as in 2A and let Φ be a convex function satisfying (1). If G carries a nonconstant harmonic function u such that $u = 0$ on C_0 and $\Phi(|u|)$ has a harmonic majorant on G, then G is hyperbolic, in the sense that there exists a nonconstant bounded harmonic function on G vanishing on C_0.

Proof. In view of the preceding result, we may suppose that $u \geqq 0$. Let U be the LHM of $\Phi(u)$. Since $\Phi(u) = 0$ on C_0, U vanishes on C_0. We define $\{G_n: n = 1, 2, \ldots\}$ as before and set $v_n = H[\phi_n; G_n]$, where $\phi_n = 0$ on C_{0n} and $= 1$ on C_{1n}. Then $v_1 \geqq v_2 \geqq \cdots$. Take any $a \in G_1$ and denote by $\omega_a^{(n)}$ the harmonic measure for the region

G_n at the point a. Then

$$u(a) = \int_{C_{1n}} u\,d\omega_a^{(n)}, \quad U(a) = \int_{C_{1n}} U\,d\omega_a^{(n)}, \quad v_n(a) = \int_{C_{1n}} d\omega_a^{(n)}.$$

Since Φ is convex and increasing, Jensen's inequality (Rudin [59], p. 63) implies that

$$\Phi\left(\int_{C_{1n}} u\,d\omega_a^{(n)} \Big/ \int_{C_{1n}} d\omega_a^{(n)}\right) \leqq \int_{C_{1n}} \Phi(u)\,d\omega_a^{(n)} \Big/ \int_{C_{1n}} d\omega_a^{(n)}$$

$$\leqq \int_{C_{1n}} U\,d\omega_a^{(n)} \Big/ \int_{C_{1n}} d\omega_a^{(n)}.$$

Namely,

$$(2) \qquad v_n(a)\Phi(u(a)/v_n(a)) \leqq U(a)$$

for $n = 1, 2, \ldots$. This means that $\{v_n(a)\}$ is bounded away from 0; for, otherwise, the left-hand side of (2) would be unbounded as $n \to \infty$ in view of (1). So the limit, say v, of $\{v_n\}$ is a nonconstant bounded harmonic function on G which vanishes on C_0, as was to be shown. □

2C. <u>Theorem</u>. Let S be a region in $\overline{\mathbb{C}}$ and let Φ be a convex function satisfying (1). If v is a harmonic function on S such that $\Phi(|v|)$ has a harmonic majorant on S, then both v and $\mathrm{LHM}(\Phi(|v|))$ are quasibounded.

<u>Proof</u>. By Lemma 2A we may assume that $v \geqq 0$. If v is bounded, then it is trivially quasibounded, so that we may assume v unbounded. Thus there exists a constant $c > 0$ such that $\{z \in S: v(z) > c\}$ is a nonvoid proper open subset of S. If G denotes one of its components, then G satisfies the condition mentioned in 2A. We set $u = v - c$ on G. Since u satisfies the hypothesis of Theorem 2B, G is hyperbolic; namely, there exists a nonconstant harmonic function u_0 on G such that $u_0 = 0$ on C_0 and $0 \leqq u_0 \leqq 1$ on G. We extend u_0 to the whole S by setting identically zero in $S \setminus G$. The resulting function, say u_1, is subharmonic on S and is majorized by v. The LHM of u_1 is thus bounded and is majorized by v. This means that the quasibounded part $\mathrm{pr}_Q(v)$ of v is strictly positive. Since this is true for any $v > 0$ for which $\Phi(v)$ has a harmonic majorant on G, the inner part $\mathrm{pr}_I(v)$ should be zero. Hence, such a harmonic function v should be quasibounded. □

2D. Theorem. Let S be a region in $\overline{\mathbb{C}}$ and let F be a compact polar set (Ch. I, 6C) in S. Set $G = S \setminus F$. If u is a harmonic function on G such that $\Phi(|u|)$ has a harmonic majorant U on G for some convex function Φ satisfying (1), then u is extended to F as a harmonic function.

Proof. Let S_0 be a regular subregion of S which contains F and let v be the solution of the Dirichlet problem for S_0 with the boundary data equal to u. Since u is bounded on ∂S_0, v is also bounded on S_0. We set $u_0 = (u - v)/2$ on $S_0 \setminus F$. Then u_0 is harmonic on $S_0 \setminus F$, $u_0 = 0$ on ∂S_0 and $\Phi(|u_0|) \leqq \Phi((|u| + |v|)/2) \leqq (\Phi(|u|) + \Phi(|v|))/2 \leqq (U + \Phi(\|v\|_\infty))/2$ on G. If u_0 were nonconstant, then Theorem 2B would imply that $S_0 \setminus F$ should be hyperbolic. This is impossible, for F is a polar set. Hence $u_0 \equiv 0$ on S_0 and therefore u is extended to F as a harmonic function. □

§2. NULL SETS OF CLASS N_Φ

3. Preliminary Lemmas

3A. Lemma. Let Φ be a convex function satisfying (1). Then there exists a convex function Ψ such that

(A1) $\Psi(t + \log 2) \leqq 2\Psi(t)$ for all large t;

(A2) $\Psi(t)$ satisfies (1);

(A3) $\lim_{t\to\infty} \Psi(t)/\Phi(t) = 0$ and $\lim_{t\to\infty} \Psi(t)/t^2 = 0$.

Proof. Since $\Phi(t)$ is convex, it has the right derivative $\Phi'(t + 0)$ for all t, which tends to ∞ as $t \to \infty$ nondecreasingly, in view of condition (1). Thus there exists a number $t_0 > 0$ such that $\Phi(t)/t$ is nondecreasing for $t \geqq t_0$. We write $\Xi(t) = \min\{\Phi(t)/t, t\}$ and $\Xi_1(t) = \min\{\Phi'(t + 0), 2t\}$ for $t \geqq t_0$, so that both $\Xi(t)$ and $\Xi_1(t)$ tend to ∞ as $t \to \infty$ in a nondecreasing way.

We now construct our $\Psi(t)$ by induction. Let t_1 be a real number such that $t_1 \geqq \max\{t_0, \log 2\}$, $\Xi(t_1) \geqq 4$ and $\Xi_1(t_1) \geqq 2$. We set

$$\Psi(t) = t_1 + 2(t - t_1) \quad \text{for} \quad t_1 \leqq t \leqq t_2,$$

where t_2 is a real number with the property: $t_2 \geqq t_1 + 1$, $\Psi(t_2) \geqq \max\{t_2, 4\log 2\}$, $\Xi(t_2) \geqq 3^2$ and $\Xi_1(t_2) \geqq 3$. When $\Psi(t)$ is determined for $t_{n-2} \leqq t \leqq t_{n-1}$, $n \geqq 3$, with the property: $t_{n-1} \geqq t_{n-2} + 1$, $\Psi(t_{n-1}) \geqq \max\{(n-2)t_{n-1}, (n+1)\log 2\}$, $\Xi(t_{n-1}) \geqq n^2$ and $\Xi_1(t_{n-1}) \geqq n$, we define

$$\Psi(t) = \Psi(t_{n-1}) + n(t - t_{n-1}) \quad \text{for} \quad t_{n-1} \leqq t \leqq t_n,$$

where t_n is a real number with the property: $t_n \geqq t_{n-1} + 1$, $\Psi(t_n) \geqq \max\{(n-1)t, (n+2)\log 2\}$, $\Xi(t_n) \geqq (n+1)^2$ and $\Xi_1(t_n) \geqq n+1$. In view of the monotonicity property of Ξ and Ξ_1, these operations are easily seen to be possible, so that $\Psi(t)$ is defined for all $t \geqq t_1$. Obviously, $\Psi(t)$ is convex for $t \geqq t_1$ and it is easy to continue $\Psi(t)$ for all $0 \leqq t \leqq t_1$ so as to have a nondecreasing convex function on $[0, \infty)$ with $\Psi(0) = 0$.

We now have only to show that $\Psi(t)$ possesses the desired properties for $t \geqq t_1$. Let $t_{n-1} \leqq t \leqq t_n$. Then $t + \log 2 \leqq t_n + 1 \leqq t_{n+1}$ and therefore

$$\begin{aligned}
\Psi(t + \log 2) &\leqq \Psi(t_{n-1}) + (n+1)(t + \log 2 - t_{n-1}) \\
&= \Psi(t_{n-1}) + (n+1)(t - t_{n-1}) + (n+1)\log 2 \\
&\leqq 2\Psi(t_{n-1}) + 2n(t - t_{n-1}) \\
&= 2\Psi(t).
\end{aligned}$$

The condition (A1) is thus satisfied. For the same t we have

$$\begin{aligned}
n(t - t_1) + t_1 \geqq \Psi(t) &= \Psi(t_{n-1}) + n(t - t_{n-1}) \\
&\geqq (n-2)t_{n-1} + n(t - t_{n-1}) > (n-2)t.
\end{aligned}$$

As we know that $\Xi(t) \geqq n^2$, we have $\Phi(t) \geqq n^2 t$ and $t^2 \geqq n^2 t$. Thus $\Psi(t)/\Phi(t) \to 0$ and $\Psi(t)/t^2 \to 0$ as $t \to \infty$. The above inequality also shows that $t/\Psi(t) \to 0$ as $t \to \infty$. □

3B. The following result shows the importance of the condition (A1) or, more generally,

$$\frac{\Phi(t + \log 2)}{\Phi(t)} = O(1), \quad t \to \infty, \tag{3}$$

which is equivalent to the Δ_2-condition in the theory of Orlicz spaces (cf. Krasnosel'skii and Rutickii [41]).

Lemma. Let Φ be a convex function which satisfies (1) and (3) and let E be a totally disconnected compact set in $\overline{\mathbb{C}}$. Then $E \in N_\Phi$ if and only if $\overline{\mathbb{C}} \setminus E \in O_\Phi$.

Proof. Since the necessity is trivial, we only show that $\overline{\mathbb{C}} \setminus E \in O_\Phi$

implies $E \in N_\Phi$. So we suppose that $\overline{\mathbb{C}} \setminus E \in \mathcal{O}_\Phi$. We take any region $V \supseteqq E$ and any $f \in H^\Phi(V \setminus E)$. We may suppose of course that f does not vanish identically. Let U be the LHM of $\Phi(\log^+|f|)$ on $V \setminus E$ and u the LHM of $\log^+|f|$ on $V \setminus E$ (see 1B). We easily see that $\Phi(u) \leqq U$ on $V \setminus E$ and in fact U is the LHM of $\Phi(u)$. The proof is now divided into three parts.

(i) First suppose that E is a polar set. Then, by Theorem 2C, both u and U are quasibounded on $V \setminus E$. Thus, by Theorem 5C, Ch. II, $u = \lim_{n\to\infty} u \wedge n$ and $U = \lim_{n\to\infty} U \wedge n$ on $V \setminus E$. Since both $u \wedge n$ and $U \wedge n$ are bounded, these are extended to V as harmonic functions by means of Theorem 2D or otherwise. It follows that u and U are harmonic on V and that $\Phi(u) \leqq U$ on V, as seen by continuity. Consequently, f is bounded in a neighborhood of E and therefore is holomorphic even on E. Hence, $f \in H^\Phi(V)$, as was to be proved.

(ii) So we suppose, in what follows, that E is not a polar set. In this part, we suppose that $V \in \mathcal{O}_G$, i.e. $F = \overline{\mathbb{C}} \setminus V$ is a polar set. We set $S = \overline{\mathbb{C}} \setminus E$. Then S is a region including F and $f \in H^\Phi(S \setminus F)$. Since F is a polar set, our proof in (i) shows that $f \in H^\Phi(S)$. Our original assumption $S \in \mathcal{O}_\Phi$ now implies that f is a constant function. Hence it is trivial that $f \in H^\Phi(V)$.

(iii) Finally, suppose that E is not a polar set and $V \notin \mathcal{O}_G$. We may assume without loss of generality that $\infty \notin V$. We begin with the case in which V is a Jordan region bounded by a finite number of analytic curves and f is continuous up to the boundary of V. Let $V \setminus E_n$, $n \geqq 1$, be an exhaustion of $V \setminus E$ by Jordan regions with analytic boundaries, where $E \subseteqq E_{n+1} \subseteqq E_n \subseteqq V$, $n \geqq 1$. For any $z \in V \setminus E_n$ we have

$$f(z) = \frac{1}{2\pi i}\int_{\partial V} \frac{f(\zeta)}{\zeta - z}\,d\zeta - \frac{1}{2\pi i}\int_{\partial E_n} \frac{f(\zeta)}{\zeta - z}\,d\zeta.$$

The first and the second members of the right-hand side are denoted respectively by $g(z)$ and $h_n(z)$, which are defined on V and on $\overline{\mathbb{C}} \setminus E_n$, respectively. From the fact $f(z) = g(z) - h_n(z)$ on $V \setminus E_n$ follows that $g(z)$ is continuous up to the boundary of V and so is bounded in modulus by a constant $M > 0$. We see also that all $h_n(z)$ coincide on $V \setminus E_1$ and therefore define a single holomorphic function, $h(z)$, on $\overline{\mathbb{C}} \setminus E$. We thus have $f(z) = g(z) - h(z)$ on $V \setminus E$ and consequently $|h(z)| \leqq M + |f(z)|$ there. By use of the property (3) of Φ we find positive constants C_1 and C_2 in such a way that $\Phi(\log^+|h|) \leqq C_1 U + C_2$ on $V \setminus E$.

Let us fix a point $z_0 \in V \setminus E_1$ and denote by μ_n (resp. ν_n) the harmonic measure at the point z_0 with respect to the region $\overline{\mathbb{C}} \setminus E_n$ (resp. $V \setminus E_n$) for $n = 1, 2, \ldots$. Since E is not a polar set, we can find a constant $C \geqq 1$ such that $d\mu_n \leqq C d\nu_n$ on ∂E_n for all n. To see this, take a Jordan region Ω with analytic boundary such that $E_1 \subsetneqq \Omega \subsetneqq \mathrm{Cl}(\Omega) \subsetneqq V \setminus \{z_0\}$. Denoting by $g(a,z:D)$ the Green function for a region D, we set

$$m_n = \max_{w \in \partial\Omega} \frac{g(w,z_0;\overline{\mathbb{C}} \setminus E_n)}{g(w,z_0;V \setminus E_n)} < \infty$$

for $n = 1, 2, \ldots$. Since E is not a polar set, both $V \setminus E$ and $\overline{\mathbb{C}} \setminus E$ have Green functions. We know by assumption that $V \setminus E_1 \subsetneqq V \setminus E_2 \subsetneqq \cdots$ and $V \setminus E = \cup_{n=1}^{\infty} V \setminus E_n$. So, by use of Harnack's theorem, we see that the sequence $\{g(w,z_0;V \setminus E_n) : n = 1, 2, \ldots\}$ tends to $g(w,z_0;V \setminus E)$ almost uniformly in $V \setminus E$. In particular, $g(w,z_0;V \setminus E_n) \to g(w,z_0;V \setminus E)$ uniformly on $\partial\Omega$. Similarly, $g(w,z_0;\overline{\mathbb{C}} \setminus E_n) \to g(w,z_0;\overline{\mathbb{C}} \setminus E)$ uniformly on $\partial\Omega$. Consequently, the sequence $\{m_n\}$ converges to

$$\max_{w \in \partial\Omega} \frac{g(w,z_0;\overline{\mathbb{C}} \setminus E)}{g(w,z_0;V \setminus E)} < \infty.$$

Hence there exists a constant $1 \leqq C < \infty$ such that $m_n \leqq C$ for all n, i.e. $g(w,z_0;\overline{\mathbb{C}} \setminus E_n) \leqq C g(w,z_0;V \setminus E_n)$ for all n and all $w \in \partial\Omega$. For each fixed n, the maximum principle applied to the region $\Omega \setminus E_n$ shows that

$$(4) \qquad g(w,z_0;\overline{\mathbb{C}} \setminus E_n) \leqq C g(w,z_0;V \setminus E_n)$$

for all $w \in \Omega \setminus E_n$. On the other hand, we have along ∂E_n

$$d\mu_n(w) = -\frac{1}{2\pi i} \frac{\partial g}{\partial n_w}(w,z_0;\overline{\mathbb{C}} \setminus E_n)|dw|$$

and

$$d\nu_n(w) = -\frac{1}{2\pi i} \frac{\partial g}{\partial n_w}(w,z_0;V \setminus E_n)|dw|,$$

where $\partial/\partial n_w$ denotes the derivative at $w \in \partial E_n$ in the direction of the outward normal relative to $\overline{\mathbb{C}} \setminus E_n$ or $V \setminus E_n$, and $|dw|$ denotes the arc-length element along ∂E_n. Combining these with (4), we have the desired result.

Since $|h(z)|$ is bounded on $\overline{\mathbb{C}} \setminus E_n$, we can define the LHM, v_n, of $\Phi(\log^+|h|)$ on $\overline{\mathbb{C}} \setminus E_n$ and in fact

$$v_n(z_0) = \int_{\partial E_n} \Phi(\log^+|h(z)|) d\mu_n(z)$$

$$\leqq C \int_{\partial E_n} \Phi(\log^+|h(z)|)d\nu_n(z)$$

$$\leqq C \int_{\partial E_n} (C_1 U(z) + C_2)d\nu_n(z)$$

$$\leqq C \int_{\partial(V\setminus E_n)} (C_1 U(z) + C_2)d\nu_n(z)$$

$$= C(C_1 U(z_0) + C_2).$$

Since $\{v_n: n = 1, 2,\ldots\}$ is a monotonically increasing sequence which is bounded at the point z_0, Harnack's theorem implies that it converges to a harmonic function, say v, on $\overline{\mathbb{C}} \setminus E$, which clearly majorizes the function $\Phi(\log^+|h|)$ on $\overline{\mathbb{C}} \setminus E$. So $h \in H^\Phi(\overline{\mathbb{C}} \setminus E)$ and our assumption on E implies that h should be a constant function. As h vanishes at infinity, we conclude that $h \equiv 0$. Thus f, which is equal to g, is a bounded analytic function on V and hence belongs to $H^\Phi(V)$.

Now let us consider any region $V \notin \mathcal{O}_G$ such that $E \in V$ but $\infty \notin V$. Let $\{V_n: n = 1, 2,\ldots\}$ be a regular exhaustion of V with $E \subsetneqq V_1$. Since the restriction of f to $V_n \setminus E$ satisfies the assumption made previously, we see that $f \in H^\Phi(V_n)$ for $n = 1, 2,\ldots$. Let u_n and v_n be the LHM's of $\Phi(\log^+|f|)$ on $V_n \setminus E$ and V_n, respectively, for $n = 1, 2,\ldots$. Fix a point $z_0 \in V_1 \setminus E$ and let μ_n and ν_n with $n = 1, 2,\ldots$ be the harmonic measures at the point z_0 with respect to the regions $V_n \setminus E$ and V_n, respectively. By an argument similar to that used for (4), we find a positive constant C independent of n such that $d\nu_n \leqq C d\mu_n$ on ∂V_n for all n. So

$$U(z_0) \geqq u_n(z_0) \geqq \int_{\partial V_n} \Phi(\log^+|f(z)|)d\mu_n(z)$$

$$\geqq C^{-1}\int_{\partial V_n} \Phi(\log^+|f(z)|)d\nu_n(z)$$

$$= C^{-1} v_n(z_0)$$

for all n. This shows that the sequence $\{v_n\}$ tends to a harmonic function, v, on V, which majorizes $\Phi(\log^+|f|)$ on V. Hence f belongs to $H^\Phi(V)$, as was to be proved. □

3C. We denote by $\Gamma(z_0;r_0)$, $|z_0| < \infty$ and $0 < r_0 < \infty$, the circumference $\{|z - z_0| = r_0\}$. When $z_0 = 0$, we write simply $\Gamma(r_0)$. In the following, we always deal with the extended complex plane $\overline{\mathbb{C}}$. So, when we write, for example, $\{|z| > a\}$, we include the point at infinity ∞ in the set.

<u>Lemma</u>. Let $0 < a < b < \infty$ and let F be a bounded closed set included in $\{|z| \geq b\}$ such that $D_0 = \{|z| > a\} \setminus F$ is a region. Denote by μ the harmonic measure at the point ∞ with respect to D_0. Then

$$\max\{\frac{ds}{d\mu}(z): z \in \Gamma(a)\} \leqq A(a/b) \cdot \frac{2\pi a}{\mu(\Gamma(a))},$$

where ds denotes the arc-length element on $\Gamma(a)$ and $A(t)$, $t > 0$, is a finite, positive, nondecreasing function in t.

<u>Proof</u>. Let $D_1 = \{a < |z| < b\}$, $D_2 = \{|z| > a\}$ and $c = (ab)^{1/2}$. By Harnack's inequality there exists a constant $A' = A'(a/b)$, depending only on the ratio a/b in a nondecreasing way, such that $U(z_1) \leqq A'U(z_2)$ for any $z_1, z_2 \in \Gamma(c)$ and any nonnegative harmonic function U on D_1. For any arc e on $\Gamma(a)$ we denote by $U_j(e;z)$, $0 \leqq j \leqq 2$, the bounded harmonic function on D_j whose boundary values are equal to 1 on e and to 0 elsewhere. Clearly,

$$U_1(e;z) \leqq U_0(e;z) \leqq U_2(e;z)$$

on D_1. By use of Harnack's inequality we see for $|w| = c$

$$\frac{|e|}{4\pi a} \leqq A'U_1(e;w), \quad U_2(e;w) \leqq \frac{|e|}{2\pi a} \frac{1 + (a/b)^{1/2}}{1 - (a/b)^{1/2}},$$

where $|e|$ denotes the arc length of e. So

(5) $$\max\{U_2(e;w): |w| = c\} \leqq A \cdot \min\{U_1(e;w): |w| = c\},$$

where

$$A = A(a/b) = 2A'(a/b) \cdot \frac{1 + (a/b)^{1/2}}{1 - (a/b)^{1/2}}.$$

From (5) follows at once that

$$\max\{U_0(e_1;w): |w| = c\} \leqq A \cdot \min\{U_0(e_2;w): |w| = c\}$$

for any arcs e_1, e_2 on $\Gamma(a)$ with the same length. This shows that $\mu(e_1) \leqq A\mu(e_2)$ if $|e_1| = |e_2|$, because $\mu(e_j) = U_0(e_j;\infty)$, $j = 1, 2$, and the function $U_0(e_j;z)$, $j = 1, 2$, is the solution of the Dirichlet problem for the region $\{|z| > c\} \setminus F$ with the boundary data equal to $U_0(e_j;z)$ on $\Gamma(c)$ and to 0 elsewhere. Hence

$$\max\{\frac{ds}{d\mu}(z): z \in \Gamma(a)\} \leqq A \cdot \min\{\frac{ds}{d\mu}(z): z \in \Gamma(a)\} \leqq A \cdot \frac{2\pi a}{\mu(\Gamma(a))}. \quad \square$$

3D. The next lemma is the key to our construction of null sets of class N_Φ.

Lemma. Let $0 < a^2/b < a_0 < a < b < \infty$ and let $F_1, \ldots, F_k$ be a finite number of bounded closed subsets of $\{|z| > b\}$ such that each $\overline{\mathbb{C}} \setminus F_i$, $1 \leq i \leq k$, is a region including all F_j with $j \neq i$. Let $\{\ell(n): n = 1, 2, \ldots\}$ be an increasing sequence of positive numbers such that $\ell(n)/n = o(1)$, $n \to \infty$, and $\ell(n) \geq n^{1/2}$ for all large n. Let $w_{n,j} = a_0 \exp(2\pi ji/n)$ and $K_{n,j} = \{|z - w_{n,j}| \leq a_0 e^{-\ell(n)}\}$ for $j = 1, \ldots, n$. We set $K_n = \cup_{j=1}^n K_{n,j}$. Suppose that each F_j, $1 \leq j \leq k$, is not a polar set and that we consider only large n's so that K_n is included in the disk $\{|z| < a\}$ and $K_{n,j}$'s, $1 \leq j \leq n$, are mutually disjoint. Let μ and μ_n be the harmonic measures at the point ∞ with respect to the regions $\{|z| > a_0\} \setminus F$ and $\overline{\mathbb{C}} \setminus (K_n \cup F)$, respectively, with $F = \cup_{j=1}^k F_j$. Then, for any $\varepsilon > 0$, there exists an interger $N = N(\varepsilon) > 0$ such that for $n \geq N$

$$|\mu(\Gamma(a_0)) - \mu_n(\partial K_n)| < \varepsilon,$$

$$|\mu(F_j) - \mu_n(F_j)| < \varepsilon, \quad 1 \leq j \leq k,$$

$$\mu_n(\partial K_{n,j}) \geq (Bn)^{-1}\mu(\Gamma(a_0)), \quad 1 \leq j \leq n,$$

where $B = B(a/b)$ is a constant depending only on the ratio a/b in a nondecreasing fashion.

Proof. We may assume without loss of generality that $b = 1$. Let $\varepsilon > 0$ be given. We first choose $\varepsilon' > 0$ with

$$\max\{(1+\varepsilon') - (1+\varepsilon')^{-3}, \varepsilon'\} < 2^{-1}\mu(\Gamma(a_0))^{-1}\varepsilon \tag{6}$$

and then choose $0 < a' < a_0 < a'' < a$ so close to a_0 as to have

$$(1+\varepsilon')^{-1}\mu(\Gamma(a_0)) < \mu'(\Gamma(a')) < \mu''(\Gamma(a'')) < (1+\varepsilon')\mu(\Gamma(a_0)), \tag{7}$$

where μ' (resp. μ'') denotes the harmonic measure at the point ∞ with respect to the region $\{|z| > a'\} \setminus F$ (resp. $\{|z| > a''\} \setminus F$).

Let $g(z,w) = \log(|1 - z\bar{w}|/|z - w|)$ be the Green function for the open unit disk with pole at w. We denote by ν_n the positive measure of mass 1 with support on ∂K_n and with uniform density on ∂K_n, and by $U_n(z)$ the Green potential of ν_n, i.e.

$$U_n(z) = \int_{\partial K_n} g(z,w)\,d\nu_n(w)$$

on the disk $\{|z| \leq 1\}$. Clearly, $U_n(z)$ is continuous on $\{|z| \leq 1\}$, harmonic on $\{|z| < 1\} \setminus K_n$, and vanishes on $\Gamma(1)$. An explicit computation shows that

$$U_n(z) = \begin{cases} \frac{1}{n}\sum_{j=1}^{n} g(z,w_{n,j}) & \text{if } z \notin K_n, \\ \frac{1}{n}\sum_{j\neq i} g(z,w_{n,j}) + \frac{\ell(n)}{n} + \frac{1}{n}R_{n,i}(z) & \text{if } z \in K_{n,i}, \end{cases}$$

where $R_{n,i}(z) = \log(|1 - z\bar{w}_{n,i}|/a_0) \leq \log(2/a_0)$. By use of the property of $\{\ell(n)\}$, it is easily seen that (i) $U_n(z)$ converges uniformly on the circle $\Gamma(a')$ to the value

$$\frac{1}{2\pi}\int_0^{2\pi} g(a',a_0e^{it})dt = \log\frac{1}{a_0},$$

and (ii) there exists an integer $N' = N'(\varepsilon') > 0$ such that

$$(1+\varepsilon')^{-1}\log\frac{1}{a_0} < U_n(z) < (1+\varepsilon')\log\frac{1}{a_0}$$

on K_n for $n \geq N'$.

Let u' (resp. u_n) be the solution of the Dirichlet problem for the region $\{|z| > a'\} \setminus F$ (resp. $\overline{\mathbb{C}} \setminus (K_n \cup F)$) with the boundary data equal to 1 on $\Gamma(a')$ (resp. ∂K_n) and to 0 elsewhere. Take $N'' = N''(\varepsilon')$ so large that, for $n \geq N''$, K_n lies in the annulus $\{a' < |z| < a''\}$ and $U_n(z) \geq (1+\varepsilon')^{-1}\log(1/a_0)$ on $\Gamma(a')$, the latter being possible in view of (i). Suppose $n \geq N(\varepsilon) = \max\{N', N''\}$. Then, by use of (ii),

$$u_n(z) = 1 \geq (1+\varepsilon')^{-1}(\log\frac{1}{a_0})^{-1}U_n(z)$$

on ∂K_n and therefore everywhere on $\{|z| \leq 1\}$. So on $\Gamma(a')$ we have $u_n(z) \geq (1+\varepsilon')^{-2} = (1+\varepsilon')^{-2}u'(z)$. Since $u_n(z)$ is superharmonic on $\{|z| > a'\} \setminus F$ and has the same boundary values as $u'(z)$ on ∂F, we see that $u_n(z) \geq (1+\varepsilon')^{-2}u'(z)$ everywhere on $\{|z| > a'\} \setminus F$. This, together with (7), means in particular that

$$(8) \qquad \mu_n(\partial K_n) = u_n(\infty) \geq (1+\varepsilon')^{-2}\mu'(\Gamma(a')) \geq (1+\varepsilon')^{-3}\mu(\Gamma(a_0)).$$

On the other hand, since K_n lies inside $\Gamma(a'')$, we have

$$(9) \qquad \mu_n(\partial K_n) \leq \mu''(\Gamma(a'')) \leq (1+\varepsilon')\mu(\Gamma(a_0)).$$

Combining the inequalities (8), (9) and using (6), we get the first inequality in the lemma. To show the second, we take any j with $1 \leq$

$j \leqq k$. Then, by (6), (7), (8) and (9)

$$\begin{aligned} |\mu(F_j) - \mu_n(F_j)| &\leqq \sum_{i=1}^{k} |\mu(F_i) - \mu_n(F_i)| \\ &\leqq \sum_{i=1}^{k} (\mu(F_i) - \mu''(F_i)) + \sum_{i=1}^{k} (\mu_n(F_i) - \mu''(F_i)) \\ &= \mu''(\Gamma(a'')) - \mu(\Gamma(a_0)) + \mu''(\Gamma(a'')) - \mu_n(\partial K_n) \\ &\leqq \varepsilon'\mu(\Gamma(a_0)) + ((1+\varepsilon') - (1+\varepsilon')^{-3})\mu(\Gamma(a_0)) < \varepsilon. \end{aligned}$$

Now we have only to prove the last bunch of inequalities. As our previous observation shows, $U_n(z)$ converges uniformly on the circle $\Gamma(a)$ to

$$\frac{1}{2\pi}\int_0^{2\pi} g(a,a_0e^{it})dt = \log\frac{1}{a},$$

in view of the fact $a_0 < a$. So there exists a positive integer $N_1 \geqq N''$ such that, for $n \geqq N_1$, we have

$$U_n(z) \geqq 2^{-1}\log\frac{1}{a} \tag{10}$$

on $\Gamma(a)$. We take any such n, set $c = a^{1/2}$ and define u_j (resp. u_{1j}, u_{2j}), $1 \leqq j \leqq n$, to be the solution of the Dirichlet problem for the region $\overline{\mathbb{C}} \setminus (K_n \cup F)$ (resp. $\overline{\mathbb{C}} \setminus K_n$, $\{|z| < 1\} \setminus K_n$) with the boundary data equal to 1 on $\partial K_{n,j}$ and to 0 elsewhere. Because of symmetry, $m_i = \min\{u_{ij}(z)\colon |z| = c\}$ and $M_i = \max\{u_{ij}(z)\colon |z| = c\}$, $i = 1, 2$, are independent of j and we have $M_i \leqq A'(a)m_i$ for $i = 1, 2$, where the constant A' has already appeared in the proof of the preceding lemma. Since $\sum_{j=1}^n u_{1j}(z) \equiv 1$ on $\overline{\mathbb{C}} \setminus K_n$, we see that $1 \leqq nM_1 \leqq A'nm_1 \leqq A'$ and therefore $M_1 \leqq A'/n$. On the other hand, $\sum_{j=1}^n u_{2j}$ is harmonic in $\{|z| < 1\} \setminus K_n$ and is equal to 1 on ∂K_n. So, by letting $\varepsilon' < 1$ and using (ii), we have

$$\sum_{j=1}^{n} u_{2j}(z) \geqq ((1+\varepsilon')\log\frac{1}{a_0})^{-1}U_n(z) \geqq \frac{1}{2}(\log\frac{1}{a_0})^{-1}U_n(z)$$

on ∂K_n and thus everywhere on $\{|z| < 1\} \setminus K_n$. In view of (10), we thus get

$$\sum_{j=1}^{n} u_{2j}(z) \geqq \frac{1}{2}(\log\frac{1}{a_0})^{-1}U_n(z) \geqq \frac{1}{4}\frac{\log a}{\log a_0} \geqq \frac{1}{8}$$

on $\Gamma(a)$, because $a^2 < a_0 < a$. It follows that

$$A'nm_2 \geqq nM_2 \geqq \sum_{j=1}^{n} u_{2j}(z) \geqq \frac{1}{16}$$

on $\Gamma(c)$. So we have $A'm_2 \geqq 1/16n \geqq M_1/16A'$. Since $u_{2j}(z) \leqq u_j(z) \leqq u_{1j}(z)$, $1 \leqq j \leqq n$, on the annulus $\{a < |z| < 1\}$, the above inequality shows that

$$u_i(z) \leqq 16A'^2 u_j(z)$$

on $\Gamma(c)$ for any i and j. The same inequalities clearly hold on $\{|z| > c\} \setminus F$, so that

$$\mu_n(\partial K_{n,i}) = u_i(\infty) \leqq 16A'^2 u_j(\infty) = 16A'^2 \mu_n(\partial K_{n,j})$$

for any $i, j = 1, \ldots, n$. This implies that

$$\mu_n(\partial K_{n,j}) \geqq (16A'^2 n)^{-1} \mu_n(\partial K_n)$$

for $j = 1, \ldots, n$. Since $\mu(\Gamma(a_0)) \neq 0$, we have $\mu_n(\partial K_n) \geqq \mu(\Gamma(a_0))/2$ for all sufficiently large n. Hence we have

$$\mu_n(\partial K_{n,j}) \geqq (Bn)^{-1} \mu(\Gamma(a_0))$$

for $j = 1, \ldots, n$, with $B = B(a) = 16A'^2$, provided n is sufficiently large. This finishes the proof. □

4. Existence of Null Sets

4A. We are now in a position to prove the main result of this section.

Theorem. Let Φ be a convex function which satisfies (1) and (3). Let $0 < a < b < \infty$ and let F_j, $1 \leqq j \leqq k$, be a finite number of bounded closed subsets of $\{|z| > b\}$ such that each $\overline{\mathbb{C}} \setminus F_i$ is a region containing all other F_j, $j \neq i$. Then for any positive numbers $\varepsilon > 0$ and $0 < \delta < 1$ there exists a non-polar set $E \in N_\Phi$ such that $E \subseteqq \{\delta a \leqq |z| \leqq a\}$,

$$|\mu(F_j) - \mu_E(F_j)| < \varepsilon, \quad 1 \leqq j \leqq k, \tag{11}$$

and

$$|\mu(\Gamma(a)) - \mu_E(E)| < \varepsilon, \tag{12}$$

where μ (resp. μ_E) denotes the harmonic measure at the point ∞ with respect to the region $\{|z| > a\} \setminus F$ (resp. $\overline{\mathbb{C}} \setminus (E \cup F)$) with $F =$

$\cup_{j=1}^{k} F_j$.

Proof. By applying Lemma 3A to the function $t \to \Phi(t/2)$, we get a convex function $\Lambda(t)$ such that (i) $t/\Lambda(t) = o(1)$, $t \to \infty$; (ii) $\Lambda(t) \leq t^2$ for all large t; and (iii) $\Lambda(t)/\Phi(t/2) = o(1)$, $t \to \infty$. We then find a positive number t_0 such that both $\Phi(t)$ and $\Lambda(t)$ are strictly increasing for $t \geqq t_0$. Let $t_1 = \max\{\Phi(t_0), \Lambda(t_0)\}$. Then the inverse functions $h(t)$ and $\ell(t)$ of $\Phi(t)$ and $\Lambda(t)$, respectively, are uniquely determined as strictly increasing functions in t for $t \geqq t_1$. The properties (i) - (iii) now imply the following: (i') $\ell(t)/t = o(1)$, $t \to \infty$; (ii') $\ell(t) \geqq t^{1/2}$ for all large t; and (iii') for any $\varepsilon > 0$ there exists a number $t(\varepsilon)$ ($\geqq t_1$) such that $h(t/\varepsilon) \leqq \ell(t)/2$ for $t \geqq t(\varepsilon)$. So the sequence $\{\ell(n): n \geqq t_1\}$ satisfies the conditions in Lemma 3D. In order to construct a set E with the desired property, we may assume that F is a non-polar set. If F is a polar set, then we have only to add a non-polar set, say F_{-1}, to F as a new member during our construction, which will cause no harm. We may assume also that ε is so small as to have $0 < \varepsilon < (1 - \mu(F))/\mu(F)$. We set $\rho = (b/a)^{1/4}$ and $B = B(a/b)$, the constant appearing in Lemma 3D.

By induction we construct families K_n, K_n', $n = 0, 1, \ldots,$ of closed disks included in $\{|z| \leqq a\}$. Each K_n (resp. K_n') consists of a finite number of mutually disjoint, closed disks of the same radius r_n (resp. r_n') in $\{|z| \leqq a\}$, whose union is denoted by K_n (resp. K_n'). By μ_n (resp. μ_n') we mean the harmonic measure at the point ∞ with respect to the region $\overline{\mathbb{C}} \setminus (K_n \cup F)$ (resp. $\overline{\mathbb{C}} \setminus (K_n' \cup F)$).

As the 0-th step of our induction, we define K_0 to be empty and K_0' to consist of only one member $\{|z| \leqq a\}$, so that $\mu_0' = \mu$ and $r_0' = a$. The definition of ρ says that the disk $\{|z| \leqq \rho^4 r_0'\}$ is disjoint from the set F.

Suppose that we have finished the n-th step with $n \geqq 0$. Namely, we have constructed K_n' in such a way that K_n' consists of closed disks D_α', $1 \leqq \alpha \leqq N(n)$, of center w_α and of common radius r_n' so that the disks $\{|z - w_\alpha| \leqq \rho^4 r_n'\}$ are mutually disjoint and also disjoint from F. This condition is clearly fulfilled by K_0', for $b/a = \rho^4$.

We then define K_{n+1} and K_{n+1}' as follows. Let $\max\{1/\rho, \delta\} r_n' < r_{n+1} < r_n'$ and set $D_\alpha = \{|z - w_\alpha| \leqq r_{n+1}\}$, $1 \leqq \alpha \leqq N(n)$. The family K_{n+1} consists of these disks. The value r_{n+1} is fixed so close to r_n' as to have

(13) $$\mu_{n+1}(F_j) - \mu_n'(F_j) \leqq \varepsilon\mu(F_j)/2^{n+1}, \quad 1 \leqq j \leqq k,$$

and

(14) $$\mu_{n+1}(\partial D_\alpha) \geqq \mu'_n(\partial D'_\alpha)/2, \quad 1 \leqq \alpha \leqq N(n).$$

In order to define K'_{n+1}, we take an integer $N'(n+1) \geqq \max\{n+1, t_1\}$, which will be fixed later. We set

$$w_{\alpha,j} = w_\alpha + r_{n+1}\exp[2\pi ji/N'(n+1)], \quad 1 \leqq j \leqq N'(n+1),$$

$$N(n+1) = N(n)N'(n+1), \quad r'_{n+1} = r_{n+1}\exp[-\ell(N(n+1))]$$

and

$$D'_{\alpha,j} = \{|z - w_{\alpha,j}| \leqq r'_{n+1}\}, \quad 1 \leqq j \leqq N'(n+1).$$

The disks $\{D'_{\alpha,j}: 1 \leqq \alpha \leqq N(n),\ 1 \leqq j \leqq N'(n+1)\}$ then form the family K'_{n+1}. We fix $N'(n+1)$ so large as to have the following:

(a) the disks $\{|z - w_{\alpha,j}| \leqq \rho^4 r'_{n+1}\}$ are mutually disjoint and also disjoint from F;

(b) $|\mu_{n+1}(F_j) - \mu'_{n+1}(F_j)| \leqq \varepsilon\mu(F_j)/2^{n+2}$ for $j = 1, \ldots, k$;

(c) $\mu'_{n+1}(\partial D'_{\alpha,j}) \geqq (BN'(n+1))^{-1}\mu_{n+1}(\partial D_\alpha)$ for $j = 1, \ldots, N'(n+1)$ and $\alpha = 1, \ldots, N(n)$;

(d) $r'_{n+1} \leqq 3^{-1}\min\{r'_n - r_{n+1}, r_{n+1} - \delta r'_n\}$;

(e) $h((n+1)(2B)^{n+1}N(n+1)) \leqq \ell(N(n+1))/2$.

This is trivial for (a) and (d). Statements (b) and (c) follow from Lemma 3D and (e) from the property (iii'). Hence, by induction, the families K_n and K'_n, $n \geqq 0$, are constructed.

Combining (13) and (b), we get for $n \geqq 1$

(15) $$\begin{aligned} &|\mu_n(F_j) - \mu_{n+1}(F_j)| \\ &\leqq |\mu_n(F_j) - \mu'_n(F_j)| + |\mu'_n(F_j) - \mu_{n+1}(F_j)| \\ &\leqq \varepsilon\mu(F_j)/2^{n+1} + \varepsilon\mu(F_j)/2^{n+1} \\ &= \varepsilon\mu(F_j)/2^n, \quad 1 \leqq j \leqq k. \end{aligned}$$

It follows from (14) and (c) that for any $D'_\alpha \in K'_n$ there is a $D'_\beta \in K'_{n-1}$ such that $\mu'_n(\partial D'_\alpha) \geqq (2BN'(n))^{-1}\mu'_{n-1}(\partial D'_\beta)$. Repeating this, we see that for $D'_\alpha \in K'_n$

(16) $$\begin{aligned} \mu'_n(\partial D'_\alpha) &\geqq (2B)^{-n}(N'(n)N'(n-1)\cdots N(1))^{-1}\mu'_0(\Gamma(a)) \\ &= (2B)^{-1}N(n)^{-1}\mu(\Gamma(a)). \end{aligned}$$

We set

$$E = \bigcap_{n=1}^{\infty} [\mathrm{Cl}(\bigcup_{s=n}^{\infty} K_s)]$$

and denote by μ_E the harmonic measure at the point ∞ with respect to the region $\overline{\mathbb{C}} \setminus (E \cup F)$.

Our construction and the property (d) show that $K'_{n+1} \subseteqq \mathrm{Int}(K'_n) \subseteqq K'_1 \subseteqq \{\delta a < |z| < a\}$ and $K_{n+1} \subseteqq K'_n$ for $n = 1, 2,\dots$. It follows that

$$E \subseteqq \mathrm{Cl}[\bigcup_{s=n+2}^{\infty} K_s] \subseteqq K'_{n+1} \subseteqq \mathrm{Int}(K'_n)$$

for $n = 1, 2,\dots$ and a fortiori $E \subseteqq \{\delta a < |z| < a\}$. By use of (15) we see for $n \geqq 1$

$$\begin{aligned} \mu_n(F) &= \mu(F) + \sum_{s=1}^{n} (\mu_s(F) - \mu_{s-1}(F)) \\ &= \mu(F) + \sum_{s=1}^{n} \sum_{j=1}^{k} (\mu_s(F_j) - \mu_{s-1}(F_j)) \\ &\leqq \mu(F) + \sum_{s=1}^{\infty} \sum_{j=1}^{k} \varepsilon\mu(F_j)/2^s \\ &= (1+\varepsilon)\mu(F) \end{aligned}$$

and therefore

$$\begin{aligned} \mu_n(\partial K_n) &= 1 - \mu_n(F) \geqq 1 - \mu(F) - \varepsilon\mu(F) \\ &= \mu(\Gamma(a)) - \varepsilon\mu(F). \end{aligned}$$

The last member is strictly positive, because of our assumption $\varepsilon < (1 - \mu(F))/\mu(F)$. Let u_n (resp. v_n) be the solution of the Dirichlet problem for the region $\overline{\mathbb{C}} \setminus (F \cup \mathrm{Cl}(\bigcup_{s=n}^{\infty} K_s))$ (resp. $\overline{\mathbb{C}} \setminus (F \cup K_n)$) with the boundary data equal to 1 on $\partial[\mathrm{Cl}(\bigcup_{s=n}^{\infty} K_s)]$ (resp. ∂K_n) and to 0 on ∂F. Clearly, the sequence $\{u_n: n = 1, 2,\dots\}$ is monotonically decreasing and bounded below, so that it converges, by Harnack's theorem, to a nonnegative harmonic function, $u(z)$, on $\overline{\mathbb{C}} \setminus (E \cup F)$. Obviously, we have $0 \leqq u(z) \leqq 1$ and $u(z) = 0$ on ∂F. We know that $u_n(z) \geqq v_n(z)$ on the domain of u_n and that $v_n(\infty) = \mu_n(\partial K_n) \geqq 1 - (1+\varepsilon)\mu(F) > 0$. Hence $u(\infty) > 0$. The function u is seen to be the solution of the Dirichlet problem with the boundary data equal to 1 on E and to 0 on ∂F. Thus $\mu_E(E) = u(\infty) > 0$ and consequently E is not a polar set. Moreover we have

$$(17) \qquad \mu_E(E) = u(\infty) = \lim_{n\to\infty} u_n(\infty) \geqq \limsup_{n\to\infty} v_n(\infty)$$
$$= \limsup_{n\to\infty} \mu_n(\partial K_n) \geqq \mu(\Gamma(a)) - \varepsilon\mu(F).$$

On the other hand, since E is included in $\{|z| \leqq a\}$, $\mu_E(F_j) \geqq \mu(F_j)$ and therefore $\mu_E(F) \geqq \mu(F)$. Thus $\mu(\Gamma(a)) \geqq \mu_E(E)$, which, together with (17), proves (12). We also have

$$\mu_E(F_j) - \mu(F_j) \leqq \mu_E(F) - \mu(F) = \mu(\Gamma(a)) - \mu_E(E) \leqq \varepsilon\mu(F),$$

which shows (11).

We will show finally that $E \in N_\Phi$. Since Φ satisfies (3), we have only to prove, in view of Lemma 3B, that $\overline{\mathbb{C}} \setminus E \in \mathcal{O}_\Phi$. Let f be a nonzero element in $H^\Phi(\overline{\mathbb{C}} \setminus E)$ and v a harmonic majorant of $\Phi(\log^+|f|)$ on $\overline{\mathbb{C}} \setminus E$. To show that f is a constant function, we first note that E is included in the interior of K_n' for $n = 0, 1, \ldots$. Let us fix $n \geqq 0$ and let $D' = \{|z - w'| \leqq r'\}$ be one of the members in $\mathcal{K}_n'$. Then the property (a) says that the annulus $\{r' \leqq |z - w'| \leqq \rho^4 r'\}$ does not meet E, F and $K_n' \setminus D'$. Let ξ_w be the harmonic measure with respect to the annulus $\{r' < |z - w'| < \rho^2 r'\}$ at a point w in it. If w is on the circle $\Gamma(w';\rho r')$, then Harnack's inequality implies that $d\xi_w/ds$ is bounded by $A'/(4\pi r')$ on $\Gamma(w';r')$ and by $A'/(4\pi\rho^2 r')$ on $\Gamma(w';\rho^2 r')$, where $A' = A'(\rho^{-2})$ is the constant appearing in the proof of Lemma 3C. Since f is bounded and analytic on the annulus $\{r' < |z - w'| < \rho^2 r'\}$, we have for any w in this annulus

$$\log^+|f(w)| \leqq \int \log^+|f(\zeta)|\,d\xi_w(\zeta).$$

Applying the convex function Φ to both sides and using Jensen's inequality, we get

$$\Phi(\log^+|f(w)|) \leqq \Phi\left(\int \log^+|f(\zeta)|\,d\xi_w(\zeta)\right)$$
$$\leqq \int \Phi(\log^+|f(\zeta)|)\,d\xi_w(\zeta))$$
$$= \int_{\Gamma(w';r')} + \int_{\Gamma(w';\rho^2 r')}$$
$$= I_1 + I_2.$$

We now assume that $w \in \Gamma(w';\rho r')$. Let η, η' denote the harmonic measures at ∞ with respect to the regions $G' = \{|z - w'| > r'\} \setminus E$ and

$G'' = \{|z - w'| > \rho^2 r'\} \setminus E$, respectively. Since $E \subseteq K_n'$ and $\Gamma(w';r') = \partial D'$, we see that

$$\eta'(\Gamma(w';\rho^2 r')) \geq \eta(\Gamma(w';r')) \geq \mu_n'(\partial D').$$

By use of (16) we also have $\mu_n'(\partial D') \geq (2B)^{-n} N(n)^{-1} \mu(\Gamma(a))$. By applying Lemma 3C to the annuli $\{r' < |z - w'| < \rho^2 r'\}$ and $\{\rho^2 r' < |z - w'| < \rho^4 r'\}$, we see that $ds/d\eta$ is bounded by $2\pi A r'/\eta(\Gamma(w';r'))$ on the circle $\Gamma(w';r')$ and $ds/d\eta'$ is bounded by $2\pi A\rho^2 r'/\eta'(\Gamma(w';\rho^2 r'))$ on $\Gamma(w';\rho^2 r')$, where $A = A(\rho^{-2})$ is the constant given in Lemma 3C. So we have

$$\begin{aligned} I_1 &= \int_{\Gamma(w';r')} \Phi(\log^+|f(\zeta)|) \frac{d\xi_w(\zeta)}{ds(\zeta)} \frac{ds(\zeta)}{d\eta(\zeta)} \, d\eta(\zeta) \\ &\leqq \frac{A'}{4\pi r'} \frac{2\pi A r'}{\eta(\Gamma(w';r'))} \int_{\Gamma(w';r')} \Phi(\log^+|f(\zeta)|) d\eta(\zeta) \\ &\leqq \frac{AA'}{2\eta(\Gamma(w';r'))} \int_{\partial G'} \Phi(\log^+|f(\zeta)|) d\eta(\zeta) \\ &\leqq C(2B)^n N(n)/2, \end{aligned}$$

where $C = AA'v(\infty)/\mu(\Gamma(a))$. Similarly, we have $I_2 \leqq C(2B)^n N(n)/2$ by using η', G'' and the corresponding estimates. It follows that

$$\Phi(\log^+|f(w)|) \leqq C(2B)^n N(n)$$

for $w \in \Gamma(w';\rho r')$. As we may always assume that $v(\infty) \neq 0$, the right-hand side exceeds t_1 for all sufficiently large $n \geqq n_0$, say. So we can apply the inverse function $h(t)$ and get the following inequality:

$$|f(w)| \leqq \exp[h(C(2B)^n N(n))].$$

We integrate this along the circle $\Gamma(w';\rho r')$. Since $r' = r_n' = r_n \exp[-\ell(N(n))]$, we have, by use of the property (e),

$$\text{(18)} \quad \begin{aligned} \int_{\Gamma(w';\rho r')} |f(w)| ds(w) &\leqq 2\pi\rho r_n \exp[-\ell(N(n)) + h(C(2B)^n N(n))] \\ &\leqq 2\pi\rho r_n \exp[-\ell(N(n))/2], \end{aligned}$$

if $n \geqq \max\{n_0, C\}$. We now look at the union Γ_n of paths $\Gamma(w';\rho r')$ corresponding to all $D' \in K_n'$. Then, as we have seen, the set E is included in the inside of Γ_n, $n \geqq 1$. Since the function f is holomorphic at ∞, it has an expansion $f(z) = \sum_{j=0}^{\infty} c_j z^{-j}$. Since we have

$\Phi(t+\log 2)/\Phi(t) = O(1)$, $t \to \infty$, it is easily seen that $f(z) - c_0$ also belongs to $H^\Phi(\overline{\mathbb{C}} \setminus E)$; so we may assume $f(\infty) = 0$. If f does not vanish identically, let c_p, $p \geqq 1$, be the first nonzero coefficient of f. If n is sufficiently large, then we can use the inequality (18) and get

$$|c_p| \leqq \frac{1}{2\pi} \int_{\Gamma_n} |f(w)||w|^{p-1} ds(w)$$

$$\leqq \rho r_n b^{p-1} N(n) \exp[-N(n)^{1/2}/2].$$

This is a contradiction, for the last member tends to zero as $n \to \infty$. Hence f must be a constant function, as was to be proved. □

4B. In the preceding theorem we constructed a null set of class N_Φ in an annulus with center at the origin. Trivially, the same construction can be performed for an annulus with arbitrary center. In the theorem we assumed that $\delta < 1$. When Φ satisfies a stronger condition, the value $\delta = 1$ is possible, i.e. E is linear. This was observed by Heins [31] and is here stated in the following form.

<u>Theorem</u>. Every closed subset of a circle having zero linear measure belongs to the class N_Φ if the convex function Φ satisfies the condition $e^t/\Phi(t) = O(1)$, $t \to \infty$.

<u>Proof</u>. Let F be any closed set of zero linear measure on a circle, say $\Gamma(1)$, and V a region in $\overline{\mathbb{C}}$ including F. The hypothesis on Φ implies that $H^\Phi(V \setminus F) \subseteqq H^1(V \setminus F)$. As Heins shows, every $f \in H^1(V \setminus F)$ is holomorphic throughout V. Then it is easy to verify that such an f belongs to $H^\Phi(V)$. □

For a general convex function Φ we do not know whether $\delta = 1$ is possible or not. But we can take δ arbitrarily close to 1 so that there will be no practical difference.

§3. CLASSIFICATION OF PLANE REGIONS

5. <u>Lemmas</u>

5A. <u>Lemma</u>. Let $0 < a < b < c < \infty$ and let F be a bounded closed subset of $\{|z| \geqq c\}$ such that $\{|z| > b\} \setminus F$ is a region. Let μ and ν be the harmonic measures at the point ∞ with respect to the regions $\{|z| > b\} \setminus F$ and $\{|z| > a\} \setminus F$, respectively. Then

$$\frac{\log(c/b)}{\log(c/a)}\,\mu(\Gamma(b)) \leqq \nu(\Gamma(a)).$$

Proof. Let u be the solution of the Dirichlet problem for the region $\{|z| > a\} \setminus F$ with the boundary data equal to 1 on $\Gamma(a)$ and to 0 elsewhere. Since the restriction of u to the region $\{|z| > b\} \setminus F$ is equal to the solution of the Dirichlet problem for this region with the boundary data equal to u on $\Gamma(b)$ and to 0 elsewhere, we have

$$\nu(\Gamma(a)) = u(\infty) = \int_{\Gamma(b)} u(z)d\mu(z).$$

Since F lies in $\{|z| \geqq c\}$, we see that $u(z) \geqq \log(|z|/c)/\log(a/c)$ for any $a < |z| < c$, from which we get

$$\int_{\Gamma(b)} u(z)d\mu(z) \geqq \frac{\log(c/b)}{\log(c/a)}\,\mu(\Gamma(b)).$$

This shows the desired inequality. □

5B. Lemma. Let F_j, $j = 1,\ldots, k$, be a finite number of bounded closed sets such that $\overline{\mathbb{C}} \setminus F_i$ for each i is a region including all F_j with $j \neq i$, and let $z_0 \in \overline{\mathbb{C}} \setminus F$ with $F = \cup_{j=0}^{k} F_j$ be any finite point. Suppose that each F_i is a non-polar set. Let μ be the harmonic measure at the point ∞ with respect to the region $\overline{\mathbb{C}} \setminus F$ and, for each $0 < a < \inf\{|z - z_0| : z \in F\}$, let μ_a be the harmonic measure at the point ∞ with respect to the region $\{|z - z_0| > a\} \setminus F$. Then there is, for any $\varepsilon > 0$, a number a with the above property such that

$$\mu(F_j) - \mu_a(F_j) < \varepsilon, \quad 0 \leqq j \leqq k.$$

Proof. Take any index j, $0 \leqq j \leqq k$, which is held fixed. Let $u(a;z)$ be the solution of the Dirichlet problem for $D_a = \{|z - z_0| > a\} \setminus F$ with the boundary data equal to 1 on ∂F_j and to 0 elsewhere, so that $\mu_a(F_j) = u(a;\infty)$. Since F_j is assumed to be non-polar, we have $u(a;z) > 0$ on D_a. Take any sequence $\{a_n : n = 1, 2,\ldots\}$ of positive numbers strictly decreasing to zero and consider the corresponding sequence $\{u(a_n;z)\}$ of harmonic functions. Since this sequence is monotonically increasing and is bounded above, it converges by Harnack's theorem to a harmonic function, say $v(z)$, uniformly on any compact set in $\overline{\mathbb{C}} \setminus (F \cup \{z_0\})$. Since v is bounded, the point z_0 is a removable singularity of v, so that v is harmonic on $\overline{\mathbb{C}} \setminus F$. In fact, it is easy to see that v is the solution of the Dirichlet problem for $\overline{\mathbb{C}} \setminus F$ with the boundary data equal to 1 on ∂F_j and to 0 elsewhere.

Thus $v(\infty) = \mu(F_j)$. Since $u(a_n;\infty)$ tend increasingly to $v(\infty)$ and since $u(a_n;\infty) \leqq u(a;\infty)$ for any $0 < a < a_n$, we have $\mu(F_j) - \mu_a(F_j) < \varepsilon$ for all sufficiently small $a > 0$. As the index j was arbitrarily chosen, the lemma is proved. □

5C. Lemma. Let Φ be a convex function satisfying (1) in 1A, $\{b_n : n = 0, 1, \ldots\}$ a sequence of positive numbers, $0 < \rho < \delta < 1$, and $0 < d < 1$. Then there exist a sequence $\{a_n : n = 0, 1, \ldots\}$ of positive numbers and a sequence $\{E_n : n = 0, 1, \ldots\}$ in N_Φ such that $na_n < 1$, $a_{n+1}/a_n \leqq \rho$, $a_n \leqq b_n$, $E_n \subseteqq \{\delta a_n \leqq |z| \leqq a_n\}$ for $n = 0, 1, \ldots$, and

(19) $$d \leqq \Phi(-\log(na_n))m(E_n) \leqq 1$$

for $n = 1, 2, \ldots$, where m is the harmonic measure at the point ∞ with respect to the region $\overline{\mathbb{C}} \setminus E$ with $E = \cup_{n=0}^{\infty} E_n \cup \{0\}$.

Proof. Let $\{d_n : n = 0, 1, \ldots\}$, $0 < d_n < 1$, be a strictly decreasing sequence with limit d. We denote by μ_n (resp. ν_n) the harmonic measure at the point ∞ with respect to the region $\overline{\mathbb{C}} \setminus (\cup_{k=0}^{n} E_k)$ (resp. $\{|z| > a_{n+1}\} \setminus (\cup_{k=0}^{n} E_k)$). In the following construction we shall take a_n so small that $na_n < 1$ and $\Phi(-\log(na_n)) > 0$. We write $\alpha_n = 1/\Phi(-\log(na_n))$.

In order to construct our set E by induction, we first choose any a_0, satisfying $0 < a_0 < \min\{b_0, 1\}$ and $\Phi(-\log a_0) > 0$, and also a non-polar $E_0 \in N_\Phi$ which is included in the annulus $\{\delta a_0 \leqq |z| \leqq a_0\}$. This is possible by Lemma 3A and Theorem 4A. Suppose that we have chosen non-polar $E_k \in N_\Phi$ in $\{\delta a_k \leqq |z| \leqq a_k\}$, $0 \leqq k \leqq n$, with the following property:

(C_n) $0 < a_k < \min\{b_k, 1/k\}$ and $\Phi(-\log(na_n)) > 0$ for $0 \leqq k \leqq n$; $a_{k+1}/a_k \leqq \rho$ for $0 \leqq k \leqq n-1$; and

(20) $$d_n\alpha_k \leqq \mu_n(E_k) \leqq \alpha_k, \quad 1 \leqq k \leqq n.$$

Then we choose a_{n+1} so small that

(21) $$0 < a_{n+1} < \min\{b_{n+1}, (n+1)^{-1}\}, \quad 0 < a_{n+1}/a_n \leqq \rho;$$

(22) $$d_{n+1}\alpha_k \leqq \nu_n(E_k) \leqq \alpha_k, \quad 1 \leqq k \leqq n;$$

(23) $$\nu_n(\Gamma(a_{n+1})) > 1/\Phi(-\log(n+1)a_{n+1})) = \alpha_{n+1}.$$

There is no problem about (21). Statement (23) is clear from Lemma 5A and the fact $t/\Phi(t) = o(1)$, $t \to \infty$, for $\nu_n(\Gamma(a_{n+1}))$ decreases as $a_{n+1} \to 0$ no faster than $(-\log a_{n+1})^{-1}$ while $1/\Phi(-\log((n+1)a_{n+1}))$

decreases much faster than this. For (22) we have only to set, in Lemma 5B, $F_j = E_j$ for $j = 0,\ldots, n$, $\mu = \mu_n$ and $\mu_a = \nu_n$ with $a = a_{n+1}$, and then use (20). By use of Theorem 4A we find a set $K \in N_\Phi$ in $\{\delta a_{n+1} \leqq |z| \leqq a_{n+1}\}$ in such a way that

$$\nu_n(\Gamma(a_{n+1})) \geqq \mu_K(K) > \alpha_{n+1},$$

where μ_K denotes the harmonic measure at the point ∞ with respect to $\overline{\mathbb{C}} \setminus (\cup_{k=0}^{n} E_k \cup K)$. Then it is not hard to find a closed subset E_{n+1} of K such that

$$d_{n+1}\alpha_{n+1} \leqq \mu_{n+1}(E_{n+1}) \leqq \alpha_{n+1}.$$

Since $\nu_n(E_k) \leqq \mu_{n+1}(E_k) \leqq \mu_n(E_k)$ for $k = 1,\ldots, n$, the property (C_{n+1}) is satisfied. By induction we can construct $\{E_n: n = 0, 1,\ldots\}$ which satisfies (C_n) for all $n \geqq 1$. Set $E = \cup_{n=0}^{\infty} E_n \cup \{0\}$. It is then easy to see that, for each fixed k, $\mu_n(E_k) \to m(E_k)$ as $n \to \infty$. So, (22) implies (19), for $d_n \to d$. □

6. Classification Theorem

6A. Let $\{a_n: n = 0, 1,\ldots\}$ be a sequence of positive numbers and let $0 < \rho < \delta < 1$ be constants. Suppose that $a_{n+1}/a_n \leqq \rho$ for $n \geqq 0$. For each n let E_n be a closed, totally disconnected, non-polar set included in $\{\delta a_n \leqq |z| \leqq a_n\}$. We set $E = \cup_{n=0}^{\infty} E_n \cup \{0\}$, so that E is also bounded, closed and totally disconnected. In the following we call such a set E a circular set with center at the origin. The definition of circular sets with center at an arbitrary point is obvious. By m and m_n, $n \geqq 1$, we denote the harmonic measures at the point ∞ with respect to the regions $\overline{\mathbb{C}} \setminus E$ and $\{|z| > a_n\} \setminus E$, respectively. We use various circular sets in order to obtain our classification result. Namely we have the following

Theorem. Let Φ and Ψ be convex functions satisfying

(A1) $\Psi(t)/\Phi(t-s) = o(1)$, $t \to \infty$, for any fixed $s > 0$.

Then there exists a circular set E with center at the origin and with $E_n \in N_\Phi$, $n \geqq 0$, such that the function z^{-1} belongs to $H^\Psi(\overline{\mathbb{C}} \setminus E)$, while $H^\Phi(\overline{\mathbb{C}} \setminus E)$ contains only constant functions.

Proof. Let $0 < \rho < \delta < 1$ be fixed. By use of (A1) we can find a sequence $\{b_n: n = 0, 1,\ldots\}$ of positive numbers such that $b_0 = 1$, $b_n < 1/n$ and

$$0 < \Psi(\log(\delta^{-1}t)) \leqq 2^{-n}\Phi(\log(n^{-1}t)), \quad t \geqq 1/b_n,$$

for $n = 1, 2, \ldots$. By Lemma 5C we get a sequence $\{a_n : n = 0, 1, \ldots\}$ of positive numbers and a sequence $\{E_n : n = 0, 1, \ldots\}$ in N_Φ such that $a_{n+1}/a_n \leqq \rho$, $a_n \leqq b_n$ and $E_n \subsetneq \{\delta a_n \leqq |z| \leqq a_n\}$ for $n \geqq 0$, and $d \leqq \Phi(-\log(na_n))m(E_n) \leqq 1$ for $n \geqq 1$, where m and d have the same meaning as in Lemma 5C.

We first show that $z^{-1} \in H^\Psi(\overline{\mathbb{C}} \setminus E)$. Since $a_n \leqq b_n$, we have

$$\Psi(-\log(\delta a_n)) \leqq 2^{-n}\Phi(-\log(na_n)), \quad n \geqq 1.$$

As E_n is included in $\{\delta a_n \leqq |z| \leqq a_n\}$, so we have

$$\int_E \Psi(\log|z^{-1}|)dm(z) = \sum_{n=0}^{\infty} \int_{E_n} \Psi(\log|z^{-1}|)dm(z)$$

$$\leqq \sum_{n=0}^{\infty} \psi(-\log(\delta a_n))m(E_n)$$

$$\leqq \Psi(-\log(\delta a_0))m(E_0) + \sum_{n=1}^{\infty} \Psi(-\log(\delta a_n))/\Phi(-\log(na_n))$$

$$\leqq \Psi(-\log(\delta a_0))m(E_0) + \sum_{n=1}^{\infty} 2^{-n} < \infty.$$

This shows that z^{-1} belongs to $H^\Psi(\overline{\mathbb{C}} \setminus E)$.

The latter half of the theorem can be shown by an argument similar to the one already used in the proof of Theorem 4A. We look at any a_n with $n \geqq 1$. Since $0 < \rho < \delta < 1$, the annulus $\{a_n < |z| < \delta\rho^{-1}a_n\}$ is included in $\overline{\mathbb{C}} \setminus E$. We set $\sigma = (\delta\rho^{-1})^{1/4}$. Let ξ_w be the harmonic measure with respect to the annulus $\{a_n < |z| < \sigma^2 a_n\}$ at the point w on the circle $\Gamma(\sigma a_n)$. Harnack's inequality then implies that

$$d\xi_w/ds \leqq A'/(4\pi a_n) \quad \text{on} \quad \Gamma(a_n), \tag{24}$$

$$d\xi_w/ds \leqq A'/(4\pi\sigma^2 a_n) \quad \text{on} \quad \Gamma(\sigma^2 a_n), \tag{25}$$

where $A' = A'(\sigma^{-2})$ is the constant appearing in the proof of Lemma 3C. We take any $f \in H^\Phi(\overline{\mathbb{C}} \setminus E)$ and a harmonic majorant u of $\Phi(\log^+|f|)$ on $\overline{\mathbb{C}} \setminus E$. Our objective is to show that f is constant. So we may assume that f and u do not vanish identically. Since each E_n belongs to the class N_Φ, we see that f can only be singular at the origin. Namely, we have the following expansion: $f(z) = \sum_{j=0}^{\infty} c_j z^{-j}$ for $0 < |z| < \infty$. Let $n \geqq 1$. Since $f(z)$ is a bounded analytic function on any annulus $\{a_n < |z| < \sigma^2 a_n\}$, we have

$$\text{(26)} \qquad \log^+|f(w)| \leq \int \log^+|f(\zeta)|d\xi_w(\zeta)$$

for any $w \in \Gamma(\sigma a_n)$, where ξ_w is the harmonic measure defined above. Applying $\Phi(t)$ to both sides and using Jensen's inequality, we get

$$\Phi(\log^+|f(w)|) \leq \left(\int_{\Gamma(a_n)} + \int_{\Gamma(\sigma^2 a_n)}\right) \Phi(\log^+|f(\zeta)|)d\xi_w(\zeta).$$

Assume now that $w \in \Gamma(\sigma a_n)$. Using (24), (25), (26), Lemma 3C and the fact $m_n(\Gamma(a_n)) \geq m(E_n) \geq d\Phi(-\log(na_n))^{-1}$, and computing as in the proof of Theorem 4A, we see that the right-hand side of the above inequality is majorized by $C\Phi(-\log a_n)$, where $C = A'(\sigma^{-2})A(\sigma^{-1})u(\infty)/d$. Thus we have

$$\text{(27)} \qquad \Phi(\log^+|f(w)|)/\Phi(-\log a_n) \leq C$$

for any $w \in \Gamma(\sigma a_n)$. Take a positive number t_0 so large that $\Phi(t)/t$ is positive and nondecreasing for $t \geq t_0$. Then we have $t_2/t_1 \leq 1 + \Phi(t_2)/\Phi(t_1)$ for any $t_1 \geq t_0$ and any $t_2 \geq 0$. We choose an integer $N \geq 1$ satisfying $-\log a_N \geq t_0$. Then (27) implies

$$|f(w)| \leq a_n^{-C-1}$$

for any $w \in \Gamma(\sigma a_n)$ and any $n \geq N$. Letting $n \geq N$ and writing $r = \sigma a_n$, we have

$$|c_k| = \left|\frac{1}{2\pi i}\int_{\Gamma(r)} f(z)z^{k-1}dz\right| \leq r^k a_n^{-C-1} = \sigma^k a_n^{k-C-1}.$$

Letting $n \to \infty$, we see that $c_k = 0$ for $k > C+1$ and therefore $f(z)$ is a polynomial in z^{-1}. Let c_p be its highest nonzero coefficient. If $p \geq 1$, then we would have $|f(z)| \geq 2^{-1}|c_p||z|^{-p} \geq 1$ for all sufficiently small z, say $|z| \leq a_{N'}$, with $N' \geq N$. Consequently,

$$\int_E \Phi(\log^+|f(z)|)dm(z) \geq \sum_{n=N'}^{\infty} \Phi(\log(2^{-1}|c_p|a_n^{-p}))m(E_n)$$

$$\geq d\sum_{n=N'}^{\infty} \Phi(\log(2^{-1}|c_p|a_n^{-1}))/\Phi(-\log(na_n)) = \infty.$$

This contradiction shows that $p = 0$. Hence $H^\Phi(\overline{\mathbb{C}} \setminus E)$ contains only constant functions. □

6B. In terms of null classes O_Φ the preceding theorem can be expressed as follows:

Theorem. Let Φ and Ψ be convex functions satisfying (A1) above. Then O_Φ strictly includes O_Ψ.

We set $O_p^- = \cup\{O_q : 0 < q < p\}$ and $O_p^+ = \cap\{O_q : p < q < \infty\}$. Then we have the following, where the inequality sign $<$ means the strict inclusion.

Corollary. (a) $O_p^- < O_p < O_p^+$ for $0 < p < \infty$.

(b) $O_{AB^*} < \cap\{O_q : 0 < q < \infty\}$, $\cup\{O_q : 0 < q < \infty\} < O_{AB}$.

Proof. We first note that the condition (A1) is equivalent to a simpler one: $\Psi(t)/\Phi(t) = o(1)$, $t \to \infty$, if either Φ or Ψ satisfies the condition (3) in 3B.

(a) Let $0 < p < \infty$. In order to show that $O_p^- < O_p$, we have only to set $\Phi(t) = e^{pt} - 1$ and

$$\Psi(t) = \begin{cases} p^2e^2t/4, & 0 \leqq t \leqq 2/p, \\ e^{pt}/t, & t \geqq 2/p. \end{cases}$$

Indeed, it is easy to check that Ψ is a convex function and the condition (A1) is satisfied. So, by the theorem, $O_\Psi < O_\Phi = O_p$. If $0 < q < p$, then $e^{qt}/\Psi(t) = o(1)$, $t \to \infty$, and therefore $O_q < O_\Psi$ again by the theorem. Hence $O_p^- = \cup\{O_q : 0 < q < p\} \subseteqq O_\Psi < O_p$. In order to show that $O_p < O_p^+$, we set $\Phi(t) = te^{pt}$ and $\Psi(t) = e^{pt} - 1$. Then these Φ and Ψ satisfy (A1) and therefore $O_p < O_\Phi$. On the other hand, we have $\Phi(t)/e^{qt} = o(1)$, $t \to \infty$, for any $q > p$, so that $O_\Phi < O_q$. Hence, $O_p < O_\Phi \subseteqq O_p^+$.

(b) To show the first half, we set $\Phi(t) = t^2$ and $\Psi(t) = t$. Φ and Ψ are then convex functions and $O_\Psi = O_{AB^*}$, as mentioned in 1A. Since $\Psi(t)/\Phi(t) = o(1)$, $t \to \infty$, we have $O_{AB^*} < O_\Phi$ by the theorem. On the other hand, $\Phi(t)/e^{qt} = o(1)$, $t \to \infty$, for any $q > 0$, so that $O_\Phi < O_q$. Hence $O_{AB^*} < O_\Phi \subseteqq \cap\{O_q : 0 < q < \infty\}$.

The latter half is seen by taking $\Phi(t) = \exp(e^{2t}) - e$ and $\Psi(t) = \exp(e^t) - e$. In fact, Φ and Ψ are seen to satisfy the condition (A1), so that $O_\Psi < O_\Phi$. As $e^{qt}/\Psi(t) = o(1)$, $t \to \infty$, so we have $O_q < O_\Psi$. Consequently, $\cup\{O_q : 0 < q < \infty\} \subseteqq O_\Psi < O_\Phi \subseteqq O_{AB}$. □

6C. We remark that, if two null classes of regions, e.g. O_p and O_q, are distinct, their difference is very wide. Here we state the following without proof: If the convex functions Φ and Ψ satisfy the condition (A1), then there exists a region $D \subseteqq \overline{\mathbb{C}}$ such that $H^\Phi(D)$ contains only constant functions, while $H^\Psi(D)$ is infinite dimensional and contains functions having essential singularities. A proof is outlined in Hasumi [21].

7. Majoration by Quasibounded Harmonic Functions

7A. First we have the following, which can be derived easily from Theorem 2C.

Lemma. Let f be an analytic function on a plane region and let Φ be a convex function such that (1) holds. If $\Phi(\log^+|f|)$ has a harmonic majorant, then its LHM is quasibounded.

7B. For a region D in $\overline{\mathbb{C}}$ let $AS(D)$ denote the set of analytic functions f on D such that $\log^+|f(z)|$ is majorized by a quasibounded harmonic function. We denote by $\mathcal{O}_{AS}$ the class of regions D in $\overline{\mathbb{C}}$ for which $AS(D)$ contains only constant functions.

Theorem. $\mathcal{O}_{AS} = \mathcal{O}_G$.

Proof. We have only to show that $\mathcal{O}_{AS} \subseteqq \mathcal{O}_G$. So, let D be a region in $\overline{\mathbb{C}}$, which possesses a Green function, i.e. $D \notin \mathcal{O}_G$. If $\overline{\mathbb{C}} \setminus D$ has a connected component which is a continuum, then D clearly carries a nonconstant bounded analytic function and thus $D \notin \mathcal{O}_{AS}$. We now assume that $\overline{\mathbb{C}} \setminus D$ (= K, say) is totally disconnected. Since $D \notin \mathcal{O}_G$, K is a non-polar set. Thus it is possible to divide K into two closed sets F and F', each being a non-polar set. We set $D' = \overline{\mathbb{C}} \setminus F'$ and so $D = D' \setminus F$. Since F is not a polar set, Theorem 6D, Ch. I, implies that F contains a point z_0 which is a regular boundary point of the region D. Let $g'(z_0,z)$ be the Green function for D' with pole z_0 and set $F_n = \{z \in F: g'(z_0,z) \leq n\}$ for $n = 1, 2, \ldots$. Finally we define functions u_x, with $x \in D'$, on the region D' by the formula $u_x(z) = g'(x,z)$.

Let $v_n = (u_{z_0})^{D'}_{F_n}$ be the balayaged function of u_{z_0} relative to F_n (Ch. I, 6E). Clearly, v_n are bounded harmonic functions on D and are majorized there by $g'(z_0,z)$. Since $\{F_n\}$ is increasing with n, $\{v_n\}$ is increasing and therefore converges to a quasibounded harmonic function, say v, on D. Next we take any $y \in D$, which is held fixed, and set $w_n = (u_y)^{D'}_{F_n}$ for $n = 1, 2, \ldots$ and $w_0 = (u_y)^{D'}_F$. Then, as above, $\{w_n: n = 1, 2, \ldots\}$ is increasing and $w_n \leq w_0$ for all n. So $\{w_n\}$ tends to a superharmonic function, say w, on D', satisfying $w \leqq w_0$. In view of Theorem 6E, Ch. I, we have $w_n = u_y$ on F_n except for a polar set. Since the union of countably many polar sets is again a polar set, we see that $w = u_y = g'(y,\cdot)$ on F except for a polar set. The definition of balayaged functions shows that $w = w_0$ on D. Let $g_n'(z_0,z)$ be the Green function for $D' \setminus F_n$ with pole z_0. Again by Theorem 6E, Ch. I, we have

$$\begin{aligned} v_n(y) &= H[u_{z_0}; D' \setminus F_n](y) \\ &= g'(z_0) - g_n'(z_0,y) \\ &= H[u_y; D' \setminus F_n](z_0) \\ &= w_n(z_0) \to w(z_0). \end{aligned}$$

Since z_0 is a regular boundary point of D, $w(z_0) = w_0(z_0) = u_y(z_0) = g'(y,z_0)$. Hence $v_n(y) \to g'(y,z_0)$ on D. Since each v_n is a bounded harmonic function on D, we conclude that $g'(z_0,\cdot)$ is a quasibounded harmonic function on D.

Finally, consider the function $f(z) = (z - z_0)^{-1}$ on $\overline{\mathbb{C}}$. Then,

$$\log^+|f(z)| = \log^+\frac{1}{|z - z_0|} \leqq g'(z_0,z) + C$$

for a sufficiently large constant C. This means that $\log^+|f(z)|$ on D is majorized by a quasibounded harmonic function. As f is not constant, we conclude that $D \notin \mathcal{O}_{AS}$, as was to be proved. □

<u>Corollary</u>. $\mathcal{O}_G = \mathcal{O}_{AB*} = \mathcal{O}_{AS}$.

NOTES

Heins' classification theorem is in his lecture note [31; Chapter III]. For the case of plane regions he only showed $\mathcal{O}_{AB*} < \mathcal{O}_1$ and proposed further investigation. Afterwards, interesting contributions were made by Hejhal [32], Kobayashi [38, 39] and Obrock [49]. The results mentioned above are due to Hasumi [21]. See Hasumi [22] for some related results.

Lemma 3B is in Hejhal [32]. The fact stated in 6C is proved in Hasumi [21]. Theorem 7B is adapted from Segawa [63].

APPENDICES

A.1. The Classical Fatou Theorem

A.1.1. We begin with a theorem on differentiation of measures.

Theorem. Let μ be a finite complex measure on an interval I and set $F(x) = \mu(I \cap (-\infty, x))$ for every $x \in I$. Then

(a) $F'(x)$ exists a.e. on I and is summable; and

(b) if we set $\mu_c(A) = \int_A F'(t)dt$ for $A \subseteqq I$ and $\mu_s = \mu - \mu_c$, then μ_c and μ_s are the absolutely continuous and the singular parts of μ, respectively, with respect to the Lebesgue measure dx on I.

For a proof, see any standard textbook on real analysis, e.g. Rudin [59], Ch. 8.

A.1.2. Let ζ_0 be any point on the unit circle $\mathbb{T}$. By a Stolz region in the open unit disk $\mathbb{D}$ with vertex ζ_0 and angular measure 2α we mean the set

$$S(\zeta_0;\alpha) = \{z \in \mathbb{D}: \ -\alpha < \arg(1 - z\bar{\zeta}_0) < \alpha\},$$

where we assume $0 < \alpha < \pi/2$. We set

$$P(r,\theta) = (1 - r^2)/(1 - 2r\cos\theta + r^2), \quad 0 \leqq r < 1.$$

This is called the Poisson kernel and we have the following

Theorem. Let μ be a finite complex Borel measure on $\mathbb{T}$ and set

$$(1) \qquad u(z) = u(re^{i\theta}) = \int_{\mathbb{T}} P(r, \theta - t)d\mu(e^{it}).$$

Then u is harmonic on $\mathbb{D}$. Take any real t_0 and set

$$(2) \qquad F(t) = \mu(\{e^{i\theta}: \ t_0 \leqq \theta < t\})$$

for $t_0 \leqq t \leqq t_0 + 2\pi$. Then $F'(\theta)$ exists for almost every θ in the interval $(t_0, t_0 + 2\pi)$ and, for any such θ, $u(z)$ tends to $2\pi F'(\theta)$ provided that z tends to $e^{i\theta}$ through any fixed Stolz region $S(e^{i\theta};\alpha)$ with $0 < \alpha < \pi/2$.

Proof. First we see easily that $u(z)$ is harmonic in $\mathbb{D}$. By the preceding theorem, the function F is differentiable a.e. So we have only to show the existence of limits through Stolz regions.

If μ is the normalized Lebesgue measure $d\sigma(t) = dt/2\pi$ on $\mathbb{T}$, then $u(z) \equiv 1$ on $\mathbb{D}$ and $F(t) = (t - t_0)/2\pi$, so that the theorem is trivially true for any θ. So the theorem can be proven for any given μ if it is proven for the modified measure $\mu - \mu(\mathbb{T})\sigma$. We may therefore assume without loss of generality that $\mu(\mathbb{T}) = 0$. In order to simplify the notation slightly, it is also assumed that $t_0 = -\pi$ and $\theta = 0$.

The definition of F implies that $F(-\pi) = F(\pi) = 0$ and so

$$u(re^{i\theta}) = \int_{-\pi}^{\pi} P(r,\theta - t)dF(t) = - \int_{-\pi}^{\pi} \frac{\partial}{\partial t} P(r,\theta - t)F(t)dt.$$

The Poisson integral formula for the harmonic function $r\cos\theta$ states:

$$r\cos\theta = \int_{-\pi}^{\pi} P(r,\theta - t)\cos t\, d\sigma(t) = - \int_{-\pi}^{\pi} \frac{\partial}{\partial t} P(r,\theta - t)\sin t\, d\sigma(t).$$

Thus we have

$$\begin{aligned} u(re^{i\theta}) - 2\pi F'(0)r\cos\theta &= - \int_{-\pi}^{\pi} \frac{\partial}{\partial t} P(r,\theta - t)(F(t) - F'(0)\sin t)\, dt \\ &= \int_{-\pi}^{\pi} (-\sin t \frac{\partial}{\partial t} P(r,\theta - t)) \left(\frac{F(t)}{\sin t} - F'(0)\right) dt \\ &= \int_{-\delta}^{\delta} + \int_{\delta}^{\pi/2} + \int_{\pi/2}^{\pi} + \int_{-\pi/2}^{-\delta} + \int_{-\pi}^{-\pi/2} \\ &= I_0 + I_1 + I_2 + I_1' + I_2', \text{ say,} \end{aligned}$$

with $0 < \delta < \pi$. Since I_1 and I_1' (resp. I_2 and I_2') are similar, we have only to estimate I_0, I_1 and I_2.

Since F is differentiable at 0,

$$\max\left\{\left|\frac{F(t)}{\sin t} - F'(0)\right|, \left|\frac{F(-t)}{-\sin t} - F'(0)\right|\right\} \tag{3}$$

tends to zero as $t \to 0$. So there exists a continuous monotonically increasing function $\varepsilon(t)$, $0 \leqq t \leqq \pi/2$, with $\varepsilon(0) = 0$, which majorizes (3) for $0 < t \leqq \pi/2$.

After this preparation, take any Stolz region $S(1;\alpha)$ with vertex 1 and angular measure 2α, $0 < \alpha < \pi/2$. If $z = re^{i\theta}$ with $|\theta| < \min\{\pi/2, \pi - 2\alpha\}$ belongs to $S(1;\alpha)$, then $r|\sin\theta|/(1 - r\cos\theta) \leqq \tan\alpha$ and therefore

$$|\theta| \leqq A(1 - r), \tag{4}$$

where A is a constant depending only on α. Since $\varepsilon(\alpha)$ is monotonically increasing, so is $\delta\varepsilon(\delta)^{1/4}$. Let δ_0 be chosen so that

$$(5) \qquad 0 < \delta_0 < 1 \quad \text{and} \quad \varepsilon(\delta_0) < \min\{(2A)^{-4},(2\delta_0)^{-4},1\}.$$

Suppose now that $z = re^{i\theta} \in \mathbb{D}$ satisfies (4) and $1-r \leqq \delta_0\varepsilon(\delta_0)^{1/4}$. Then there exists a positive number δ with $0 < \delta < \delta_0$ such that $1-r = \delta\varepsilon(\delta)^{1/4}$. It follows from (4) and (5) that

$$|\theta| \leqq A(1-r) = A\delta\varepsilon(\delta)^{1/4} \leqq \delta/2.$$

If $|t| \leqq \delta$, then

$$|-\sin t \frac{\partial}{\partial t} P(r,\theta-t)| = \frac{2r(1-r^2)|\sin t \sin(\theta-t)|}{((1-r)^2 + 4r\sin^2((\theta-t)/2))^2}$$

$$\leqq 4(1-r)^{-3}\delta(|\theta|+\delta) \leqq 4(\delta\varepsilon(\delta)^{1/4})^{-3}\cdot\delta\cdot 3\delta/2$$

$$= 6\delta^{-1}\varepsilon(\delta)^{-3/4}$$

and thus

$$|I_0| \leqq \int_{-\delta}^{\delta} |-\sin t \frac{\partial}{\partial t} P(r,\theta-t)| \left|\frac{F(t)}{\sin t} - F'(0)\right| dt$$

$$\leqq 6\delta^{-1}\varepsilon(\delta)^{-3/4}\cdot 2\delta\varepsilon(\delta) = 12\varepsilon(\delta)^{1/4}.$$

If $\delta \leqq t \leqq \pi$, then $3\pi/2 \geqq |t-\theta| \geqq |t| - |\theta| \geqq |t|/2$ and therefore there exists a constant $B > 0$ with

$$(6) \qquad \sin^2 \frac{\theta - t}{2} \geqq Bt^2.$$

So, for $\delta \leqq t \leqq \pi/2$,

$$|-\sin t \frac{\partial}{\partial t} P(r,\theta-t)| \leqq \frac{4(1-r)t}{\sin^3((t-\theta)/2)} \leqq \text{Const. } t^{-2}\delta\varepsilon(\delta)^{1/4}$$

and thus

$$|I_1| \leqq \int_{\delta}^{\pi/2} |-\sin t \frac{\partial}{\partial t} P(r,\theta-t)| \left|\frac{F(t)}{\sin t} - F'(0)\right| dt$$

$$\leqq \text{Const.}\,\delta\varepsilon(\delta)^{1/4}\int_{\delta}^{\pi/2} t^{-2}dt = \text{Const.}\,\varepsilon(\delta)^{1/4}.$$

Finally, let $\pi/2 \leqq t \leqq \pi$. By (5), $1-r \leqq \delta_0\varepsilon(\delta_0)^{1/4} \leqq 1/2$ and so, in view of (6),

$$|\frac{\partial}{\partial t} P(r,\theta-t)| \leqq \frac{4(1-r)|\sin((t-\theta)/2)|}{16\sin^4((t-\theta)/2)} = \frac{1-r}{|\sin^3((t-\theta)/2)|}$$

$$\leqq \text{Const.}(1-r) \leqq \text{Const.}\,\varepsilon(\delta)^{1/4}.$$

This implies that

$$|I_2| \leq \int_{\pi/2}^{\pi} \left|\frac{\partial}{\partial t} P(r,\theta - t)\right| |F(t) - F'(0)\sin t|\, dt$$

$$\leq \text{Const.}\varepsilon(\delta)^{1/4}(\max_t |F(t)| + |F'(0)|) = \text{Const.}\varepsilon(\delta)^{1/4}.$$

Combining these estimates, we get

$$|u(re^{i\theta}) - 2\pi F'(0) r \cos\theta| \leq \text{Const.}\varepsilon(\delta)^{1/4}.$$

If z tends to 1 through $S(1;\alpha)$, then $\delta \to 0$ and thus $\varepsilon(\delta) \to 0$; so $u(re^{i\theta}) \to 2\pi F'(0)$, as desired. □

A.1.3. Corollary. Under the same assumption as in Theorem A.1.2, let

$$u^*(e^{i\theta}) = \lim_{r \to 1-0} u(re^{i\theta}),$$

which exists a.e. on $\mathbb{T}$ as a summable function. Then

$$\int_{\mathbb{T}} P(r,\theta - t) u^*(e^{it})\, d\sigma(t), \tag{7}$$

for $z = re^{i\theta} \in \mathbb{D}$, is the quasibounded part of $u(z)$.

Proof. Since $u^*(e^{i\theta}) = 2\pi F'(\theta)$ a.e. and $F'(\theta)$ is summable, (7) defines a harmonic function on $\mathbb{D}$, which is denoted by v. v is easily seen to be quasibounded. To show that v is the quasibounded part of u, it is enough to see that $u - v$ is an inner harmonic function. For this purpose we may assume that μ is nonnegative. Since Theorem A.1.1 implies that $u^*(e^{it})d\sigma(t) = F'(t)dt$ is the absolutely continuous part μ_c of μ,

$$(u - v)(re^{i\theta}) = \int_{\mathbb{T}} P(r,\theta - t)\, d\mu_s(e^{it}),$$

where $\mu_s = \mu - \mu_c$ is the singular part of μ. Since μ is nonnegative, μ_s is also nonnegative and therefore $u - v$ is a nonnegative harmonic function. So it is sufficient to show that $(u - v) \wedge n = 0$ for any positive integer n. Set $v_n = (u - v) \wedge n$. Since v_n is a bounded harmonic function, the first paragraph of the proof of Theorem 2F, Ch. IV shows that

$$v_n(re^{i\theta}) = \int_{\mathbb{T}} P(r,\theta - t) v_n^*(e^{it})\, d\sigma(t) \tag{8}$$

for $re^{i\theta} \in \mathbb{D}$, where v_n^* denotes a bounded measurable function on $\mathbb{T}$. Applying Theorem A.1.2 to the expression (8), we find that v_n^* is the radial boundary function for v_n. Since $0 \leqq v_n \leqq u - v$ on $\mathbb{D}$, $0 \leqq v_n^*(e^{it}) \leqq u^*(e^{it}) - v^*(e^{it}) = 0$ a.e. and thus $v_n \equiv 0$ by (8), as was to be proved. □

A.1.4. Let $P(\mathbb{T})$ be the set of analytic trigonometric polynomials $\sum_{k=0}^n c_k e^{ik\theta}$ on $\mathbb{T}$. $P(\mathbb{T}) + \overline{P(\mathbb{T})}$ is thus the set of trigonometric polynomials on $\mathbb{T}$. Then we have the following

Corollary. For any $f \in L^\infty(d\sigma)$ there exists a sequence $\{f_n\}$ in $P(\mathbb{T}) + \overline{P(\mathbb{T})}$ such that $\|f_n\|_\infty \leqq \|f\|_\infty$ and $f_n \to f$ a.e. If $f \in H^\infty(d\sigma)$ (for the definition, see Ch. IV, 2A), then f_n can be taken from $P(\mathbb{T})$.

Proof. We may suppose that $\|f\|_\infty > 0$. Set

$$u(re^{i\theta}) = \int_{\mathbb{T}} P(r, \theta - t) f(e^{it}) d\sigma(t) \tag{9}$$

for $re^{i\theta} \in \mathbb{D}$. By the Fatou theorem A.2.1 and Theorem A.1.1, $u(re^{i\theta})$ tends to $f(e^{i\theta})$ a.e. as $r \to 1-0$. Let $0 < r_1 < r_2 < \cdots < 1$ be an ascending sequence tending to 1 and let $h_n(e^{i\theta}) = u(r_n e^{i\theta})$, so that $\|h_n\|_\infty \leqq \|f\|_\infty$ and $h_n \to f$ a.e. Using the expansion

$$P(r, \theta - t) = \sum_{j=-\infty}^{\infty} r^{|j|} e^{ij(\theta - t)},$$

we have

$$h_n(e^{i\theta}) = \sum_{j=-\infty}^{\infty} c_j r_n^{|j|} e^{ij\theta} \tag{10}$$

with

$$c_j = \int_{\mathbb{T}} f(e^{it}) e^{-ijt} d\sigma(t)$$

for $j = 0, \pm 1, \ldots$. Since the right-hand side of (10) converges uniformly on $\mathbb{T}$, one can find an integer $N = N(n) > 0$ such that

$$f_n(e^{i\theta}) = \frac{n-1}{n} \sum_{j=-N}^{N} c_j r_n^{|j|} e^{ij}$$

satisfies the conditions $\|h_n - f_n\|_\infty \leqq 2\|f\|_\infty / n$ and $\|f_n\|_\infty \leqq \|f\|_\infty$. The sequence $\{f_n\}$ thus fulfills the requirement. When $f \in H^\infty(d\sigma)$, we have $c_j = 0$ for $j < 0$ and therefore $f_n \in P(\mathbb{T})$. □

A.1.5. For continuous functions we have

Theorem. Let f be a real continuous function on $\mathbb{T}$. Then there is a sequence $\{f_n\}$ in $\mathrm{Re}(P(\mathbb{T}))$ such that $f \leqq f_n$ and $f_n \to f$ uniformly.

Proof. We may suppose that $\|f\|_\infty > 0$. Let $u(re^{i\theta})$ be defined by (9). The characteristic properties of the Poisson kernel then show that u is continuous up to the boundary $\mathbb{T}$ and $u(e^{i\theta}) = f(e^{i\theta})$ everywhere on $\mathbb{T}$. Using the same notation as in the proof of Corollary A.1.4, we find that $h_n \to f$ uniformly. When $\{r_n\}$ tends to 1 rapidly enough, we can assume moreover that $|f - h_n| < 1/n$. We then take $N = N(n)$ so large that $u_n = \sum_{j=-N}^{N} c_j r_n^{|j|} e^{ij\theta}$ is real and satisfies the inequality $|h_n - u_n| < 1/n$. Setting $f_n = u_n + 2/n$, we see that $f_n \in \mathrm{Re}(P(\mathbb{T}))$, $f \leqq h_n + 1/n \leqq f_n$ and also $f_n = u_n + 2/n \leqq h_n + 3/n \leqq f + 4/n$. Hence, $f_n \to f$ uniformly. □

A.2. Kolmogorov's Theorem on Conjugate Functions

A.2.1. To any $u \in \mathrm{Re}(R(\mathbb{T}))$ we associate a unique $\tilde{u} \in \mathrm{Re}(R(\mathbb{T}))$ such that $u + i\tilde{u} \in P(\mathbb{T})$ and $\int \tilde{u}\,d\sigma = 0$. $\tilde{u}$ is called the harmonic conjugate of u.

Lemma. Let $0 < s < 1$. Then we have the following:

(a) for $u \in \mathrm{Re}(P(\mathbb{T}))$ with $u \geqq 0$,

$$\|\tilde{u}\|_s \leqq \left(\cos\frac{\pi s}{2}\right)^{-1/s} \|u\|_1; \tag{11}$$

(b) for any $u \in \mathrm{Re}(P(\mathbb{T}))$

$$\|\tilde{u}\|_s \leqq 2^{(s+1)/s}\left(\cos\frac{\pi s}{2}\right)^{-1/s} \|u\|_1. \tag{12}$$

Proof. (a) First suppose $u > 0$. Then $(u + i\tilde{u})^s$ has a single-valued branch on $\mathbb{D}$ with values in the sector $|\arg z| < \pi s/2$. So

$$\mathrm{Re}\{(u + i\tilde{u})^s\} \geqq (u^2 + \tilde{u}^2)^{s/2}\cos\frac{\pi s}{2} \geqq |\tilde{u}|^s \cos\frac{\pi s}{2}$$

and therefore

$$\cos\frac{\pi s}{2}\int |\tilde{u}|^s d\sigma \leqq \int \mathrm{Re}\{(u + i\tilde{u})^s\}d\sigma = \mathrm{Re}\int (u + i\tilde{u})^s d\sigma = \left(\int u\,d\sigma\right)^s,$$

which shows the inequality (11) for $u > 0$. In case $u \in \mathrm{Re}(P(\mathbb{T}))$ with $u \geqq 0$, we have only to consider $u + \varepsilon$ with $\varepsilon > 0$ and then let ε tend to 0.

(b) For each $n = 1, 2, \ldots$ there exists, by Theorem A.1.5, a function $u_n \in \mathrm{Re}(P(\mathbb{T}))$ with $u^+ \leqq u_n \leqq u^+ + 1/n$, where $u^+(e^{it}) = \max\{u(e^{it}),0\}$. Then $u_n - u \in \mathrm{Re}(P(\mathbb{T}))$ and $u_n - u \geqq 0$. So by (11) we have $\|\tilde{u}_n\|_s \leqq c_s\|u_n\|_1 \leqq c_s(\|u^+\|_1 + 1/n)$ and also $\|\tilde{u}_n - \tilde{u}\|_s \leqq c_s\|u_n - u\|_1 \leqq c_s(\|u^-\|_1 + 1/n)$, where $c_s = (\cos(\pi s/2))^{-1/s}$. Therefore,

$$\begin{aligned}\|\tilde{u}\|_s^s &\leqq 2^s(\|\tilde{u} - \tilde{u}_n\|_s^s + \|\tilde{u}_n\|_s^s)\\ &\leqq 2^s c_s^s((\|u^+\|_1 + 1/n)^s + (\|u^-\|_1 + 1/n)^s).\end{aligned}$$

As n is arbitrary,

$$\|\tilde{u}\|_s^s \leqq 2^s c_s^s(\|u^+\|_1^s + \|u^-\|_1^s) \leqq 2^{s+1}c_s^s\|u\|_1^s,$$

which shows the inequality (12). □

A.2.2. <u>Theorem</u>. Let $u \in L^1(d\sigma)$ be real and set

$$\Phi(z) = \int P(r,\theta - t)u(e^{it})d\sigma(t)$$

for $z = re^{i\theta} \in \mathbb{D}$. Let $\Psi(z)$ be the harmonic conjugate of $\Phi(z)$ normalized by $\Psi(0) = 0$. Then $\lim_{r\to 1-0} \Psi(re^{i\theta})$ (= $v(e^{i\theta})$, say) exists a.e. on $\mathbb{T}$ and

$$\|v\|_s \leqq 2^{(s+1)/s}(\cos \tfrac{\pi s}{2})^{-1/s}\|u\|_1. \tag{13}$$

<u>Proof</u>. Suppose first that $u \geqq 0$. By Theorem A.1.5 and the density of continuous functions in $L^1(d\sigma)$, we can find $u_n \in \mathrm{Re}(P(\mathbb{T}))$, $u_n \geqq 0$, such that $u_n \to u$ in $L^1(d\sigma)$. By Lemma A.2.1 $\tilde{u}_n$ converge in $L^s(d\sigma)$, $0 < s < 1$, to some element, say $u^\# \in L^s(d\sigma)$. By taking a subsequence if necessary, we may assume that $u_n \to u$ and $\tilde{u}_n \to u^\#$ a.e. We set $f_n = \exp(-u_n - i\tilde{u}_n)$ and $f = \exp(-u - iu^\#)$. Then, $|f_n| \leqq 1$, $|f| \leqq 1$, $f_n \in H^\infty(d\sigma)$ and $f_n \to f$ a.e. From this follows that $f_n \to f$ weakly* and so $f \in H^\infty(d\sigma)$. If we denote the analytic extensions, into $\mathbb{D}$, of f_n and f by the same symbols, then $f_n \to f$ almost uniformly in $\mathbb{D}$. Since $f_n(0)$ are real, so is $f(0)$. We have $-\log|f| = u$ on $\mathbb{T}$ and therefore $-\log|f| = \Phi$ in $\mathbb{D}$. Since $\Psi(0) = 0$ and $\Phi + i\Psi$ is holomorphic in $\mathbb{D}$, we see that $f = \exp(-\Phi - i\Psi)$ in $\mathbb{D}$. As $f(z) \in H^\infty(\mathbb{D})$, we have $f(re^{i\theta}) \to f(e^{i\theta}) = \exp(-u(e^{i\theta}) - iu^\#(e^{i\theta}))$ as $r \to 1-0$ for almost all θ. For each such θ, $\exp(-i\Psi(re^{i\theta})) \to \exp(-iu^\#(e^{i\theta}))$. Since $\Psi(re^{i\theta})$, with fixed θ, varies continuously as $r \to 1-0$, we conclude that $v(e^{i\theta}) = \lim_{r\to 1-0} \Psi(re^{i\theta})$ exists a.e. The same can be seen for any real $u \in L^1(d\sigma)$ by applying the above argument to $\max\{u,0\}$ and $\max\{-u,0\}$.

To show the inequality (13), we take any $0 < \rho_1 < \rho_2 < 1$ and look at the analytic functions $\Phi(\rho_j z) + i\Psi(\rho_j z)$, $j = 1, 2$. Let $u_n \in \mathrm{Re}(P(\mathbb{T}))$ be chosen so that $u_n \to u$ in $L^1(d\sigma)$. Then $\Phi_n \to \Phi$ and $\Psi_n \to \Psi$ almost uniformly in $\mathbb{D}$, where Φ_n (resp. Ψ_n) denotes the harmonic extension of u_n (resp. $\tilde{u}_n$) to $\mathbb{D}$. Thus $\Phi_n(\rho_j e^{i\theta}) \to \Phi(\rho_j e^{i\theta})$ and $\Psi_n(\rho_j e^{i\theta}) \to \Psi(\rho_j e^{i\theta})$ uniformly on $\mathbb{T}$ for $j = 1, 2$. So

$$\begin{aligned}
&\|\Psi(\rho_1 e^{i\theta}) - \Psi(\rho_2 e^{i\theta})\|_s^s \\
&\leqq 3^s(\|\Psi(\rho_1 e^{i\theta}) - \Psi_n(\rho_1 e^{i\theta})\|_s^s + \|\Psi_n(\rho_1 e^{i\theta}) - \Psi_n(\rho_2 e^{i\theta})\|_s^s \\
&\qquad + \|\Psi_n(\rho_2 e^{i\theta}) - \Psi(\rho_2 e^{i\theta})\|_s^s).
\end{aligned}$$

By applying (12) to the central term of the right-hand side and letting $n \to \infty$, we get

$$\|\Psi(\rho_1 e^{i\theta}) - \Psi(\rho_2 e^{i\theta})\|_s \leqq 3\cdot 2^{(s+1)/s}(\cos\frac{\pi s}{2})^{-1/s}\|\Phi(\rho_1 e^{i\theta}) - \Phi(\rho_2 e^{i\theta})\|_1.$$

We set $\Phi_\rho(e^{i\theta}) = \Phi(\rho e^{i\theta})$ and $\Psi_\rho(e^{i\theta}) = \Psi(\rho e^{i\theta})$, $0 < \rho < 1$. Since $\{\Phi_\rho\}$ forms a Cauchy net in $L^1(d\sigma)$ tending to u as $\rho \to 1$, we have seen that $\{\Psi_\rho\}$ forms a Cauchy net in $L^s(d\sigma)$ as $\rho \to 1$. Since $\Psi_\rho(e^{i\theta}) \to v(e^{i\theta})$ a.e. as $\rho \to 1$, $v \in L^s(d\sigma)$ and $\Psi_\rho \to v$ in $L^s(d\sigma)$. The inequality (12) then implies the desired inequality (13), for $\Phi_\rho \to u$ in $L^1(d\sigma)$ as $\rho \to 1$. □

A.2.3. As a consequence of the preceding theorem, we see that, if $u_n \in \mathrm{Re}(P(\mathbb{T}))$ tend to u in $L^1(d\sigma)$, $\tilde{u}_n$ tend in $L^s(d\sigma)$, $0 < s < 1$, to v, which is given by A.2.2 in relation to u. v is called the <u>harmonic conjugate</u> of $u \in L^1(d\sigma)$ and is denoted by $\tilde{u}$. As our discussion shows, $\exp(-u - i\tilde{u}) \in H^\infty(d\sigma)$ provided that $u \in L^1(d\sigma)$ is bounded below.

A.3. The F. and M. Riesz Theorem

A.3.1. We will give Øksendal's proof of the F. and M. Riesz theorem. We first note the following, which can easily be verified by induction.

<u>Lemma</u>. Let $w_k \in \mathbb{C}$, $k = 1, \ldots, n$, satisfy $|w_k| < 1$ and $\mathrm{Re}(w_k) < 0$. Then

$$|1 - \prod_{k=1}^{n} (1 + w_k)| \leqq \prod_{k=1}^{n} (1 + |w_k|) - 1.$$

A.3.2. Theorem. Let μ be a complex Borel measure on the unit circle $\mathbb{T}$ in the complex plane such that

$$(14) \qquad \int_{\mathbb{T}} e^{int} d\mu(e^{it}) = 0$$

for $n = 1, 2, \ldots$. Then μ is absolutely continuous with respect to the Lebesgue measure on $\mathbb{T}$.

Proof. By considering $\mu - \mu(\mathbb{T})\sigma$ instead of μ, we may suppose that (14) holds for $n = 0$, too. Let F be a compact set in $\mathbb{T}$ with length zero, i.e. $|F| = 0$. So there exists, for each positive integer n, a finite number of disks $\Delta_1, \ldots, \Delta_N$ with centers $z_1, \ldots, z_N$ in F and with radii $r_1, \ldots, r_N$, respectively, such that these cover the set F and $\sum_{j=1}^N r_j < 1/n^2$. We set $\rho_j = nr_j$ and define

$$g_n(z) = 1 - \prod_{j=1}^{N} \frac{z - z_j}{z - (1 + \rho_j)z_j}.$$

Then, g_n is analytic in $|z| < 1 + \min\{\rho_j : j = 1, \ldots, N\}$, a neighborhood of the closed unit disk $\overline{\mathbb{D}}$. By the Runge theorem there exists a sequence of polynomials converging to g_n uniformly on the closed unit disk. Since μ is orthogonal to every polynomial, it follows that $\int_{\mathbb{T}} g_n(e^{it})d\mu(e^{it}) = 0$.

We claim that (a) $|g_n(z)| \leqq 2$ for $z \in \mathbb{T}$ and $n = 1, 2, \ldots$; and (b) the sequence $\{g_n\}$ converges pointwise on $\mathbb{T}$ to the characteristic function of F. In fact, for each $z \in \mathbb{T}$ we clearly have $|z - z_j| < |z - (1 + \rho_j)z_j|$ for all $j = 1, \ldots, N$. So $|g_n(z)| \leqq 2$, which shows the claim (a). If $z \in F$, then $z \in \Delta_k$ for some k and thus

$$\begin{aligned} |g_n(z) - 1| &= \prod_{j=1}^{N} \left| \frac{z - z_j}{z - (1 + \rho_j)z_j} \right| \\ &\leqq \frac{|z - z_k|}{|z - (1 + \rho_k)z_k|} \leqq \frac{r_k}{\rho_k} = \frac{1}{n}. \end{aligned}$$

Thus $g_n(z) \to 1$ on F. On the other hand, consider any $z \in \mathbb{T}$ with $\mathrm{dist}(z,F) \geqq \delta$ for any fixed $\delta > 0$. Then, $|\rho_j z_j/(z - (1 + \rho_j)z_j)| < 1$ and $\mathrm{Re}\{\rho_j z_j/(z - (1 + \rho_j)z_j)\} < 0$ for $j = 1, \ldots, N$. By applying Lemma A.3.1, we have

$$\begin{aligned} |g_n(z)| &= \left| 1 - \prod_{j=1}^{N} \left(1 + \frac{\rho_j z_j}{z - (1 + \rho_j)z_j} \right) \right| \\ &\leqq \prod_{j=1}^{N} \left(1 + \left| \frac{\rho_j z_j}{z - (1 + \rho_j)z_j} \right| \right) - 1 \end{aligned}$$

$$= \prod_{j=1}^{N} \left(1 + \frac{\rho_j}{|z - (1+\rho_j)z_j|}\right) - 1$$

$$\leq \exp\left(\sum_{j=1}^{N} \log(1 + \frac{\rho_j}{\delta})\right) - 1$$

$$= \exp\left(\sum_{j=1}^{N} \frac{\rho_j}{\delta}\right) - 1 = \exp\left(\sum_{j=1}^{N} \frac{nr_j}{\delta}\right) - 1$$

$$< \exp(\frac{1}{n\delta}) - 1,$$

for $\sum_{j=1}^{N} r_j < 1/n^2$. So, $g_n(z) \to 0$. Hence, the statement (b) is verified.

Consequently, the dominated convergence theorem shows that $\mu(F) = \int_F d\mu = \lim_{n\to\infty} \int g_n d\mu = 0$. As this holds for any compact set F in $\mathbb{T}$ of length zero, we conclude that μ is absolutely continuous with respect to the Lebesgue measure dt on $\mathbb{T}$. □

A.3.3. Let μ be as above. Then, by Theorem A.3.2,

$$d\mu(e^{it}) = f(e^{it})d\sigma(t)$$

for some $f \in L^1(d\sigma)$. Thus, the condition (14) states that

$$\int_{\mathbb{T}} f(e^{it})e^{int}d\sigma(t) = 0$$

for $n = 1, 2, \ldots$. This means that f belongs to $H^1(d\sigma)$.

REFERENCES

[AS] Ahlfors, L. V. and Sario, L., "Riemann Surfaces," Princeton Univ. Press, Princeton, N.J., 1960.

[CC] Constantinescu, C. and Cornea, A., "Ideale Ränder Riemannscher Flächen," Springer-Verlag, Berlin, 1963.

[1] Behrens, M. F., The maximal ideal space of algebras of bounded analytic functions on infinitely connected domains, Trans. Amer. Math. Soc. 161 (1971), 359-379.

[2] Beurling, A., On two problems concerning linear transformations in Hilbert space, Acta Math. 81 (1949), 239-255.

[3] Bishop, E., Subalgebras of functions on a Riemann surface, Pacific J. Math. 8 (1958), 29-50.

[4] Bishop, E., Analyticity in certain Banach algebras, Trans. Amer. Math. Soc. 102 (1962), 507-544.

[5] Brelot, M., Topology of R. S. Martin and Green lines, in "Lectures on Functions of a Complex Variable" (W. Kaplan, ed.), pp. 105-121, Univ. of Michigan Press, Ann Arbor, 1955.

[6] Brelot, M. and Choquet, G., Espaces et lignes de Green, Ann. Inst. Fourier (Grenoble) 3 (1952), 199-264.

[7] Carleson, L., Interpolation by bounded analytic functions and the corona problem, Ann. of Math. 76 (1962), 542-559.

[8] Dunford, N. and Schwartz, J. T., "Linear Operators," Part I, Wiley (Interscience), New York, 1958.

[9] Forelli, F., Bounded holomorphic functions and projections, Illinois J. Math. 10 (1966), 367-380.

[10] Gamelin, T. W., "Uniform Algebras," Prentice-Hall, Englewood Cliffs, N.J., 1969.

[11] Gamelin, T. W., "Uniform Algebras and Jensen Measures," Cambridge Univ. Press, London and New York, 1978.

[12] Garnett, J., "Bounded Analytic Functions," Academic Press, New York, 1981.

[13] Gunning, R. C., "Lectures on Riemann Surfaces," Princeton Univ. Press, Princeton, N.J., 1966.

[14] Hara, M. On Gamelin constants, preprint.

[15] Hardy, G. H., The mean value of the modulus of an analytic function, Proc. London Math. Soc. ser. 2, 14 (1915), 269-277.

[16] Hasumi, M., Invariant subspace theorems for finite Riemann surfaces, Cand. J. Math. 18 (1966), 240-255.

[17] Hasumi, M., Invariant subspaces on open Riemann surfaces, Ann. Inst. Fourier (Grenoble) 24, 4 (1974), 241-286.

[18] Hasumi, M., Invariant subspaces on open Riemann surfaces, II, Ann. Inst. Fourier (Grenoble) 26, 2 (1976), 273-299.

[19] Hasumi, M., The Fatou theorem for open Riemann surfaces, Duke Math. J. 43 (1976), 731-746.

[20] Hasumi, M., Weak-star maximality of H^{∞} on surfaces of Parreau-Widom type, unpublished note, 1977.

[21] Hasumi, M., Hardy classes on plane domains, Arkiv för Mat. 16 (1978), 213-227.

[22] Hasumi, M., Remarks on Hardy classes on plane domains, Bull. Fac. Sci. Ibaraki Univ. Ser. A. Math. 11 (1979), 65-71.

[23] Hasumi, M. and Srinivasan, T. P., Doubly invariant subspaces, II, Pacific J. Math. 14 (1964), 525-535.

[24] Hasumi, M. and Srinivasan, T. P., Invariant subspaces of continuous functions, Canad. J. Math. 17 (1965), 643-651.

[25] Hasumi, M. and Rubel, L. A., Multiplication isometries of Hardy spaces and of double Hardy spaces, Hokkaido Math. J. 10 (1981) Sp., 221-241.

[26] Hayashi, M., Hardy classes on Riemann surfaces, Thesis, UCLA (1979).

[27] Hayashi, M., Invariant subspaces on Riemann surface of Parreau-Widom type, preprint, 1980.

[28] Hayashi, M., Characterization of Riemann surfaces possessing invariant subspace theorem, preprint, 1981.

[29] Hayashi, M., Szegö's theorem on Riemann surfaces, Hokkaido Math. J. 10 (1981) Sp., 242-254.

[30] Hayashi, M., Seminar talk (February, 1982).

[31] Heins, M. H., "Hardy Classes on Riemann Surfaces," Lecture Notes in Math. Vol. 98, Springer-Verlag, Berlin, 1969.

[32] Hejhal, D. A., Classification theory for Hardy classes of analytic functions, Ann. Acad. Sci. Fenn. Ser. A. I. no. 566 (1973), 1-28.

[33] Helson, H., "Lectures on Invariant Subspaces," Academic Press, New York, 1964.

[34] Hoffman, K., "Banach Spaces of Analytic Functions," Prentice-Hall, Englewood Cliffs, N.J., 1962.

[35] Hurwitz, A. and Courant, R., "Vorlesungen über allgemeine Funktionentheorie and elliptische Funktionen. Geometrische Funktionentheorie," (4th ed.) Springer-Verlag, Berlin, 1964.

[36] Kaplan, W., Approximation by entire functions, Michigan Math. J. 3 (1955), 43-52.

[37] Kelley, J. L. and Namioka, I., "Linear Topological Spaces," D. Van Nostrand, Princeton, N.J., 1963.

[38] Kobayashi, S., On H_p classification of plane domains, Kodai Math. Sem. Rep. 27 (1976), 458-463.

[39] Kobayashi, S., On a classification of plane domains for Hardy classes, Proc. Amer. Math. Soc. 68 (1978), 79-82.

[40] Koosis, P., "Lectures on H_p Spaces," Cambridge Univ. Press, Cambridge, 1980.

[41] Krasnosel'skii, M. A. and Rutickii, Ya. B., "Convex Functions and Orlicz Spaces," Noordhoff, Gronigen, 1961.

[42] Kusunoki, Y., "Theory of Functions--Riemann Surfaces and Conformal Mappings," Asakura Shoten, Tokyo, 1973. (Japanese)

[43] Nakai, M., Corona problem for Riemann surfaces of Parreau-Widom type, Pacific J. Math. 103 (1982), 103-109.

[44] Nevanlinna, R., "Analytic Functions," Springer-Verlag, Berlin, 1970.

[45] Neville, C. W., Ideals and submodules of analytic functions on infinitely connected plane domains, Thesis, Univ. of Illinois at Urbana-Champaign, 1972.

[46] Neville, C. W., Invariant subspaces of Hardy classes on infinitely connected plane domains, Bull. Amer. Math. Soc. 78 (1972), 857-860.

[47] Neville, C.W., Invariant subspaces of Hardy classes on infinitely connected open surfaces, Mem. Amer. Math. Soc. no. 160 (1975), 151 pp.

[48] Niimura, M., Asymptotic value sets of bounded holomorphic functions along Green lines, Arkiv för Mat. 20 (1982), 125-128.

[49] Obrock, A., Null Orlicz classes of Riemann surfaces, Ann. Acad. Sci. Fenn. Ser. A. I. no. 498 (1972), 22 pp.

[50] Øksendal, B., A short proof of the F. and M. Riesz theorem, Proc. Amer. Math. Soc. 30 (1971), 204.

[51] Parreau, M., Sur les moyennes des fonctions harmoniques et analytiques et la classification des surfaces de Riemann, Ann. Inst. Fourier (Grenoble) 3 (1951), 103-197.

[52] Parreau, M., Théorème de Fatou et problème de Dirichlet pour les lignes de Green de certaines surfaces de Riemann, Ann. Acad. Sci. Fenn. Ser. A. I. no. 250/25 (1958), 8 pp.

[53] Phelps, R., "Lectures on Choquet's Theorem," D. Van Nostrand, Princeton, N.J., 1966.

[54] Pommerenke, Ch., On the Green's function of Fuchsian groups, Ann. Acad. Sci. Fenn. Ser. A. I. 2 (1976), 409-427.

[55] Pranger, W., Bounded sections on a Riemann surface, Proc. Amer. Math. Soc. 69 (1978), 77-80.

[56] Pranger, W., Riemann surfaces and bounded holomorphic functions, Trans. Amer. Math. Soc. 259 (1980), 393-400.

[57] Rubel, L. A. and Shields, A. L., The space of bounded analytic functions, Ann. Inst. Fourier (Grenoble) 16 (1966), 235-277.

[58] Rudin, W., Essential boundary points, Bull. Amer. Math. Soc. 70 (1964), 321-324.

[59] Rudin, W., "Real and Complex Analysis," McGraw-Hill, New York, 1966.

[60] Sarason, D., The H^p spaces of an annulus, Mem. Amer. Math. Soc. no. 56 (1965), 78 pp.

[61] Sarason, D., "Function Theory on the Unit Circle," Virginia Poly. Inst. and State Univ., Blacksburg, Virginia, 1978.

[62] Sario, L. and Nakai, M., "Classification Theory of Riemann Surfaces," Springer-Verlag, Berlin, 1970.

[63] Segawa, S., Harmonic majoration of quasi-bounded type, Pacific J. Math. 84 (1979), 199-200.

[64] Stanton, C. M., Bounded analytic functions on a class of open Riemann surfaces, Pacific J. Math. 59 (1975), 557-565.

[65] Tsuji, M., "Potential Theory in Modern Function Theory," Maruzen, Tokyo, 1959.

[66] Voichick, M., Ideals and invariant subspaces of analytic functions, Trans. Amer. Math. Soc. 111 (1964), 493-512.

[67] Voichick, M., Invariant subspaces on Riemann surfaces, Canad. J. Math. 18 (1966), 399-403.

[68] Weyl, H., "Die Idee der Riemannschen Flache," (3rd ed.) Teubner, Stuttgart, 1958.

[69] Widom, H., The maximum principle for multiple-valued analytic functions, Acta Math. 126 (1971), 63-82.

[70] Widom, H., $\mathcal{H}_p$ sections of vector bundles over Riemann surfaces, Ann. of Math. 94 (1971), 304-324.

[71] Zygmund, A., "Trigonometric Series" (2nd ed.), 2 vols., Cambridge Univ. Press, London, 1959.

INDEX OF NOTATIONS

$A(a/b)$, $A'(a/b)$: XI.3C

$B(\alpha,a)$: V.1A

$C_\kappa(R,\mathbb{R})$: III.1A
$C_0(R)$: IV.5B

$\mathbb{D}$: IV.1A
$D(a)$: VI.3A
$\mathcal{D}(f)$: III.4B
$d\chi$: III.2C
$d\sigma$: IV.1A
$\Delta(R)$ $(=\Delta)$: III.1B
$\Delta_1(G)$: III.4B
$\Delta_1(R)$ $(=\Delta_1)$: III.2B

$\mathbb{E}_0(R,a)$ $(=\mathbb{E}_0(a))$: VI.1A

$F_0(R)$: I.3A
Φ (conformal map): VI.3A
Φ (convex function): XI.1A

$g(a,z)$: I.6A
$g^{(a)}(z)$: VII.1A
$G[f;G(a)]$: VI.1B
$\mathbb{G}(R,a)$ $(=\mathbb{G}(a))$: VI.1A
$\mathbb{G}'(R,a)$ $(=\mathbb{G}'(a))$: VI.1A
$G(b)$: III.4A
Γ, Γ_c, Γ_{c0}, Γ_e, Γ_{e0}, Γ_h: I.8D
Γ_{h0}: I.9B; V.6D
Γ^1, Γ^1_a, Γ^1_c, Γ^1_e, Γ^1_h: I.8A
$\Gamma M(Cl(G),\xi)$: V.4B
$\Gamma(r)$, $\Gamma(z;r)$: XI.3C

$h[f]$: III.5A
$H[f]$: III.3A
$H[f;D]$: I.5A
$h^p(\mathbb{D})$, $h^\infty(\mathbb{D})$, $h^1_Q(\mathbb{D})$: IV.1C
$H^p(\mathbb{D})$, $H^\infty(\mathbb{D})$: IV.1B
$H^p(d\sigma)$: IV.2A
$H^\Phi(D)$: XI.1A
$H^p(R)$, $H^\infty(R)$, $h^p(R)$, $h^1_Q(R)$, $H^p(d\chi)$: IV.3A
$H^p_0(R)$, $H^p_0(d\chi)$: VIII.1A
$h^\infty_\beta(R)$: IV.5C
$H^p(\mathbb{D},\xi)$, $H^p d\sigma,\xi)$, $H^p(d\sigma,Q)$: IX.1A
$\mathcal{H}^p(R,\xi)$:V.2A
$\tilde{\mathcal{H}}^p(G,\xi)$: V.4B
$HP(a)$, $HP(R)$: II.4A
$HP'(R)$: II.4B
$H_1(R)$: I.2A
$H^1(R,\mathbb{T})$: II.1A

I_0: IX.2A
$I(R)$: II.5A

$k(b,z)$ $(=k_b(z))$: III.1B
$K'(a)$, $K_n(a)$: VII.7A
χ_R $(=d\chi)$: III.2C
ξ_A: IX.2A
ξ_C: IX.2D

$\mathbb{L}(R,a)$ $(=\mathbb{L}(a))$: VI.1A
$\Lambda(R,a)$ $(=\Lambda(a))$: VI.5B

$m(a)$, $m(\xi,a)$: V.8A
$m^p(a)$, $m^p(\xi,a)$: IX.1D

$M(\Delta_1)$: III.2C
$M(H^\infty(R))$:X.5A
M^Δ: IX.3C
$\{M\}_p$: VIII.1B

N_p, N_Φ: XI.1A

$\mathcal{O}_{AB}$, $\mathcal{O}_{AB*}$, $\mathcal{O}_\Phi$, $\mathcal{O}_G$, $\mathcal{O}_p$: XI.1A
$\mathcal{O}_{AS}$: XI.7B
ω_a^G: I.5D
Ω_0: VI.3A

$P(a.a';z)$: VI.5B
$P(r,\theta)$: IV.1A; A.1.2
pr_I, pr_Q: II.5A
Ψ (conformal map): VI.3A

q_C: IX.3A
Q_A: IX.2A
Q_C: IX.2D
Q_g: IX.2D
Q_M: III.1B
$Q(R)$: II.5A

R^*: III.1B
$R^\dagger$: V.3A
$r(a;z)$: VI.1A
$R(\alpha,a)$: V.1A
u_E^R: I.6E

$S(\zeta;\alpha)$: A.1.2
$S(\zeta:\alpha,\rho)$: VI.3D
$S(b;\alpha,\rho)$: VI.7B
$SP'(R)$: II.4D
$\sigma(c)$: I.9B
$\theta(a;z)$: VI.1A

T_R: III.6B

$w(E,E')$, $w(E',E)$: IV.5A
$W(R)$: III.5A

$z(\ell;\alpha)$: VI.1A
$Z(a,R)$: V.1C

$\vee$, $\wedge$ (lattice operations): I.4C
^ (fine limit): III.4B
ˇ (radial limit): VI.4A
~ (harmonic conjugation): A.2.1
$\|\cdot\|_p$ (norms): IV.1B; IV.1C; V.2A

INDEX

β-dual: IV.5C
β-topology: IV.5B
Behrens theorem: X.7B
Band: II.5A
Betti number: V.1A
Bishop theorem: X.6C
Blaschke product: IV.2C
Boundary m-function: VIII.4A
Bounded characteristic: II.5D; XI.1B
Brelot-Choquet problem: VI.5A
Bundle (= unitary flat complex line bundle): II.2A
— of a multiplicative meromorphic function: II.2D
— of an l.m.m.: II.2C

Canonical basis: I.2B
Canonical measure: III.2B
Cauchy theorem
direct — —: VII.3C; IX.4A
direct — — (weak type): VII.4B
inverse — —: VII.1B; VII.1C; IX.1C
Cauchy kernel: I.11B
Cauchy-Read theorem: V.4A
Character: I.3B
— group: I.3B
— of a multiplicative meromorphic function: II.2D
— of an l.m.m.: II.2C
— of an m-function: VIII.4A; IX.1A
Choquet theorem: II.4C
Circular set: XI.6A
Cluster value theorem: VI:7D; VI.7E
Cocyle: II.1A
Cohomology group: II.1A
Convex function: XI.1A
Corona problem: X.5A
Cover transformation: III.6B
group of — —: III.6B
Covering map: III.6A
Covering triple: III.6A
Critical point: V.1C

(DCT) (= (DCT_a)): VII.3C; IX.4A
Differential: I.7A
analytic —: I.8A
conjugate —: I.7A
harmonic —: I.8A
reproducing — (σ(c)): I.9B
— with singularity: I.9A
Dirichlet problem: I.5A; III.3A; VI.1B
Divisor: I.10B

End: VI.5A
Exhaustion: I.1A
canonical —: I.1C
regular —: I.1A

Fatou theorem: A.1.2
Fine limit: III.4B
Fine boundary function: III.4B
Function
balayaged —: I.6E
boundary m-—: VIII.4A

Function (continued)
convex —: XI.1A
i-—: IX.1A
m-—: VIII.4A; IX.1A
Martin —: III.1B
Wiener —: III.5A
Fundamental group: I.3A

Gelfand transform: X.5A
Genus: I.2B
Green function: I.6A
modified — —: V.7D
Green line: VI.1A
convergent — —: VI.5A; VI.5B
regular — —: VI.1A
Green measure: VI.1A
Green star region: VI.1A

Hardy class: IV.1B; IV.3A
Hardy-Orlicz class: XI.1A
Harmonic
— conjugate: A.2.1
— differential: I.8A
least — majorant: I.4C
— measure: I.5D; III.2C
minimal — function: III.2B
quasibounded — function: II.5A
singular — function: III.2C
Harmonizable: III.5A
Hayashi theorem: IX.1C; IX.5D; X.2B; X.4A
Hejhal theorem: XI.3B
Homology group: I.2A
Hyperbolic region: I.5C; V.7A
Hyperbolically rare: X.7B

Inner: II.5A
common — factor: II.5D
— factor of an l.a.m.: II.5D
greatest common — factor: II.5D
— part: II.5A
Integral representation: III.2B
Invariant subspace: VIII.1A; VIII.2A
doubly — —: VIII.1A; VIII.2C
simply — —: VIII.1A; VIII.2C
— — theorem: VIII.2C; VIII.3B; VIII.4C
Irregular boundary point: I.5B

Kolmogorov theorem: A.2.2

LHM (= least harmonic majorant): I.4C
Line bundle (= bundle): II.2A
L.a.m. (= locally analytic modulus): II.2C
L.m.m. (= locally meromorphic modulus): II.2C
— of bounded characteristic: II.5D
inner —: II.5D
quasibounded —: II.5D

Martin boundary: III.1B
Martin compactification: III.1B
Martin function: III.1B
pole of a — —: III.1B
Maximal ideal space: X.5A
Mean value theorem: IX.2D
Minimal point: II.4C; III.2B
Modulus of a section: II.2A
Multiplicative meromorphic function: II.2D
inner factor of — — —: II.5D
outer factor of — — —: II.5D
— — — of bounded characteristic: II.5D

Nakai theorem: X.8A; X.8C
Normal operator: V.6B
Null set: XI.1A

Origin: II.2B
Orthogonal decomposition: II.5A
Outer character: IX.1A
  p-— —: IX.1A

Parreau theorem: VI.4A; VI.4B; XI.2B; XI.2C; XI.2D
Partition
  identity —: V.6A
  — of the ideal boundary: V.6A
Poisson kernel: IV.1A; A.1.2
Polar set: I.6C
Potential: I.6B
Pranger theorem: X.6A
Principal branch: II.2D
Principal operator: V.6D
PWS (= surface of Parreau-Widom type): V.1A

Quasibounded: II.5A; IV.1C
  — part: II.5A
Quasi-everywhere (= q.e.): I.6C

Radial limit: VI.4A
Region (= subregion): I.1A
  canonical —: I.1C
  curvilinear Stolz —: VI.3D
  Green star —: VI.1A
  hyperbolic —: I.5C; V.7A
  regular —: I.1A
  Stolz —: VI.3D; VI.7B; A.1.2
Regular boundary point: I.5B
Regularization: V.3B
Riemann-Roch theorem: I.10C
Riemann surface
  — of Myrberg type: X.1A
  — of Parreau-Widom type: V.1A
  regular — —: V.1C
Riesz theorem: I.6F; III.3B
  F. and M. — —: A.3.2.

Section: II.2A
  holomorphic —: II.2A; V.2A
  meromorphic —: II.2A
  meromorphic-differential —: II.2E; V.4B
Segawa theorem: XI.7B
Singularity: I.9A
Standard: IX.1A
Stanton theorem: X.1A; X.5B
Stolz region: VI.3D; VI.7B; A.1.2
Subharmonic function: I.4A
Subregion = region
Superharmonic function: I.4A
Support: VIII.1A

Thin: III.4A

Universal covering surface: III.6A

Weak-star maximality: VII.5A
Weak topology: IV.5A
Widom theorem: V.2B; V.4C; V.5C; V.9A; V.9B
Wiener function: III.5A